Vulnerability and Adaptation to Climate Change

Interim Results from the U.S. Country Studies Program

Environmental Science and Technology Library

VOLUME 8

The titles published in this series are listed at the end of this volume.

Vulnerability and Adaptation to Climate Change

Interim Results from the U.S. Country Studies Program

Edited by

Joel B. Smith
Hagler Bailly Consulting, Inc., Boulder, CO, U.S.A.

Saleemul Huq
Bangladesh Centre for Advanced Studies, Dhaka, Bangladesh

Stephanie Lenhart
Hagler Bailly Consulting, Inc., Boulder, CO, U.S.A.

Luis Jose Mata
Fluid Mechanic Institute, Universidad Central de Venezuela, Caracas, Venezuela

Ivana Nemešová
Institute of Atmospheric Physics, Czech Academy of Sciences, Prague, Czech Republic

and

Sekou Toure
Ecole Nationale Supérieure des Travaux Publics, Abidjan, Cote d'Ivoîre

U.S. Country Studies Program

Participating U.S. Government Agencies:
Environmental Protection Agency
Department of Energy
Agency for International Development
Department of State
National Oceanic and Atmospheric Administration
Department of Agriculture
National Aeronautics and Space Administration
National Science Foundation
Department of the Interior
Department of Health and Human Services

SPRINGER-SCIENCE+BUSINESS MEDIA, B.V.

A C.I.P. Catalogue record for this book is available from the Library of Congress.

ISBN 978-0-7923-4141-3 ISBN 978-94-017-3653-4 (eBook)
DOI 10.1007/978-94-017-3653-4

Printed on acid-free paper

The information presented here covers preliminary work carried out by the countries and has not necessarily been endorsed by either the respective Governments or the U.S. Government. Each country chapter was subject to peer review before editing. Additionally, the entire document is currently undergoing further peer review.

Contents

Figures

Tables

Foreword

Martin Parry

University College, London, UK

The 13 country studies collected in this report represent the first of what is likely to become a worldwide, country-by-country estimate of the likely impacts of, and appropriate adaptations to, greenhouse-gas-induced global climate change.

Under the U.N. Framework Convention on Climate Change (UNFCCC), signatories agreed to two near-term actions and one major subsequent one. The two near-term actions are to make annual estimates of the emissions and sinks of greenhouse gases, which are now being reported as part of a country-by-country inventory developed by the U.N. Environment Programme, the Organization for Economic Cooperation and Development, the U.S. Country Studies Program, and the U.N. Development Program; and to take steps to reduce the impact of climate change by reducing economic and social vulnerability to climate change and by evaluating the range of appropriate responses. The subsequent action will be internationally agreed-upon reductions in greenhouse-gas emissions to avoid changes of climate that might threaten food security, natural ecosystems, and sustainable development (Article 2 of the UNFCCC).

The U.S. Country Studies Program is a significant contribution to the second of the near-term actions. It has built on the *Technical Guidelines for Climate Impacts and Adaptations Assessment* published by the Intergovernmental Panel on Climate Change (Carter et al., 1994) as an agreed technical set of scientific methods for climate impact assessment and has written its own guidance document, *Guidance for Vulnerability and Adaptation Assessment* (U.S. CSP, 1994). The U.S. Country Studies Program developed the Guidance and other reviews of methodology into a nonspecialist set of workbooks for use at the country level, which, backed up by advice from experienced scientists from the United States and other countries, enabled local scientists to conduct their own vulnerability and adaptation assessments.

This has enabled two important objectives to be attained. The first is to make available, for the first time, reliable and compatible country-by-country studies. The second is to build the capacity for country-level assessments by training local scientists in both developed and developing countries who are adept at vulnerability and adaptation assessment.

On the first objective — that of reliable and compatible country-level studies — the 13 reports collected in this volume represent a milestone toward the ultimate objective of a complete set of 165 national studies of the likely impact of climate change that enables both the compilation of a global picture of possible damage and between-country comparisons of impacts and adaptations. This is necessary if we are to be able to assess accurately which regions will be more, or less, affected and where different policies of response will be required. An initiative by the U.N. Environment Programme, stimulated by the U.S. Country Studies Program, is currently concerned with developing an internationally agreed-upon set of *Workbooks on Climate Impacts and Adaptations Assessment* to enable this country-by-country assessment.

Achievement of the second objective — building scientific capacity in climate impact assessment at the national level — is a prerequisite for a robust global strategy to combat the effects of climate change. Ten years ago there were fewer than five completed regional or country studies of the possible impacts of climate change. Today, there are at least 40.

Yet, these studies must still be regarded as preliminary, and three-quarters of the world's nation states have no such study at all. There is, therefore, still a long way to go. But, the U.S. Country Studies Program has pointed the way and has taken the first step. The U.S. Country Studies Program and the country scientists deserve both our congratulations on their achievement and our encouragement for their next efforts. This work, and other work like it, may prove to be extremely important in reducing both vulnerability to climate change and the potential suffering this could involve.

REFERENCES

Benioff, R., S. Guill, and J. Lee (eds.) (In press). *Vulnerability and Adaptation Assessments: An International Guidebook.* Dordrecht, The Netherlands: Kluwer Academic Publishers.

Carter, T.R., M.L. Parry, H. Harasawa, and S. Nishioka. 1994. *IPCC Technical Guidelines for Assessing Climate Change Impacts and Adaptations.* London: Department of Geography, University College London.

Acknowledgments

This document is the result of work by many researchers within the contributing countries. The dedication of these individuals in performing analyses and reporting results was essential for development of this interim report. The authors of the chapters in this report received extensive editorial assistance from Christina Thomas, Greg Pitts, Eloise Castillo, and production support was provided by Sylvia Luedeking, Janice Pagel, Ashley Frazee, Sylvia Northey, Wynne Cougill, and Brett Sanders. The attention to detail and dedication of this staff contributed greatly to the quality of the final report. The assistance and review of the chapters provided by the technical editors was also greatly appreciated. We would also like to acknowledge Shannon Ragland for developing the summary of results by vulnerability sector.

Technical Editors

Joel Smith, Hagler Bailly Consulting, Inc.
Scenarios and Adaptation

Cynthia Rosenzweig, Goddard Institute for Space Studies and Anna Iglesias, Instituto Nacional de Investigaciones Agrarias
Agriculture

Barry Baker, Ecosystems Research International
Grasslands/Livestock

Tom Smith, University of Virginia, Department of Environmental Sciences
Forests

Kenneth Strzepek, Strzepek & Associates
Water Resources

Stephen Leatherman, University of Maryland, Laboratory for Coastal Research
Coastal Resources

Ihor Hlohowskyj, Argonne National Laboratory
Fisheries

Kirk LaGory, Argonne National Laboratory and Jay Malcolm, University of Florida
Wildlife

John Balbus, George Washington University
Human Health

Executive Summary

Overview

Fifty-five developing countries and countries with economies in transition are undertaking studies of greenhouse gas emissions, mitigation options, and vulnerability and adaptation to climate change with financial and technical support from the U.S. Country Studies Program. This report presents preliminary work from 13 of these countries that are prepared to share their progress on their vulnerability and adaptation assessments. This report is designed to contribute to scientific understanding of the potential impacts of climate change and effective adaptation strategies. In addition, this report is designed to promote the sharing of methodologies and results with other countries that are conducting assessments of vulnerability and adaptation.

The information presented here covers preliminary results and work in progress (Table S.1). This work has not necessarily been endorsed by the respective governments or the U.S. Country Studies Program. In a few cases, work supported from other sources is presented here.

Methods for Vulnerability and Adaptation Assessment

The 55 countries participating in the U.S. Country Studies Program are using a variety of methods to assess their vulnerability to climate change and the potential for implementing adaptation options. Many of the countries are using models and methodologies provided by the U.S. Country Studies, and others are using other models and methodologies (Table S.1[1]). In all cases, the countries are adapting the models and methodologies to their unique national circumstances.

Technical assistance and training have been provided on tools that may be used to assess vulnerability and adaptation, including climate and baseline scenarios, climate model data, and analytical methods for agriculture, grassland/livestock, forests, water resources, coastal resources, fisheries, wildlife, and human health. These methods are described in more detail in *Vulnerability and Adaptation Assessments: An International Guidebook* (Benioff et al., in press).

The general approach recommended by the U.S. Country Studies Program for conducting vulnerability and adaptation assessments is presented in Figure S.1. The assessments begin with selection of climate change scenarios and baseline socioeconomic changes (population and gross national product) that will happen without climate change. The climate change scenarios involve combining changes from $1\times CO_2$ to $2\times CO_2$ in average monthly climate variables from general circulation models (GCMs) with an up to 30 year observed climate data base (typically 1951-1980 or 1961-1990) or combining incremental changes in temperature and precipitation (e.g., +2°C, +4°C, +/-10% or

[1] In this document, GFDL is referred to as GFD3, GFDL Transient is referred to as GF01, and UKMO-89 is referred to as UK89.

Table S.1. Summary of Countries Contributing to the Interim Vulnerability and Adaptation Assessment Synthesis Report

Region/Country	Sector	Scenario	Vulnerability Assessment Method	Progress
Africa				
The Gambia (Round 1)	Agriculture	CCCM, GFDL, GFDL Transient, GISS, UKMO	DSSAT 3	Biophysical results
	Coastal	SLR: 0.2 m, 0.5 m, 1.0 m, 2.0 m	Common methodology	Biophysical results
Zimbabwe (Round 1)	Agriculture	CCCM, GFDL	CERES-Maize	Biophysical results
Asia-Pacific				
Bangladesh (Round 2)	Agriculture	GFDL	CERES-Rice	Methods only
	Water	GFDL	MIKE11	Methods only
	Coastal	SLR	Common methodology	Methods only
	Forest	To be selected	Water salinity projections	Methods only
Kazakhstan (Round 1)	Agriculture	GFDL, GFDL Transient, CCCM	CERES-Wheat	Biophysical results
	Grassland	GFDL, GFDL Transient, CCCM	Dynamic Productivity Model	Biophysical results
	Water	GFDL, GFDL Transient, CCCM	KazNIGMI Model	Biophysical results
Mongolia (Round 1)	Agriculture	GISS, GFDL, Historical	CERES-Wheat (spring)	Biophysical results
	Water	GISS, GFDL, CCCM, Incremental	Basin Conceptual Model	Biophysical results
	Forest	UKMO-89	Holdridge model	Biophysical results
Sri Lanka (Round 2)	Coastal	SLR: 0.3 m, 1.0 m, 2.0 m	Modified common methodology	Biophysical and socioeconomic results
Central/Eastern Europe				
Czech Republic (Round 1)	Agriculture	CCCM, GISS, Incremental	CERES-Wheat	Biophysical results
	Forestry	CCCM, GISS, Incremental	Forest Vegetation Zones	Biophysical results
	Water	CCCM, GISS, Incremental	BILAN, CLIRUN, SAC-SMA	Biophysical results
Estonia (Round 2)	Agriculture	GCMs to be selected	CERES-Barley, SPUR2	Preliminary biophysical results
	Forestry	Incremental	Gap models	Preliminary biophysical results
	Coastal	SLR: 1.0 m	Common methodology	Preliminary biophysical results
Poland (Round 1)	Coastal	SLR: 0.1 m to 2.5 m	Common methodology	Biophysical and socioeconomic results
Slovakia (Round 2)	Agriculture	CCCM, GISS, GFDL	DSSAT 3	Methods only
	Forestry	CCCM, GISS, GFDL	Holdridge model	Methods only
	Water	CCCM, GISS, GFDL	WATBAL	Methods only
Ukraine (Round 2)	Forestry	GFDL, GISS	Holdridge model	Preliminary biophysical results
	Water	GFDL, GFDL Transient, UKMO	National models	Preliminary biophysical results
	Coastal	SLR: 0.5 m, 1.0 m	Common methodology	Preliminary biophysical results
Latin America				
Venezuela (Round 1)	Forestry	GFDL, GISS, OSU, UKMO	Holdridge model	Biophysical results
	Coastal	SLR: 0.5 m and 1.0 m	Common methodology	Biophysical and Socioeconomic results
Peru (Round 1)	Coastal	SLR: 1.0 m	Common methodology	Biophysical results

Figure S.1. U.S. Country Studies Program Primary Approach to Climate Change Vulnerability and Adaptation Assessments

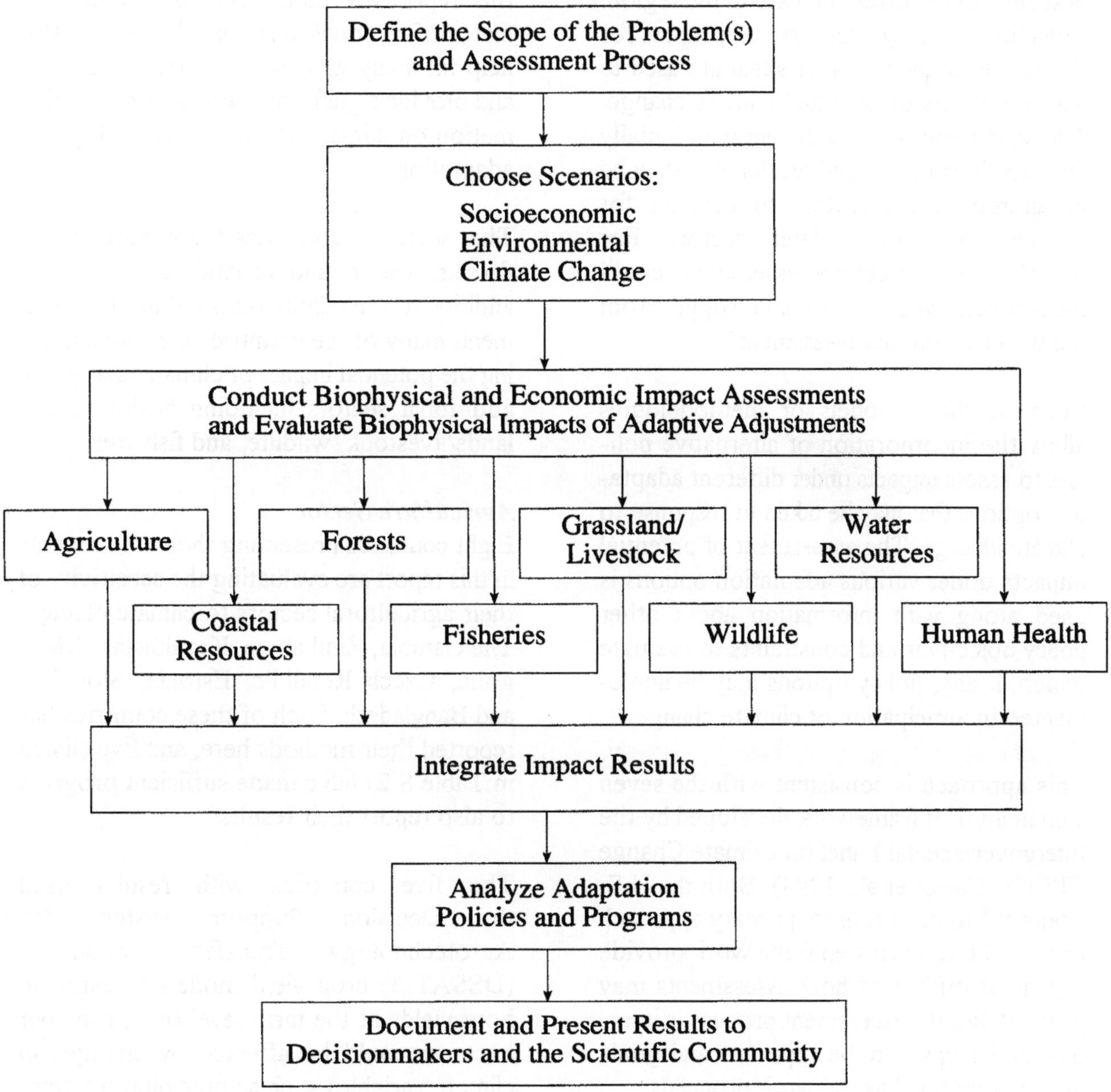

Source: Benioff et al., in press.

20% precipitation) with the observed data base. GCMs are selected after comparing their estimates of current climate with observed climate data and selecting those that best represent current climate in the region. The climate change scenarios are inputs for the models or methodologies that are used to assess impacts of potential climate change. The vulnerability of each sector is initially analyzed in isolation, and results may then be integrated across sectors to account for interactions among related sectors. For example, an agriculture assessment could incorporate changes in water supply from the water resources assessment.

Most of these models or methodologies allow the incorporation of alternative policies to assess impacts under different adaptation options that may be taken in response to climate change. The assessment of potential impacts under various adaptation options is used along with information about other policy objectives and constraints to evaluate which, if any, policy options may be implemented in anticipation of climate change.

This approach is consistent with the seven step analytical framework developed by the Intergovernmental Panel on Climate Change (IPCC) (Carter et al., 1994). Both the U.S. Country Studies Program primary approach and the IPCC seven step framework provide a general outline of how assessments may proceed, but the assessment process is iterative and steps may be repeated and goals may be redefined as the work proceeds.

Summary of Results by Sector

The following sections summarize the impacts reported in this document and compare these results to recent literature on climate change vulnerability and adaptation, including recent IPCC literature (in press). This comparison shows that the results in this report are consistent with other estimates of regional impacts. The results also help fill many gaps in the current literature and provide significant sources of new information on climate change vulnerability and adaptation.

The sectors summarized are agriculture, forests, water, and coastal resources.[2] In addition to the results reported in this document, many of the countries are also assessing the potential impact of climate change on additional sectors including health, grasslands/livestock, wildlife, and fisheries.

Agriculture Sector

Eight countries presenting their assessments in this report are evaluating the sensitivity of their agricultural sectors to climate change: The Gambia, Zimbabwe, Kazakhstan, Mongolia, Czech Republic, Estonia, Slovakia, and Bangladesh. Each of these countries has reported their methods here, and five (listed in Table S.2) have made sufficient progress to also report their results.

The five countries with results used the Decision Support System for Agrotechnology Transfer version 3 (DSSAT 3) crop yield models to estimate how yields at the farm level (e.g., tons per hectare) could be affected by changes in climate variables such as precipitation, temperature, and evapotranspiration. Their analyses typically focused on the sensitivity of important cash or staple crops such as maize. Changes in crop yields were derived

2 The grasslands/livestock sector was examined by only Kazakhstan, and thus is not summarized here.

Table S.2. Summary of Changes in Agricultural Yields

Country	Scenario	Crop	Percent Change in Farm Level Yields	Direct CO_2 Effect Included	Adaptation Included
The Gambia	CCCM, GFDL, GISS	Maize	-26% to -15%	Yes	No
		Millet (early)	-44% to -29%		
		Millet (late)	-21% to -14%		
		Groundnuts	+40% to +52%		
Zimbabwe	CCCM, GFDL	Maize (all sites)	-13.6% to -11.5%	Yes	No
		(Masvingo site)	-98%		No
		(Masvingo site)	+3% to +16%		Yes[1]
Kazakhstan	CCCM, GFDL[2] Incremental Scenarios	Wheat (spring)	-70% to -25%	Only for GCMs	No
		Wheat (winter)	-35% to +17%		
Mongolia	GFDL, GISS[3]	Wheat (spring)	-74.3% to +32.0%	No	Yes[4]
Czech Republic	Incremental Scenarios	Wheat (winter)	-3% to +16%	Yes	No

[1] Adaptation measures for Masvingo include earlier planting date and short season varieties.
[2] Also estimated impacts for 2030 using GFDL Transient climate data. Spring wheat yield declined by 27% and winter wheat yield increased by 21%.
[3] Also considered impacts if historical warming trend continues. In contrast to the GCMs, this trend predicts a cooler and wetter climate in June and July, which may lead to yield increases compared to baseline.
[4] Adaptation measure shifts planting date from May 15 to May 1 or April 25. Yield loss range is smaller for earlier planting dates (-5% to -67%).

by contrasting productivity under baseline climate conditions with productivity under altered conditions. Table S.2 summarizes the percentage changes in yields by country and crop.

These results indicate that climate change tends to decrease crop productivity in low latitude countries such as The Gambia and Zimbabwe. For example, in The Gambia, all maize and millet simulations show yield declines. In Zimbabwe, results for four study sites showed that maize yields could increase or decrease. However, notable increases in groundnut yields were predicted for The Gambia. Groundnuts are historically the country's primary cash crop, so this result has important economic consequences. Wheat crop yields in the higher latitude countries (Kazakhstan for winter wheat only, Mongolia, and the Czech Republic) are all potentially highly vulnerable to climate change, with yield changes from -74% to +32%.

Whether the yield impacts are positive or negative, they are potentially large. Given the relative importance of agriculture in some of these countries (e.g., 27% of GDP and 75% of employment in The Gambia), such impacts — if they are applicable for all areas of crop production — are likely to have significant effects on regional and national economies. Thus far, economic impacts have been qualitatively addressed in the vulnerability assessments.

Earlier studies have simulated agricultural impacts in developing countries and countries in transition (e.g., Parry et al., 1988a, b; Parry et al, 1992; Downing, 1992; Qureshi and Hobbie, 1994; Rosenzweig and Iglesias, 1994). About two-thirds of the studies summarized in IPCC (in press) indicate that yields may increase or decrease; the remaining studies show declines in yields. Yield declines appear to be most frequent in Africa. The U.S. Country Studies Program results are very similar. They show a strong trend toward maize and millet yield losses in Africa, but mixed results in other regions. The lack of much overlap by region and crop between the two tables suggests that the U.S. Country Studies Program adds a number of new study results to the existing inventory of studies.

The results in Table S.2 for Zimbabwe and The Gambia can be compared with other studies of agricultural impacts in or near those countries. Downing (1992) reported maize yield reductions for sites in Zimbabwe ranging from 5% to 27%, excluding beneficial effects of CO_2. Rosenzweig and Iglesias (1994) also showed maize decreases in Zimbabwe of 8% to 14%. These findings are consistent with the results in this report, which show reductions near the low end of this range, which include direct CO_2 effects. Downing also reported that regional millet yields in Senegal may decline by 33% to 45% throughout most of the country. These are consistent with the 14% to 43% reductions in early and late millet yields reported here by The Gambia under the Goddard Institute for Space Studies (GISS) scenario.

Forest Sector

Assessments of forest vulnerability and adaptation have been prepared by seven countries for this report: Bangladesh, Mongolia, Czech Republic, Estonia, Slovakia, Ukraine, and Venezuela. All of these assessments describe the methods used, and five of these report preliminary vulnerability results.

Mongolia, Ukraine, and Venezuela used the Holdridge Life Zone Classification to identify areas where climate change would lead to life zone shifts. GCMs were used to estimate how the location of life zones would shift from current climate to $2\times CO_2$ climate. Regions that fall into different zones between the $1\times CO_2$ and $2\times CO_2$ are likely to undergo significant change in vegetation type under climate change.

Table S.3. Summary of Forest Impacts

Country	Scenario	Model	Impact
Mongolia	UKMO	Holdridge	66 million ha shift from cool temperate to warm temperate desert/steppe
Czech Republic	GISS, CCCM	FVZ[1]	National and study area analyses show shift from Norway spruce to pine and deciduous species
Estonia	Incremental Scenarios[2]	Gap	Two sites show decline of spruce and aspen and increase in pine or deciduous species under all scenarios
Ukraine	GFDL, GISS	Holdridge	39% to 47% reduction in forested area
Venezuela	UKMO	Holdridge	About 35 million ha of subtropical forest converted to tropical forest; tropical and subtropical moist or wet forest declines by 40-50 million ha, and tropical dry and very dry forests increase by a similar amount

1 Based on a model similar to the Holdridge one that identifies forest vegetation zones (FVZ) based on climate and soil conditions. The results are simulated for temperature change in 2010 and 2030.

2 Incremental scenarios were 4°C warming with 0% and ±10% annual precipitation change. The forest gap model simulated 500 years of forest growth.

The Czech Republic uses its forest vegetation zone classification, which is similar to the Holdridge model, to study the effects of climate change on forests.

Although the Holdridge model and forest vegetation zone classification impacts indicate potential vegetation shifts, only transient analyses can show whether forest ecosystems can successfully migrate during the next 70 to 100 years. Estonia used a forest gap model to perform a transient analysis of how climate change may affect two forest sites. This approach shows how forest stand composition gradually changes over time as climate conditions change. Table S.3 summarizes the reported impacts.

Although only Ukraine reports a decline in total forested area, the other countries predict vegetation zone shifts that could cause major disruptions in forests. The estimated decline in forested area in Ukraine is between 39% and 47%, with large regions of cool temperate moist forest converting to steppe or desert scrub regions under drier conditions.[3] Furthermore, small boreal forest regions in Ukraine are converted to temperate forests. Both Mongolia and Venezuela show large shifts from one type of vegetation zone to another. Mongolia estimates that about 66 million ha of land (47 % of currently forested land) switches from cool temperate desert and steppe to warm temperate desert and steppe. In Venezuela, about 35 million ha (38% of currently forested land) change from subtropical forest to tropical forest, and between 40 million and 50 million ha (44 to 55% of currently forested land) shift from moist to dry or very dry forest. Finally, the results for both the Czech Republic and Estonia suggest that forest areas currently dominated by Norway spruce will become forests dominated by pine or deciduous trees.

Although formal economic analyses have not yet been performed, the potential economic impacts due to shifts in forest vegetation have been discussed by some of the reporting countries. For example, the Czech Republic notes the economic importance of environmental services provided by the forest, such as watershed protection and recreation, and has also estimated that annual expenditures due to "ecological destabilization" and forest management changes may be in the hundreds of thousands of U.S. dollars. Estonia reports that reduction in Norway spruce and aspen may harm the forestry industry, because these are important harvest species and they currently have relatively high growth volumes. However, it notes that more research is needed before concluding that harvest volumes will necessarily decline.

The results presented in Table S.3 can be compared with recent literature, which indicates that about one-third of global forested area will undergo major vegetation changes (IPCC, in press). Greater changes are expected in the higher latitudes, where boreal forests are expected to decline. Smaller changes are estimated for tropical and temperate forests, and the direction of change differs across the vegetation models.

Global vegetation modeling by Smith et al. (1995) showed that tropical forest areas could increase primarily through conversions from subtropical forest areas, which were

[3] These are preliminary results for the Ukraine and should be interpreted with caution; more recent analysis not completed in time for publication generated significantly different impact results.

expected to decline. The climate change results in Venezuela show this same trend from subtropical to tropical forests. Climate change impacts are also estimated to increase the potential temperate forest area as the climate zones associated with boreal forests shift poleward, and may cause significant species composition changes in many existing temperate forests (Melillo et al., 1993; Neilson, 1993; Smith et al., 1995; VEMAP, 1995). The species shifts from spruce to pine and deciduous species reported by Estonia and the Czech Republic are consistent with shifts from boreal to temperate forests. Finally, the shifts from cool temperate to warm temperate vegetation zones reported by Mongolia are consistent with results in IPCC (in press) and Smith et al. (1995).

Water Resources Sector

Of the participating countries, six included their work on effects of climate change on water resources in this report: Bangladesh, Kazakhstan, Mongolia, Czech Republic, Slovakia, and Ukraine. Of these, Kazakhstan, Mongolia, Czech Republic, and Ukraine report preliminary impacts.

The water resources assessment method recommended to participants in the U.S. Country Studies Program has three main components: (1) assessment of climate change impacts on the hydrologic resources of a country, (2) assessment of climate change impacts on the water resource management sector, and (3) assessment of the ability of the water management system to adapt to climate change impacts. The four reporting countries provide estimates of climate change impacts on their hydrologic resources obtained from four different, national water balance models. The four reporting countries have not yet completed their assessments of impacts on the water resources management sector and on the ability of the water management system to adapt to climate change impacts. Table S.4 summarizes the effects of climate change on annual runoff.

Most of the basins studied appear to be highly sensitive to climate change. Kazakhstan's two study basins show consistently lower annual runoffs and indicate that the distribution of monthly mean precipitation under climate change will probably

Table S.4. Summary of Water Resource Impacts

Country	Scenario	Basin	Runoff Change (% annual change)
Kazakhstan	CCCM, GFDL	Uba/Ulba	-23% to -29%
		Tobol	-23% to -26%
Mongolia	Incremental Scenarios,[1] GISS, GFDL, UKMO, CCCM	Bogd	-24% to +32%
		Bulgan	-53% to +79%
		Khalkhgol	-39% to +25%
Czech Republic	Incremental Scenarios,[2] GISS, GFDL, CCCM	Elbe, Zelivka, Upa, Metuje	-45% to + 15%[3]
Ukraine	GFDL, UKMO	Dnieper	-20% to +128%

[1] 2°C and 4°C warming with 0%, ±10%, and ±20% annual precipitation change.
[2] 2°C and 4°C warming with 0%, ±5%, and +10% annual precipitation change.
[3] Results not reported for individual basins.

reduce the risk of flooding events. Mongolia reports that its basins tended to be more sensitive to precipitation changes than to temperature changes. The Czech Republic found its low-to-medium runoff basins to be more sensitive to climate change than its larger basins. Ukraine's study of the Dnieper basin is particularly important because that basin provides about 80% of Ukraine's water supply. The analysis predicts a range of runoff changes from more than double the current runoff to a decline of 20%. Finally, although Slovakia has not completed its analysis, it noted that preliminary simulations through 2030 showed 20% decline in annual runoff.

Climate change is expected to have major impacts on regional water resources (IPCC, in press); however, for these country studies additional demand and supply analysis is needed to determine the socioeconomic effects of the predicted changes in runoff. Estimates of future baseline demand that include foreseeable adaptations to climate change in other sectors (e.g., energy, agriculture) are necessary to approximate the total expected shortfall or surplus. Preliminary demand estimates were reported by two countries. The Czech Republic reports preliminary estimates of baseline demand in 2075, and comments that agricultural irrigation demand may increase by as much as 40% due to climate change impacts. Ukraine compares simulated annual runoff with current demand levels for three stretches of the Dnieper River. In almost all instances, runoff exceeds demand by a large margin.

The supply and demand of water resources is quite sensitive to climate variability and is likely to be sensitive to climate change. Recent studies have found that water resources in all countries are sensitive to climate change, but whether runoff will increase or decrease is generally uncertain. The IPCC (in press) summarizes regional studies of climate change impacts on surface water and groundwater supplies, as well as water quality. Among the results for northern China, it notes that water basins located in the plains tend to be more sensitive to precipitation changes than basins located in mountain regions. This is similar to Mongolia's finding that the Bulgan basin, the lower portion of which is located in a lowland area with sandy soil, was more sensitive to climate change than the Bogd basin, which is located in a mountainous region.

Coastal Resources Sector

Eight of the 13 countries presenting their assessments in this report are studying the potential effects of sea level rise: The Gambia, Bangladesh, Sri Lanka, Estonia, Poland, Ukraine, Peru, and Venezuela. All except Bangladesh have also provided results (in some cases these results are preliminary).

These countries estimate sea level rise impacts based on several different scenarios of sea level rise for 2100. The two most common scenarios are a 0.5 m rise and a 1.0 m rise. Erosion and inundation impacts were estimated using a combination of geomorphological information and topographical maps. Coastline shifts for selected sea level rise scenarios were generated and land losses were estimated as the difference between present and predicted coastlines. Preliminary impacts were estimated for the entire coastline in The Gambia and Poland, and for selected vulnerable study sites for Sri Lanka, Estonia, Ukraine, Peru, and Venezuela. Table S.5 summarizes land loss by scenario for each country.

Table S.5. Summary of Coastal Resources Impacts

Country	Scenario (m)	Land Loss (km^2)	Other Impacts	Economic Damages	Adaptation Costs
The Gambia	0.2	5	N/A		US$3.9 million (7 km revetment)
	0.5	46			US$3.1 million (7 km seawall)
	1.0	92		US$217 million (land)	
	2.0	182			
Sri Lanka[1]	0.3	47	N/A	N/A	N/A
	1.0	106			
	2.0	197			
Estonia[2]	1.0	60	N/A	N/A	N/A
Poland	0.1 to 2.5	845 to 2,203	40,860 to 234,840 population at risk	US$10 billion to US$75 billion capital value at risk	US$2.3 billion to US$9.8 billion for full protection, US$0.01 billion to US$0.50 billion for annual maintenance
Ukraine[3]	0.5	32	N/A	N/A	N/A
	2.0	370			
Peru[4]	1.0	78.32	N/A	N/A	N/A
Venezuela[5]	0.5	85.54	N/A	US$153 million (land + buildings)	N/A
	1.0	124.74			

1 Estimate for southwest coast; combines erosion and inundation losses.
2 Estimate for two study sites only.
3 Estimate for wetland losses along the Black Sea coastline.
4 Estimate for selected vulnerable areas. Sea level rise of 0.3 m was also analyzed, but no results were reported.
5 Estimate for five vulnerable areas. Economic damages represent lost land values and building values for four study sites, excluding potential oil and gas industry losses.

The Gambia estimates that a sea level rise of 1.0 m will inundate coastal mangroves and could threaten the loss of the capital city of Banjul unless protective measures are undertaken. Other urban areas at risk include Gdańsk, Poland; Lima, Peru; and the Lake Maracaibo region of Venezuela (a key petroleum industry location). Recreational beaches or shoreside reserves and infrastructure would be lost in regions of most of the countries, including four of Venezuela's five study sites, both of Estonia's study sites, and Sri Lanka's southwest coast.

In addition to loss of land, Poland reports that between 40,860 and 234,840 people are living in areas classified as at risk from increased flooding, depending on the sea level rise scenario, and that between US$10 billion and US$75 billion in capital value is at risk under sea level rise scenarios from 0.1 m to 2.5 m. Poland's impact estimates are based on projections of future coastal development. The Gambia estimates that the value of lost land in the Banjul region from a 1.0 m sea level rise is US$217 million based on land loss estimates and market land values.

The results in Table S.5 can be compared with the sea level rise impacts reported in Nicholls and Leatherman (1995) and IPCC (in press), which generally assume a 1.0 m sea level rise and existing coastal development levels. For Poland, the land loss and at-risk population impacts of a 1.0 m rise from Nicholls and Leatherman (1995) fall within the impact ranges reported in Table S.5 for sea level rise scenarios of 0.1 m to 2.5 m. The national estimate of land loss for Venezuela is about 45 times greater than the land loss estimate for the case studies that is reported in Table S.5. However, the national damage values are of the same order of magnitude as the case study damages. If the study site results reported in Table S.5 provide better damage estimates than the national economic damage estimate reported in Nicholls and Leatherman (1995), then the national estimate probably understates damages by a significant amount.

CONCLUSIONS

The countries reporting in this document are at various stages in their assessments. A few of the reporting countries have just initiated the screening stage of their assessment, while others have completed the assessment of biophysical impacts for one or more sectors. Most of the country studies presented in this report have only begun their assessment of socioeconomic impacts and evaluation of adaptation options. Thus, while much work has been completed much work remains to be done. In addition, over 30 countries that have not reported their results here are conducting vulnerability and adaptation assessments with support from the U.S. Country Studies Program. Although the results presented in this report are preliminary and the results from many countries are still to come, a number of conclusions may be drawn:

Successful Use of Common Methodologies. Common methodologies for conducting vulnerability and adaptation assessments are widely available, and are being successfully applied by advisory groups of developing and transition country researchers to address a range of vulnerability sectors. The specific recommended methods and the U.S. Country Studies Program's primary approach build on the IPCC seven-step framework and are described in more detail in Benioff et al. (in press).

The countries participating in the program have adapted a large number of vulnerability and adaptation assessment tools to meet their unique needs. These countries have been able to develop climate change scenarios, use these scenarios and other data as input for impact assessment models, and generate impact assessment estimates for crops, livestock, forests, water resources, coastal resources, and other sectors. Participating countries have also been able to evaluate economic impacts and assess the feasibility of adapting to potential climate change. Based on the experiences of countries applying these methods, the U.S. Country Studies Program is currently assisting in the development of United Nations Environment Programme guidelines for vulnerability and adaptation assessments being prepared for IPCC.

Contribution to Scientific Understanding. This report presents results from many countries for which there was no prior research on the potential impacts of climate change or for which research has not yet been published in widely available documents. This work will not only fill gaps in the understanding of the potential impacts of climate change and the effectiveness of adaptation strategies, but will also help develop a consensus on appropriate methodologies and needs for refinement to currently available methodologies.

Sharing the information among researchers furthers the scientific understanding of both potential impacts and the usefulness of various methodologies. This document, as well as other publications, workshops, training, newsletters, and visiting analyst programs, promote the exchange of information between countries participating in the U.S. Country Studies Program.

Need for Capacity Building. Support for country studies is critical to strengthen human and institutional capacity to conduct vulnerability and adaptation assessments. However, scarcity of human and financial resources and of technical training continues to limit the ability of some countries to carry out vulnerability and adaptation assessments. For these countries, the efforts of international institutions to bring resources to these regions, improve communication, and build necessary capacity should improve the countries' abilities to conduct vulnerability and adaptation assessments. Given the complexities involved with vulnerability and adaptation assessments, in depth training and technical assistance should be a major component of such international efforts.

Development of Climate Change Plans. Impact assessments such as those reported in this document provide a basis for preparing national climate change plans, which will address adaptation measures. The process for developing climate change plans builds on assessments of vulnerability by focusing on adaptation measures for the most sensitive regions or populations within a country and by responding to the particular nature of the vulnerability. For example, the vulnerability assessment may identify the water resources in a particular region as being highly sensitive to flooding under potential climate change, thus this region could be a focus for adaptation planning and the adaptation policies considered would address the threat of flooding (with the timing and magnitude of such floods estimated in the vulnerability assessment).

Among the reasons for preparing such plans is that many measures that would be taken in response to climate change are "no regrets" measures which will produce benefits even if climate does not change, and many other measures will involve only marginal costs and be the only feasible option in anticipation of climate change. Additionally, these plans can serve as communications required by the Framework Convention on Climate Change. The U.S. Country Studies Program is providing assistance to a number of countries in preparation of national climate change plans.

Communication of Risks to Decision Makers and the Public. Country studies on vulnerability and adaptation assessments and reports such as this document play an important role in communicating the magnitude and direction of potential risks associated with climate change to decision makers and the public. Responding to climate change will require policy decisions at a number of different levels and across a variety of different sectors. Thus, a broad base of support will be necessary to comprehensively implement response measures. Effective communication of the risks associated with climate change in light of scientific uncertainties may be essential before decision makers and the public will support response measures.

REFERENCES

Benioff, R., S. Guill, and J. Lee (eds.) (In press). *Vulnerability and Adaptation Assessments: An International Guidebook.* Dordrecht, The Netherlands: Kluwer Academic Publishers.

Carter, T.R., M.L. Parry, H. Harasawa, and S. Nishioka. 1994. *IPCC Technical Guidelines for Assessing Climate Change Impacts and Adaptations*. London: Department of Geography, University College.

(IPCC) Intergovernmental Panel on Climate Change. (In press). *Scientific-Technical Analyses of Impacts, Adaptations, and Mitigation of Climate Change*. New York: Cambridge University Press.

Melillo, J.M., A.D. McGuire, D.W. Kicklighter, B. Moore, C.J. Vorosmarty, and A.L. Schloss. 1993. "Global Climate Change and Terrestrial Net Primary Productivity." *Nature* 363: 234-240.

Neilson, R.P. 1993. "Vegetation Redistribution: A Possible Biosphere Source of CO_2 during Climate Change." *Water, Air, and Soil Pollution* 70: 659-673.

Nicholls, R.J. and S.P. Leatherman. 1995. Potential Impacts of Accelerated Sea-Level Rise on Developing Countries. *Journal of Coastal Research* Special Issue No. 14, Spring.

Parry, M.L., T.R. Carter, and N.T. Konijn (eds.). 1988a. *The Impacts of Climate Variations on Agriculture: Volume 1: Assessments in Cool Temperate and Cold Regions*. Dordrecht: Kluwer Academic Publishers.

Parry, M.L., T.R. Carter, and N.T. Konijn (eds.). 1988b. *The Impacts of Climate Variations on Agriculture: Volume 2: Assessments in Semi-Arid Regions*. Dordrecht: Kluwer Academic Publishers.

Parry, M.L., M.B. de Rozari, A.L. Chong, and S. Panich. 1992. *The Potential Socio-economic Effects of Climate Change in Southeast Asia.* Nairobi: United Nations Environment Programme.

Qureshi, A. and D. Hobbie (eds.) 1994. *Climate Change in Asia: Thematic Overview.* Manila: Asian Development Bank.

Rosenzweig, C. and A. Iglesias (eds.). 1994. *Implications of Climate Change for International Agriculture: Crop Modeling Study.* EPA 230-B-94-003. Washington, DC: U.S. Environmental Protection Agency.

VEMAP Participants. 1995. "Vegetation/ Ecosystem Modeling and Analysis Project (VEMAP): Comparing Biogeography and Biogeochemistry Models in a Continental-Scale Study of Terrestrial Ecosystem Responses to Climate Change and CO_2 Doubling." *Global Biogeochemical Cycles* 9: 407-437.

Chapter 1
Introduction

The Framework Convention on Climate Change, which has been signed by more than 150 governments worldwide, calls on parties to the convention to inventory national sources and sinks of greenhouse gases (GHG) and to develop plans for responding to climate change. To assist developing countries and countries with economies in transition to meet this obligation, the U.S. Government is providing technical and financial support for the development of climate change studies.

Global climate change is an important issue for many reasons, including the significant impacts a change in the climate could have on the world economy and the world's environmental resources. Adaptation to the impacts of global climate change will be challenging because of the potential magnitude of the impacts and because the impacts often transcend national boundaries and traditional fields of study. Decision makers will not only be required to coordinate many national interests and information from many disciplines, but will also be required to make judgments that take into account the high degree of uncertainty associated with climate change impacts. These challenges are relevant for current and future policies that shape international rules and institutional arrangements.

The U.S. Country Studies Program is providing technical and financial assistance to 55 developing countries and countries with economies in transition to help them conduct studies of greenhouse gas emissions, mitigation options, and vulnerability and adaptation to climate change and help them prepare national climate change action plans. Each country forms a national committee to direct the study and identifies a lead agency or organization for the overall study and for each task. Researchers and analysts from a large number of organizations are responsible for conducting the vulnerability and adaptation assessment. The program is providing technical assistance to countries by providing opportunities to work with technical advisors, assisting in the publication of studies by countries, and by providing training and synthesis workshops.

This report presents preliminary results from the research of 13 of the countries that were ready to share their results. The countries contributing to this report from the African region are The Gambia and Zimbabwe; from the Asian-Pacific region are Bangladesh, Kazakhstan, Mongolia, and Sri Lanka; from the Central and Eastern European region are the Czech Republic, Estonia, Poland, Slovakia, and Ukraine; and from the Latin American region are Peru and Venezuela.

These countries have all undertaken research to assess their vulnerability to climate change and to evaluate various alternatives to adapt to the impacts of climate change. The U.S. Country Studies Program is sponsoring two rounds of studies. The first round of studies began in October 1993 and the second round of studies began in October 1994. The

J. B. Smith et al. (eds.), Vulnerability and Adaptation to Climate Change, 1–2.

countries submitting chapters for this interim report include countries that are participating in both rounds of the program. As such, the chapters vary widely in their scope and depth. Some of the chapters review the selection of particular methodologies or study sites, others provide results of vulnerability assessments from a number of different sectors, and others include preliminary adaptation assessments or economic analyses. In many cases, the countries chose to focus their chapter on the results from only one or two sectors, even though their study includes more sectors. This work has not necessarily been endorsed by the respective governments or the U.S. Country Studies Program. In a few cases, work supported from other sources is presented here.

The chapters included in this report represent the vulnerability and adaptation assessments prepared by the collaborative efforts of authors within each country. In addition, editors were selected from each of the four regions to prepare a chapter summarizing the results by region. All of the country chapters were reviewed and commented on by technical experts, peers in the field of climate change, and by the regional editors.

Chapter 2 of the report provides an overview of the methodologies that were recommended by the U.S. Country Studies Program and that have been used for much of the work presented in this report. This chapter is intended to be used as background describing the methods that may have been used by the countries and should be used as reference; it should be noted that many countries modified these methods or used alternative methods. Chapter 3 is a global overview that is organized by vulnerability sector. It compares the results from countries to recent literature in the field. Chapter 4, which is organized by region (Africa, Asia/Pacific, Central and Eastern Europe, Latin America), highlights significant issues in each region, summarizes results from each country that has contributed to this report, and compares the reported results to recent literature. The remaining chapters of the report are organized by region, and countries are presented alphabetically within each region.

This report is one in a series of synthesis publications that the U.S. Country Studies Program will issue. A second report synthesizing the final results of vulnerability and adaptation assessments by all participating countries is expected to be issued in 1997. It is believed that this report will not only contribute to the knowledge of climate change vulnerability and adaptation issues, but will also be useful for countries that are in the process of researching and documenting their work in this area.

Chapter 2

Methods for Vulnerability and Adaptation Assessment

Joel Smith,[1] Ana Iglesias,[2] Stephen Leatherman,[3] Kenneth Strzepek[4] Thomas Smith,[5] Barry Baker,[6] John Balbus,[7] Jay Malcolm[8] Ihor Hlohowskyj,[9] Cynthia Rosenzweig,[10] Jim Jones[8] Greg Kiker,[11] Joe J. Ritchie,[12] Ellen Hartig[13]

[1] Hagler Bailly Consulting, Inc., USA
[2] Instituto Nacional de Investigaciones Agrarias (INIA), Spain
[3] University of Maryland, USA
[4] University of Colorado, USA
[5] University of Virginia, USA
[6] Ecosystems Research International, USA
[7] George Washington University, USA
[8] University of Florida, USA
[9] Argonne National Laboratory, USA
[10] Columbia University, Goddard Institute for Space Studies, USA
[11] Cornell University, USA
[12] Michigan State University, USA
[13] Columbia University, USA

INTRODUCTION

The 55 countries participating in the U.S. Country Studies Program are using a variety of methods to assess their vulnerability to climate change and the potential for implementing adaptation options. Many of the countries are using models and methodologies provided by the U.S. Country Studies while others are using other methods and methodologies (Table 2.1). In all cases the countries are adapting the models and methodologies to their unique national circumstances. This chapter briefly describes those models and methodologies made available by the U.S. Country Studies Program. Readers will not find these models and methodologies described elsewhere in this report.

The U.S. Country Studies Program is providing technical assistance and training to countries participating in the program in assessing their vulnerability and adaptation strategies to climate change. Technical advisors were selected to assist with the climate and baseline scenarios, climate model data, and vulnerability and adaptation analytic methods in adaptation: agriculture, grassland/livestock, forests, water resources, coastal resources, fisheries, wildlife, and human health. These natural resources sectors are socioeconomically important and sensitive to climate change. All of these

J. B. Smith et al. (eds.), Vulnerability and Adaptation to Climate Change, 3–48.

Table 2.1. Summary of Countries Contributing to the Interim Vulnerability and Adaptation Assessment Synthesis Report

Region/Country	Sector	Scenario	Vulnerability Assessment Method	Progress
Africa				
The Gambia (Round 1)	Agriculture	CCCM, GFD3, GF01, GISS, UKMO	DSSAT 3	Biophysical results
	Coastal	SLR: 0.2 m, 0.5 m, 1.0 m, 2.0 m	Common methodology	Biophysical results
Zimbabwe (Round 1)	Agriculture	CCCM, GFD3	CERES-Maize	Biophysical results
Asia-Pacific				
Bangladesh (Round 2)	Agriculture	GFDL	CERES-Rice	Methods only
	Water	GFDL	MIKE11	Methods only
	Coastal	SLR	Common methodology	Methods only
	Forest	To be selected	Water salinity projections	Methods only
Kazakhstan (Round 1)	Agriculture	GFD3, GF01, CCCM	CERES-Wheat	Biophysical results
	Grassland	GFD3, GF01, CCCM	Dynamic Productivity Model	Biophysical results
	Water	GFD3, GF01, CCCM	KazNIGMI Model	Biophysical results
Mongolia (Round 1)	Agriculture	GISS, GFD3, Historical	CERES-Wheat (spring)	Biophysical results
	Water	GISS, GFD3, CCCM, Incremental	Basin Conceptual Model	Biophysical results
	Forest	UK89	Holdridge model	Biophysical results
Sri Lanka (Round 2)	Coastal	SLR: 0.3 m, 1.0 m, 2.0 m	Modified common methodology	Biophysical and socioeconomic results
Central/Eastern Europe				
Czech Republic (Round 1)	Agriculture	CCCM, GISS, Incremental	CERES-Wheat	Biophysical results
	Forestry	CCCM, GISS, Incremental	Forest Vegetation Zones	Biophysical results
	Water	CCCM, GISS, Incremental	BILAN, CLIRUN, SAC-SMA	Biophysical results
Estonia (Round 2)	Agriculture	GCMs to be selected	CERES-Barley, SPUR2	Preliminary biophysical results
	Forestry	Incremental	Gap models	Preliminary biophysical results
	Coastal	SLR: 1.0 m	Common methodology	Preliminary biophysical results
Poland (Round 1)	Coastal	SLR: 0.1 m to 2.5 m	Common methodology	Biophysical and socioeconomic results
Slovakia (Round 2)	Agriculture	CCCM, GISS, GFD3	DSSAT 3	Methods only
	Forestry	CCCM, GISS, GFD3	Holdridge model	Methods only
	Water	CCCM, GISS, GFD3	WATBAL	Methods only
Ukraine (Round 2)	Forestry	GFD3, GISS	Holdridge model	Preliminary biophysical results
	Water	GFD3, GF01, UKMO	National models	Preliminary biophysical results
	Coastal	SLR: 0.5 m, 1.0 m	Common methodology	Preliminary biophysical results
Latin America				
Venezuela (Round 1)	Forestry	GFD3, GISS, OSU, UKMO	Holdridge model	Biophysical results
	Coastal	SLR: 0.5 m and 1.0 m	Common methodology	Biophysical and Socioeconomic results
Peru (Round 1)	Coastal	SLR: 1.0 m	Common methodology	Biophysical results

sectors have been subject to past studies on climate change (IPCC, in press). The names and affiliations of the advisors for each of these areas are listed in Table 2.2.

The technical advisors provided training on specific methods for assessing vulnerability and adaptation through workshops, and in some cases through site visits. In addition, the methods for scenarios, agriculture, grasslands/livestock, forests, water resources, coastal resources, fisheries, wildlife, human health, and adaptations are described in more detail in *Vulnerability and Adaptation Assessments: An International Guidebook* (Benioff et al., in press).

The general approach recommended by the U.S. Country Studies Program for conducting vulnerability and adaptation assessments is presented in Figure 2.1. The assessments begin with selection of climate change scenarios and baseline socioeconomic (population, economic conditions) changes that will happen without climate change. The climate

Table 2.2. Vulnerability and Assessment Technical Advisors

Area	Advisors
Scenarios	Joel Smith, Hagler Bailly Consulting, Inc.
Climate Model Data	Roy Jenne, National Center for Atmospheric Research
Agriculture	Cynthia Rosenzweig, Columbia University/Goddard Institute for Space Studies
Grasslands/Livestock	Barry Baker, Ecosystems Research International
Forests	Thomas Smith, University of Virginia
Water Resources	Kenneth Strzepek, University of Colorado
Coastal Resources	Stephen Leatherman, University of Maryland
Fisheries	Ihor Hlohowskyj, Argonne National Laboratory
	Michael Brody, U.S. Environmental Protection Agency
Wildlife	Jay Malcolm, University of Florida
	Kirk LaGory, Argonne National Laboratory
	Adam Markham, World Wildlife Fund
Health	Jonathan Patz, Johns Hopkins University
	John Balbus, George Washington University
Adaptation	Joel Smith, Hagler Bailly Consulting, Inc.

Figure 2.1. U.S. Country Studies Program Primary Approach to Climate Change Vulnerability and Adaptation Assessments

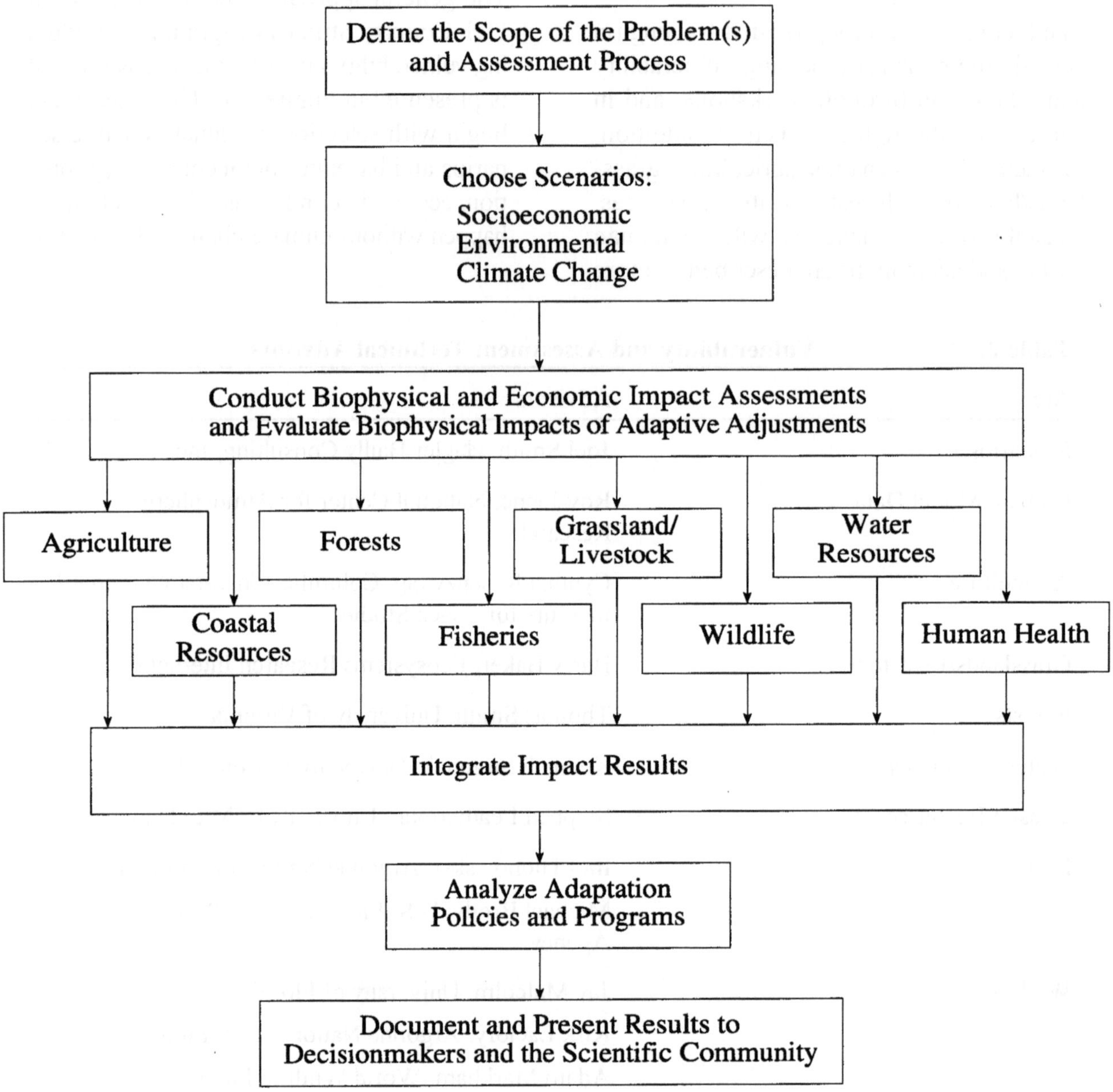

Source: Benioff et al., in press.

change scenarios are inputs for the biophysical and socioeconomic models or methodologies that are used to assess impacts of potential climate change. Most of these methods allow the incorporation of alternative policies to assess the impacts under different adaptation options that may be taken in response to climate change. The vulnerability of each sector is initially analyzed in isolation, and results may then be integrated across sectors to account for interactions among related sectors. For example, an agriculture assessment could incorporate changes in water supply from the water resources assessment. The adaptation assessment is used to evaluate which, if any, policy options may be implemented in anticipation of climate change to mitigate potential adverse climate change impacts.

This approach is consistent with the seven step analytical framework developed by the Intergovernmental Panel on Climate Change (IPCC) (Carter et al., 1994) and shown in Figure 2.2. In both approaches, the assessments begin with the definition of the scope of the problem and the assessment process. Steps 2 and 3 of the IPCC seven step framework are selecting and testing methods. Although these steps are not delineated in the primary approach outlined by the U.S. Country Studies Program., in many cases countries are presented with a choice of methods, and where appropriate, countries are encouraged to use their own models or approaches, which may provide for more thorough or accurate assessments. In addition, each of the methods made available by the U.S. Country Studies Program is accompanied by a

Figure 2.2. IPCC Seven Steps of Climate Impact Assessment

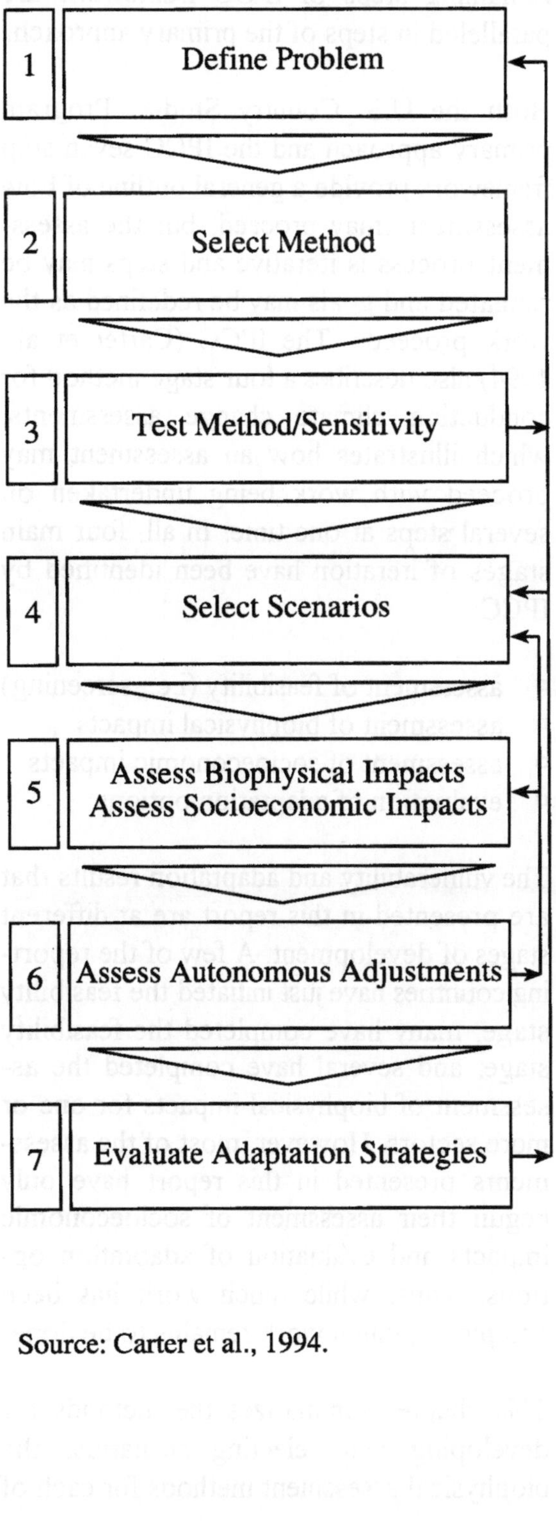

Source: Carter et al., 1994.

description of screening techniques and/or calibration and validation techniques to test the appropriateness of the methods. The remaining steps of IPCC framework are paralleled in steps of the primary approach.

Both the U.S. Country Studies Program primary approach and the IPCC seven step framework provide a general outline of how assessments may proceed, but the assessment process is iterative and steps may be repeated and goals may be redefined as the work proceeds. The IPCC (Carter et al., 1994) also describes a four stage method for conducting climate change assessments, which illustrates how an assessment may proceed with work being undertaken on several steps at one time. In all, four main stages of iteration have been identified by IPCC:

- assessment of feasibility (i.e., screening)
- assessment of biophysical impacts
- assessment of socioeconomic impacts
- evaluation of adaptation options.

The vulnerability and adaptation results that are presented in this report are at different stages of development. A few of the reporting countries have just initiated the feasibility stage, many have completed the feasibility stage, and several have completed the assessment of biophysical impacts for one or more sectors. However, most of the assessments presented in this report have only begun their assessment of socioeconomic impacts and evaluation of adaptation options. Thus, while much work has been completed, much work remains to be done.

This chapter summarizes the methods for developing and selecting scenarios, the biophysical assessment methods for each of eight vulnerability sectors, and the adaptation assessment techniques made available by the U.S. Country Studies Program. It is intended to introduce the methods and their limitations; it is not meant to be a comprehensive description. Readers who wish more information should consult the guidance document or other references in this chapter. There are many methods for assessing vulnerability and adaptation, and some countries used other approaches. For example, countries assessing water resources were trained to use a monthly water balance model called CLIRUN (Kaczmarek, 1993) for estimating changes in runoff. The Czech Republic, however, also used the BILAN and SACRAMENTO (Burnash et al., 1973) models to assess climate change impacts. To avoid repetition, the country chapters in this report describe only methods not described in this chapter.

SCENARIOS METHODS

There is significant uncertainty about how regional climate and economic conditions will change in the future. Although global climate is likely to warm, the direction and the magnitude of regional climatic changes over the next century are highly uncertain (Mitchell et al., 1990). With so much uncertainty about regional climate change, it is impossible to predict impacts with precision. In addition, socioeconomic conditions such as population, income, and technology, which in turn affect the sensitivity of sectors such as agriculture, are also likely to change (Leggett et al., 1992).

Studies of the effects of climate change usually focus on changes caused by a doubling of CO_2 levels ($2\times CO_2$) in the atmosphere (e.g., Tegart et al., 1990; Strzepek and Smith, 1995). The IPCC (Houghton et al., 1992) projected that if current emission trends remain unchanged, global temperatures will increase by 0.3°C per decade.[1] At that rate of warming, temperatures in 2075, when CO_2 levels are estimated to double, should be about 2.5°C warmer than present-day temperatures.

Vulnerability and adaptation assessments use scenarios to address the uncertainty associated with future climate. Scenarios are plausible combinations of circumstances that can be used to describe a future set of conditions.

Baseline Scenarios

Baseline Socioeconomic Scenarios

Socioeconomic conditions will most likely change significantly between now and 2075. Changes in such important variables as population, income, and technology may have a significant effect on the sensitivity of sectors to climate change. Baseline socioeconomic scenarios may be developed not to predict future socioeconomic conditions but to help measure how changes in these conditions could affect the sensitivity of sectors to climate change.

Countries that are preparing vulnerability and adaptation assessments are supplied with information on population and economic growth to assist them in preparing baseline socioeconomic scenarios. Researchers are encouraged to use expert judgment and statistics available locally in developing socioeconomic scenarios to be used in their adaptation assessments. The source of the information provided on population estimates for individual countries is Bos et al. (1994) and Bos et al. (1992). Information on regional economic growth rates, agricultural production, and deforestation is from Pepper et al. (1992) and Leggett et al. (1992). These estimates were developed by the IPCC to determine greenhouse gas emissions over the next century and the subsequent rate of climate change. If researchers use these estimates to develop baseline scenarios the regional growth rates must be adjusted to each country's circumstances.

Baseline Climate Scenarios

A baseline climate scenario may be created to examine the behavior of sectors under the current climate (e.g., crop yields under current climate). In addition, the baseline climate scenario may be compared with GCM estimates of $1{\times}CO_2$ climate conditions to select the GCMs used in the assessment.

Thirty years of observed climate data are recommended for developing the baseline climate scenario. A 30-year period is considered long enough to obtain a good representation of wet, dry, warm, or cool periods. The period 1951-80 is preferred for a baseline climate because it does not include the global warming trends of the 1980s (Carter et al., 1994). However, because the more recent 30-year period of 1961-90 may contain more complete and reliable information, many countries used it to develop their baseline climate scenario. If data in the 30-year period are missing, countries may use weather generators to fill in gaps (Benioff et al., in press).

[1] New projections by the IPCC (IPCC, 1995) are that global temperatures will increase by 0.2°C per decade.

Climate Change Scenarios

Climate change scenarios should be designed to:

- be consistent with broadly accepted predictions of climate change (e.g., Houghton et al., 1992)
- be physically plausible
- estimate a sufficient number of variables on a spatial and temporal scale for vulnerability assessment (Carter et al., 1994)
- reflect a reasonable range of potential regional climate change.

Sources of Climate Change Scenarios
Climate change scenarios for a country may be created using either estimates of regional climate change generated from general circulation models (GCMs), or a country may use incremental climate change scenarios. GCMs are mathematical representations of atmosphere, ocean, and land surface processes based on the laws of physics (Gates et al., 1990). Most GCMs have been run to estimate current climate ($1\times CO_2$) and climate resulting from a doubling of carbon dioxide levels in the atmosphere ($2\times CO_2$). In addition, some models have been run to estimate how climate may change over time assuming a gradual increase in greenhouse gas concentrations; these applications of GCMs are called transient runs. The advantage of GCMs is that they produce regional estimates of climate changes that are physically consistent with global temperature changes thought likely to happen. One of the disadvantages of using GCMs is that, in many areas, they do not accurately represent current climate.

Incremental climate change scenarios that combine temperature changes (+2°C and +4°C) with changes in precipitation (no change and ±10% and 20%) are useful for analyzing the sensitivity of sectors to a wide range of potential climate changes (such as hotter and wetter or hotter and drier scenarios) and for identifying sensitivities to changes in a specific variable, such as temperature or precipitation. The disadvantage of incremental scenarios is that uniform changes over an entire year or over large areas are not likely. GCMs and incremental scenarios can complement each other (Sulzman et al., 1995).

Selection of GCMs
GCMs vary in how well they simulate current climate. Each country should examine how well GCMs simulate current climate in their country and region. A global data set consisting of long-term means of observed data (referred to as CLIM) was provided to researchers by the National Center for Atmospheric Research (NCAR) (Möller, 1951; Taljaard et al., 1969; Crutcher and Meserve, 1970; Schutz and Gates, 1971) or observed country-specific climate data sets should be compared with GCM estimates of $1\times CO_2$ climate. Based on the results of this comparison, a country may select at least two or three GCMs that best represent current climate.

NCAR supplied GCM model output to countries participating in the U.S. Country Studies Program. Average monthly values are supplied for $1\times CO_2$, $2\times CO_2$, and the first, fourth, seventh, and tenth decades of

Table 2.3. Characteristics of GCMs Distributed to Country Studies Participants

GCM	When Calculated	Model Resolution	Δ T(°C)	% Change in Precipitation
GISS	1982	7.83° × 10.0°	4.2	11.0
OSU	1984-1985	4° × 5°	2.8	7.8
UKMO	1986	5° × 7.5°	5.2	15.0
GFD3	1989	2.22° × 3.75°	3.2	10
UK89	1989	2.50° × 3.75°	2.5	9.9
CCCM	1989	3.75° × 3.75°	3.5	3.8
GF01[1]	1991	4.44° × 7.50°	3.2[2]	7.0[2]

[1] The model is run for 100 simulated years assuming a 1% annual increase in atmospheric carbon dioxide concentrations.
[2] Transient results are changes between the first decade and the tenth decade of the transient.

the transient run. The models, their run date, spatial resolution, and estimates of changes in global temperature and precipitation are displayed in Table 2.3. The models are:

- Goddard Institute for Space Studies (GISS) (Hansen et al., 1983)
- Geophysical Fluid Dynamics Laboratory (GFDL or GFD3) (Mitchell et al., 1990)
- United Kingdom Meteorological Office (UK89) (Mitchell et al., 1990)[2]
- United Kingdom Meteorological Office (UKMO) (Wilson and Mitchell, 1987)[3]
- Canadian Climate Centre Model (CCCM) (Boer et al., 1992)
- Geophysical Fluid Dynamics Laboratory Transient (GF01) (Gates et al., 1992)
- Oregon State University (OSU) (Schlesinger and Zhao, 1988).[4]

Creation of Climate Change Scenarios

GCM climate change scenarios may be created by combining average monthly output from the models with observed climate data. For temperature, the difference between $2\times CO_2$ and $1\times CO_2$ is added to observed temperature; for precipitation and solar radiation, the ratio of $2\times CO_2$ and $1\times CO_2$ is multiplied by observed values. Thus, each day would be x° warmer or every precipitation event would have y% more or less precipitation. Incremental climate change scenarios may be created by combining the selected changes in temperature and percentage changes in rainfall with observed climate data in the same manner.

[2] The UK89 model run outputs were distributed to countries that joined the Country Studies Program in 1994 (Round 2 countries).

[3] The UKMO model run outputs were distributed to countries that joined the Country Studies Program in 1993 (Round 1 countries).

[4] This model was not widely distributed.

Sea Level Rise Scenarios

Sea level rise is a highly probable result of global warming. The sea level is likely to rise less than 1.0 m over the next century (Wigley and Raper, 1992). The U.S. Country Studies Program recommends that countries assessing sea level rise impacts examine a 0.2 m (the current rate) sea level rise and accelerated rates of sea level rise of 0.5 and 1.0 m. Local subsidence and uplift should be added to or subtracted from these sea level rise scenarios.

Atmospheric Carbon Dioxide Levels

Higher atmospheric concentrations of CO_2 can increase plant growth and water use efficiency. Thus higher CO_2 levels have the potential to offset the negative effects of climate change on vegetation. Models such as those used to estimate crop growth can account for higher atmospheric CO_2 levels. The levels of atmospheric CO_2 recommended for use in vulnerability and adaptation assessments are 355 parts per million (ppm) in 1990, 440 ppm in 2030, and 580 ppm in 2075 (Tom Wigley, NCAR, personal communication).

Limitations of Climate Change Scenarios

One of the major limitations in assessing vulnerability to climate change is uncertainty about regional climate change. None of the scenarios can be considered to be predictions of climate change.

One uncertainty is that the approach for creating GCM and incremental scenarios does not include spatial, interannual, or daily variability. GCM results are produced in grid boxes, which vary from $2.2° \times 3.75°$ of longitude and latitude to as much as $8° \times 10°$, and the approach does not indicate how climate change might vary within the grid boxes. This may be a particular problem along coastal areas and in mountains because changes in climate would probably not be uniform in these areas. Since monthly averages from the GCMs are used, the scenarios do not consider changes in year-to-year variability or day-to-day variability. Yet, not only are mean conditions likely to change, but also the variance is likely to change. A critical uncertainty, for example, is whether increased precipitation would result in more intense precipitation events, more precipitation days, or both. Such changes in variance could have a significant effect on such phenomena as droughts, storms, and floods.

Agriculture Methods

Despite technological advances such as improved crop varieties and irrigation systems, weather and climate are still key factors in agricultural productivity. Much research has been conducted in regional, national, and international assessments of the potential effects of climate change on agriculture (see, for example, Houghton et al., 1992; Rosenzweig and Parry, 1994; IPCC, 1995). Techniques for evaluating crop productivity impacts have also been developed (see Rosenzweig and Iglesias, 1994).

Figure 2.3. Flow Diagram for the Agriculture Sector Assessment

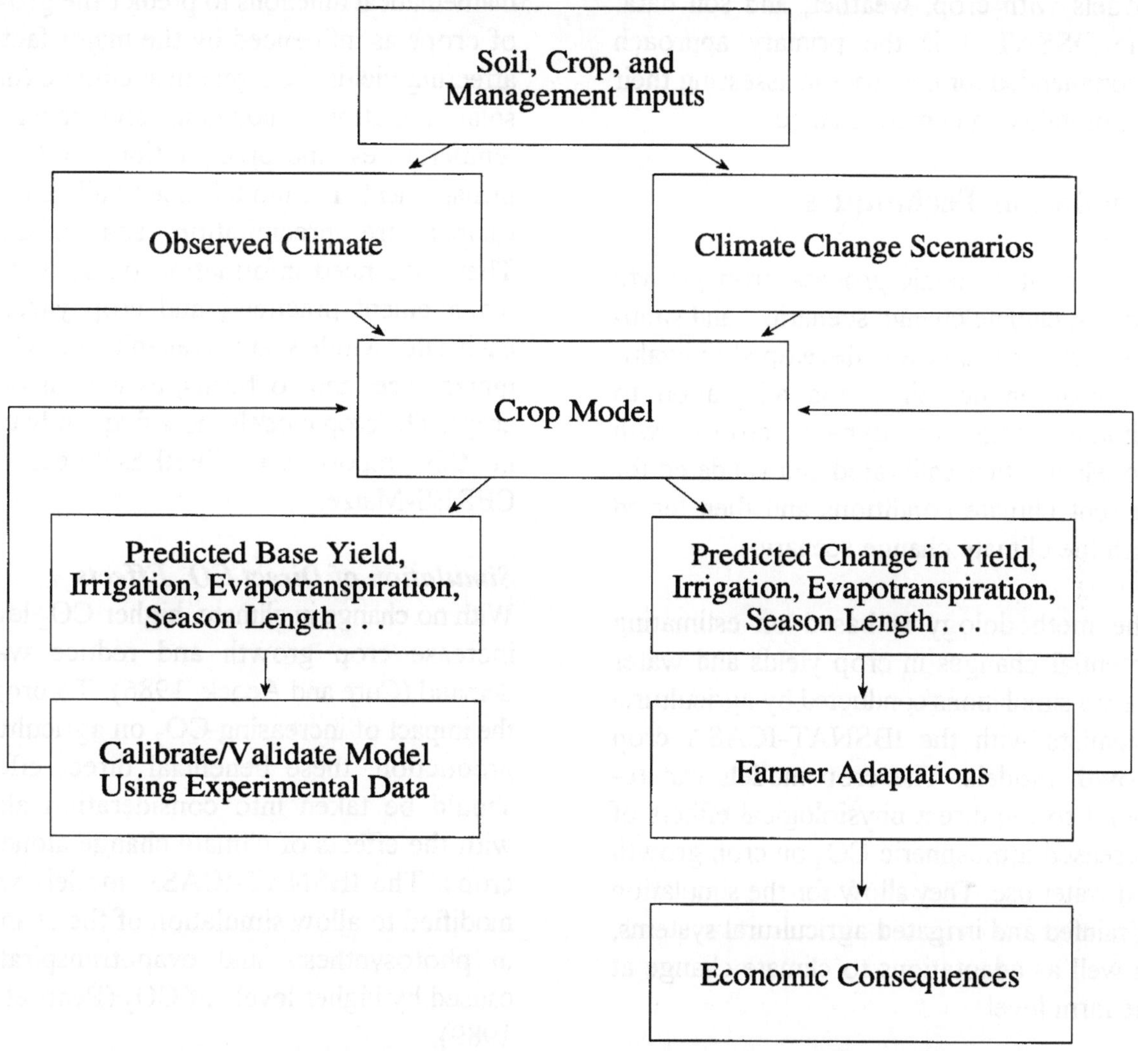

The general approach for analyzing agricultural vulnerability and adaptation to climate change is to use simulation techniques (see Figure 2.3). Simulation techniques such as crop growth models allow for a comprehensive evaluation of crop vulnerability and adaptation to climate change. Simulation permits the results to be integrated across sectors and makes it possible to compare results to other impact studies.

The International Benchmark Sites Network for Agrotechnology Transfer — International Consortium for Application of Systems Approaches to Agriculture (IBSNAT-ICASA) allows for the evaluation and dissemination of appropriate crop models and climate change impact methodologies so that participants can use site-specific information to identify climate change vulnerabilities. The IBSNAT method includes the Decision Support System for

Agrotechnology Transfer (DSSAT 3), a software system that integrates crop growth models with crop, weather, and soil data. The DSSAT 3 is the primary approach recommended for countries in assessing their vulnerability to climate change.

Simulation Techniques

A suite of dynamic-process crop growth models, climate change scenarios, and simulation experiments was developed to evaluate crop vulnerability and adaptation to climate change. The dynamic crop growth models are first calibrated and validated for current climate conditions and then tested with the climate change scenarios.

The methodology is based on estimating potential changes in crop yields and water use via simulations conducted by agricultural scientists with the IBSNAT-ICASA crop growth models. The crop models can respond to the direct physiological effects of increased atmospheric CO_2 on crop growth and water use. They allow for the simulation of rainfed and irrigated agricultural systems, as well as adaptations to climate change at the farm level.

The IBSNAT-ICASA models were selected for use because they have been validated over a wide range of environments (e.g., Otter-Nacke et al., 1986) and are not specific to any particular location or soil type. Thus they are suitable for use in international studies in which crop growing conditions vary greatly. Furthermore, because management practices can be varied in the models, they permit experiments that simulate crop management adaptations to climate change.

Crop Models

The IBSNAT-ICASA models use simplified mathematical functions to predict the growth of crops as influenced by the major factors affecting yields, i.e., genetics; climate (daily solar radiation, maximum and minimum temperatures, and precipitation); soils; and management. The models need daily data on temperature, precipitation, and radiation. They also need information on soils, crop management practices, and crop yields at each site. Models are available for wheat, maize, rice, and soybeans, as well as other crops. The crop models most frequently used in this report are CERES-Wheat and CERES-Maize.

Simulation of Direct CO_2 Effects

With no change in climate, higher CO_2 levels increase crop growth and reduce water demand (Cure and Acock, 1986). To project the impact of increasing CO_2 on agricultural production, these beneficial direct effects should be taken into consideration along with the effects of climate change alone on crops. The IBSNAT-ICASA models were modified to allow simulation of the changes in photosynthesis and evapotranspiration caused by higher levels of CO_2 (Peart et al., 1989).

Calibration and Validation of Crop Models

The validation of the IBSNAT-ICASA crop models using local experimental data from field trials is essential for assuring the credibility of the results. It is important that the crop models be validated to the fullest extent possible. Nevertheless, since the IBSNAT-ICASA crop models were created with the express purpose of applying them across a wide range of environments, validation based on any previous studies in each particular region should also be considered.

Spatial Estimates of Crop Changes

Crop model site results can be scaled up to reflect changes in regional and national crop yields. The results of the crop simulation study at different sites can thus be extended to broader agricultural regions based on agronomic characteristics, on similarities in agroecological zones, and on previous climate change impact studies.

Limitations

The limitations of the methodology for climate change scenarios are described above, but it is important to emphasize again that changes in climate variability are not incorporated into the scenarios. Changes in climate extremes for instance may affect agricultural production more than changes in mean climate variables (Mearns et al., 1992).

The IBSNAT-ICASA models contain many simple, empirically derived relationships that do not completely mimic actual plant processes. These relationships may or may not hold under differing climatic conditions, particularly the higher temperatures predicted for global warming. For example, most of the data used to derive the relationships in the crop models were obtained with temperatures below 35°C, whereas the projected temperatures under doubled CO_2 conditions are often 35°C or even 40°C during the growing period.

Other simplifications of the crop models are that weeds, diseases, and insect pests are assumed to be controlled and they do not include poor soil conditions. The crop models also do not include potential improvements in agricultural technology, although they may be used to test the effects of some potential improvements such as improved varieties and irrigation schedules.

The assessment of the relative contribution of the direct effects of CO_2 and predicted climate change on agricultural crops contains many uncertainties. The agronomic adaptation simulations have to be considered within the framework of all economic possibilities, as well as the environmental consequences of the proposed changes.

Economic Analyses and Adaptation Strategies

Once crop yield changes are estimated, it is important to examine how the market (to the extent there is one) may react to climate change and how society's welfare will be affected. Markets may shift production to areas that become relatively more productive. These economic measures are one component of the information that policy makers need in order to understand how climate change affects people. Via probability distribution, the IBSNAT-ICASA crop simulation methodology provides the primary source of information for evaluating differential changes in crop production and consumption opportunities. Methods for further economic evaluation should be derived with the assistance of resource economists.

Farmers will try to adapt to changing environmental conditions. The IBSNAT-ICASA crop models allow us to test crop management alternatives so that the effectiveness of different adaptation techniques can be evaluated for their ability to mitigate the negative effects of climate change and take advantage of the positive effects. The adaptations include irrigation and changes in fertilizers, crop varieties, crop cultivation calendars, and crop cultivation areas. The evaluation of the results of the adaptation simulation should include the assessment techniques described in the adaptation section of this chapter.

GRASSLANDS/LIVESTOCK METHODS

As precipitation patterns change, grasslands productivity and area could change. Livestock is likely to be affected directly by changes in temperature as well as indirectly by changes in grassland productivity and pests. SPUR2 (Hanson et al., 1992) provides the level of detail required for assessing the effects of climate change on grassland ecosystems. SPUR2 has been used to simulate the effects of climate change on grassland ecosystem processes and cattle production for U.S. rangelands (Baker, 1991; Baker et al., 1993; Hanson et al., 1993). The mechanistic, process-oriented structure of SPUR2 makes it well suited for examining interactions between management decisions and climatic influences on short-term ecological processes and for evaluating possible adaptive management strategies. Also, SPUR2 is the only grassland model available at the temporal and spatial resolutions required for this type of analysis that incorporates a process-oriented livestock model.

The primary recommended approach for analyzing grassland/livestock impacts for the Country Studies Program is biophysical simulation modeling using the SPUR2 grassland/livestock model.

The simulation modeling techniques should be used to gain an understanding of the effects of climatic change on ecological and biophysical processes.

SPUR2 Model

SPUR2 is a general grassland simulation model composed of five basic components: plant growth, soils/hydrology, domestic animals, wild animals, and grasshopper submodels. The model is driven by daily inputs of precipitation, maximum and minimum temperatures, solar radiation, and wind run.

The plant submodel simulates the cycling of carbon and nitrogen through several compartments, including standing green, standing dead, live roots, dead roots, seeds, litter, and soil organic matter. Competition between plant species for nutrients and the impact of grazing on vegetation are also simulated. The direct effects of increased atmospheric concentrations of CO_2 on the net photosynthetic rate are simulated assuming that net photosynthesis will increase by 35% as CO_2 concentration doubles.

The soils/hydrology component of the model calculates upland surface runoff volumes, peak flow, snow melt, upland sediment yield, and channel stream flow and sediment yield. Soil-water tensions, used to control various aspects of plant growth, are generated by using a soil-water balance equation. Surface runoff is estimated by the Soil Conservation Service curve number procedure (Hanson et al., 1980), and soil loss is computed by the modified universal soil loss equation. The snow melt routine uses an empirical relationship between air temperature and the energy flux of the snow pack.

There are two domestic herbivore submodels in SPUR2, the Colorado Beef Cattle Production Model (CBCPM) and a generalized beef steer model. CBCPM is a herd-wide, life cycle simulation model and operates at the level of the individual animal (Baker, 1991). The biological routines of CBCPM simulate animal growth, fertility, pregnancy, calving, death, and demand for nutrients. The second livestock model, the steer model, was adapted from the TAMU Beef Simulation Model and incorporated into the original SPUR model (Wight and Skiles, 1987). The model simulates the growth of a young animal approximately 12 to 24 months of age during a single grazing season. Intake of grazed forage for both CBCPM and the steer model is calculated by FORAGE, a deterministic model that interfaces with SPUR2 (Baker et al., 1992). The model is driven by weight from the animal growth curve, the animal's demand for grazed forage, and the quantity and quality of forage available for each time step of the simulation.

Data Requirements for the SPUR Model

To conduct the simulations, 30 years of continuous historical weather data are needed for each simulation site. Weather data requirements include daily measurements of maximum and minimum temperature (°C), precipitation (mm), wind run (km/day), and solar radiation (langleys). Data for the hydrology model include soil name and texture; slope; percentage of sand, clay, and silt; organic matter; parameters for universal soil loss equations; soil evaporation; bulk density; and water-holding capacity.

Plant model initial conditions include biomass estimates for green shoots, live roots, propagules, standing dead, dead roots, litter, and soil organic matter, as well as an estimate of the amount of nitrogen (g/m^2) in green shoots, live roots, propagules, standing dead, dead roots, litter, soil organic nitrogen, and soil inorganic nitrogen.

Data for the animal models include animal weight, estimate of genetic potential for gain, milk production, birth weight, weaning weight, yearling weight, mature weight, estimates of lifetime growth curve weights, reproductive capacity, and forage intake.

Limitations

The major limitation in using a biophysical simulation approach for this problem is that complete data sets for parameterizing the model rarely exist. To overcome this obstacle, default parameter values can be used initially. In addition, the model used in this approach is a site specific model, which requires making simplifying assumptions when results are aggregated to the regional level. The most limiting assumptions of this analysis are that management practices will remain constant over the simulated period and that the region of aggregation is homogenous. In reality, management strategies, microclimate, vegetation structure, and soil type may be quite heterogeneous. Therefore, some care must be exercised when describing discrete regions.

Economic Analyses and Adaptation Strategies

Economic analyses should be conducted to examine (1) the effects of climatic change on the local or producer level, (2) the effects on secondary supply and demand within a region or country, and (3) the effects on world supply and demand for meat and animal products produced from grasslands. Adaptive management strategies to be tested depend on the current and accepted management strategies within a country and on the type of livestock being studied. However, some general adaptations include spatially shifting the use of the grazing resource, shifting the genotype of grazing animals, altering the species of animals used for grazing, and using other country-dependent alternative strategies. These options can be tested using the SPUR2 model.

FOREST METHODS

Exploring the potential response of forest ecosystems to climate change is a complex topic related to a variety of plant and ecosystems processes. The range of methods for addressing these processes is likewise diverse. The objective of the methodology proposed for the U.S. Country Studies Program was to provide a framework for evaluating the potential impacts of a climate change on the distribution, composition, and productivity of forested ecosystems. Although a wide variety of computer models have been developed for specific forested regions, the purpose of the approach outlined here was to provide a methodology that could be applied to the wide variety of forested ecosystems represented by the countries participating in the program (see Figure 2.4).

Assessment Techniques

For a model to be of use in evaluating the potential response of vegetation to climate change, it must provide an explicit statement of the relationship between vegetation pattern and climate. Three methods were suggested: (1) examination of simple climate indices that are related to forest patterns, (2) the Holdridge Life Zone Classification, and (3) forest gap models.

Climate Indices and Forest Patterns

The broad-scale patterns of forest distribution and productivity have been shown to be strongly influenced by temperature and rainfall (Walter, 1985). Perhaps the simplest approach to evaluating the potential response of plant distribution to changes in climate is to examine the correlation between the current distribution of forest types (or species) and simple climate indices. Certain indices have been shown to be related to the distribution and productivity of forests: growing degree-days (the sum of daily average temperatures above some defined minimum/threshold value); absolute minimum temperature; and moisture deficit (the difference between potential evapotranspiration and precipitation). For example, the distribution of subtropical forests dominated by broadleaf evergreen species has been shown to be limited by the occurrence of frost (minimum temperatures below 0°C) (Woodward, 1987).

This method involves correlating the current distribution or productivity of forests with features of the regional climate. Potential changes in distribution are evaluated using the new climate conditions predicted by the climate change scenarios.

Holdridge Life Zone Classification

A more formal approach to examining the correlation between climate and the broad-scale distribution of terrestrial vegetation is the use of bioclimatic models or classifications. Bioclimatic models correlate multiple features of the climate with the current

Figure 2.4. Flow Diagram for the Forest Sector Assessment

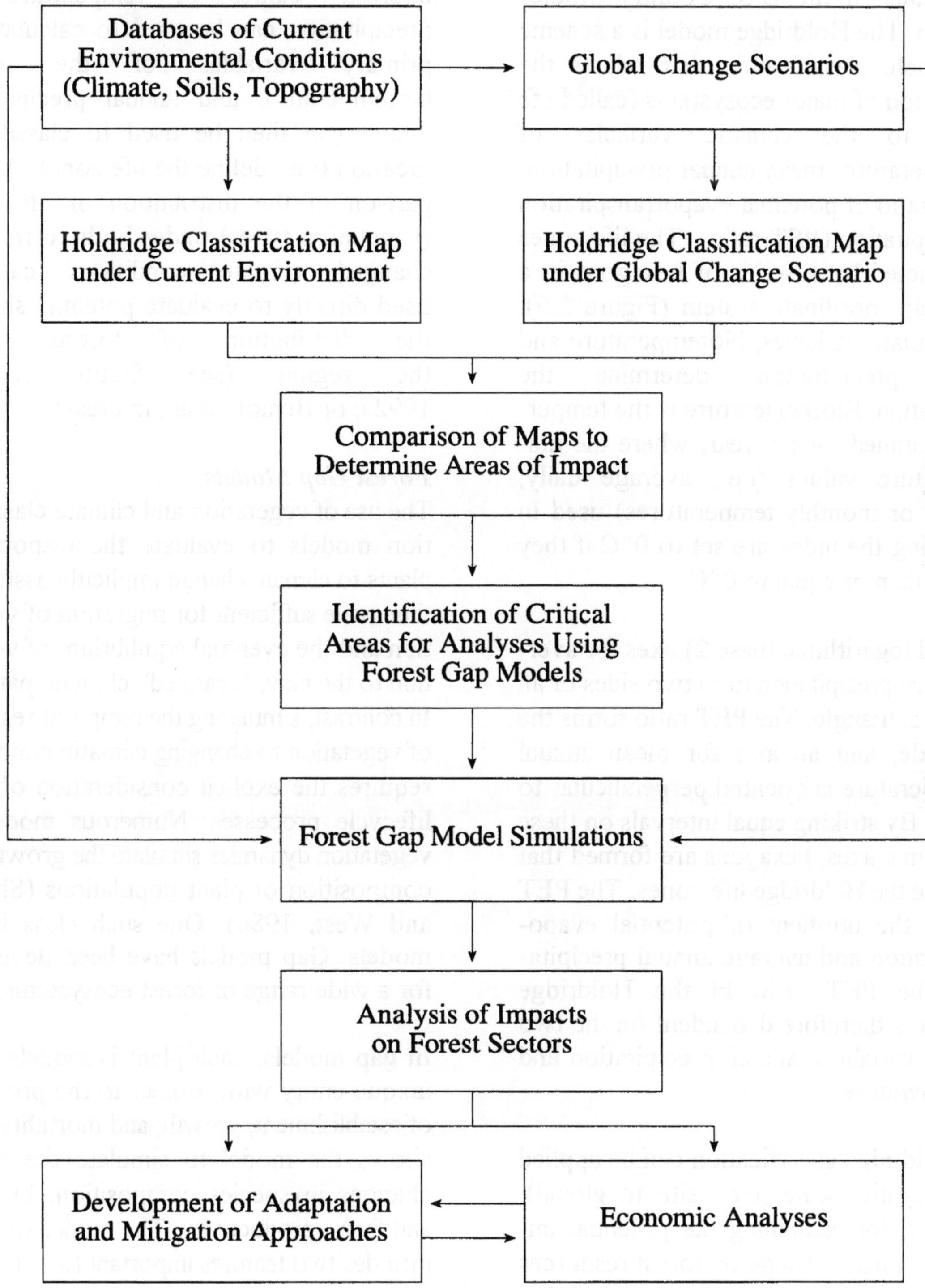

distribution of vegetation. One such model, the Holdridge Life Zone Classification (Holdridge, 1967), was suggested for use by participants in the U.S. Country Studies Program. The Holdridge model is a scheme for climate classification that relates the distribution of major ecosystems (called life zones) to the climatic variables of biotemperature, mean annual precipitation, and the ratio of potential evapotranspiration to precipitation (PET ratio). The life zones are depicted by a series of hexagons in a triangular coordinate system (Figure 2.5). Two climate variables, biotemperature and annual precipitation, determine the classification. Biotemperature is the temperature summed over a year, where the unit temperature values (i.e., average daily, weekly, or monthly temperatures) used in computing the index are set to 0°C if they are less than or equal to 0°C.

Identical logarithmic (base 2) axes for average annual precipitation form two sides of an equilateral triangle. The PET ratio forms the third side, and an axis for mean annual biotemperature is oriented perpendicular to its base. By striking equal intervals on these logarithmic axes, hexagons are formed that designate the Holdridge life zones. The PET ratio is the quotient of potential evapotranspiration and average annual precipitation. The PET ratio in the Holdridge diagram is therefore dependent on the two primary variables, annual precipitation and biotemperature.

The Holdridge classification can be applied at any spatial scale (e.g., site to global). However, for examining the potential impacts of climate change on forest resources and for subsequently developing adaptation or mitigation strategies, the model should be used in a spatial context to provide maps of potential land cover for the area (region) of interest. Values of temperature and precipitation can be used to calculate the primary variables of the model, biotemperature and annual precipitation, which can then be used to classify the location (i.e., define the life zone). A comparison of the distribution of life zones (vegetation types) under both current and changed climate conditions can be used directly to evaluate potential shifts in the distribution of forests within the region (see Smith et al., 1992a, or Benioff et al., in press).

Forest Gap Models

The use of vegetation and climate classification models to evaluate the response of plants to climate change implicitly assumes a time scale sufficient for migration of vegetation and the eventual equilibrium of vegetation to the new, "changed" climatic patterns. In contrast, simulating the temporal response of vegetation to changing climatic conditions requires the explicit consideration of plant lifecycle processes. Numerous models of vegetation dynamics simulate the growth and composition of plant populations (Shugart and West, 1980). One such class is gap models. Gap models have been developed for a wide range of forest ecosystems.

In gap models, each plant is modeled as a unique entity with respect to the processes of establishment, growth, and mortality. This allows the model to simulate the annual changes in species composition, biomass, and forest productivity. The model structure includes two features important to a dynamic

Figure 2.5. Holdridge Life Zone Classification Model

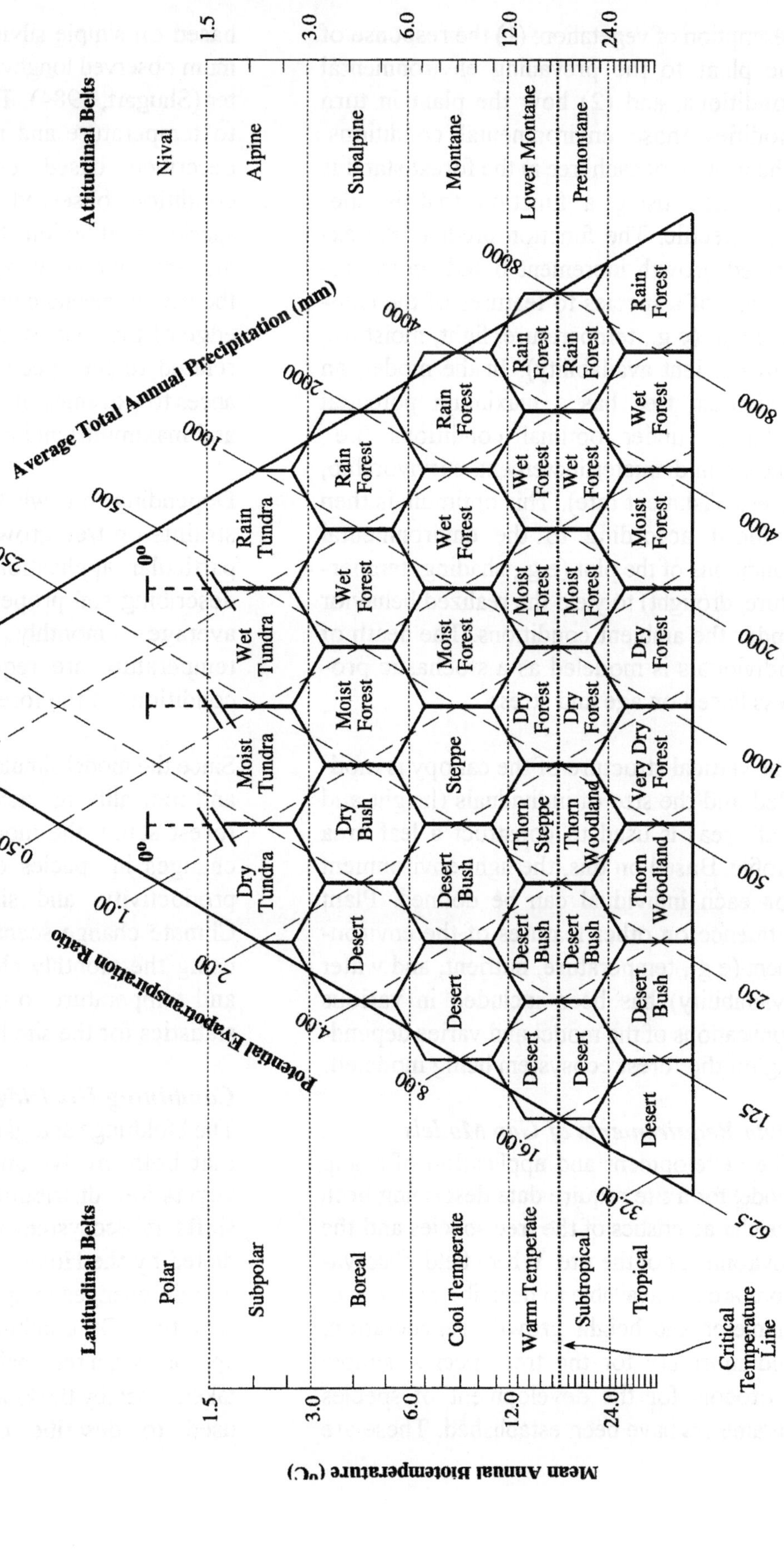

Source: Holdridge, 1967.

description of vegetation: (1) the response of the plant to the prevailing environmental conditions, and (2) how the plant in turn modifies those environmental conditions. The growth of each tree in the forest stand is calculated using a function that is species-specific. The function predicts the expected growth increment based on the response of a species to features of the environment (e.g., temperature, light, moisture, and nutrient availability). In the model, an individual tree has a maximum potential behavior under optimal conditions (i.e., maximum diameter increment, survivorship, or establishment rate). This optimum is then reduced according to the environmental conditions of the plot (e.g., shading, temperature, drought) to yield the realized behavior under the ambient conditions. The death of individuals is modeled as a stochastic process based on age and stress.

The vertical structure of the canopy is modeled and the size of individuals (height and leaf area) is used to construct a leaf area profile. Based on this, the light environment for each individual can be defined. Plant influence on other features of the environment (e.g., temperature, nutrient, and water availability) has been included in various applications of the model and varies depending on the forest ecosystem being modeled.

Data Requirements of Gap Models

The development and application of a gap model for a site require data describing both the characteristics of the tree species and the environment of the site. Where field observations are not available to describe patterns of diameter and height growth, regeneration, and mortality for the tree species, simple protocols for the development of species parameters have been established. These are based on simple silvicultural data on maximum observed longevity, height, and diameter (Shugart, 1984). Tree species' responses to temperature and moisture are generally described based on the environmental conditions observed at the boundary of the species' distribution. This approach of defining environmental responses assumes that the environmental conditions observed at the edge of the species' geographical range are related to the species' physiological tolerances to extremes of temperature (minimum and maximum) and aridity.

Depending on what environmental constraints on tree growth are included in the particular application of the model, data describing soil properties, topography, and average monthly precipitation and temperature are required to describe the conditions of the forest stand.

Since the model simulates the annual growth and mortality of each individual tree in the forest stand, the model can predict annual changes in species composition, biomass, productivity, and size class distribution. Climate change scenarios are evaluated by using the monthly changes in precipitation and temperature to modify current climate statistics for the site being simulated.

Combining Holdridge and Gap Models

The Holdridge and gap models are similar in that both involve an explicit link between vegetation distribution and climate. The shifts in ecosystem type or life zone predicted by the Holdridge classification imply certain changes in species composition and structure. By combining the two modeling approaches, a regional analysis can be undertaken whereby the Holdridge classification is used to describe broad scale shifts in ecosystems, and gap models are used to examine both the transient dynamics associated with the predicted shifts as well as possible changes in composition and structure within a given ecosystem or life zone.

This combined approach can be achieved in two steps. The Holdridge classification can be used to examine regional patterns of vegetation/forest distribution under both current climate and climate change scenarios. The results of this analysis can then be used to identify areas of potential forest impact, such as forest decline. These areas are used as focal sites for the development and application of forest gap models to examine the potential changes in forest composition and productivity. In this way, the Holdridge classification can be used as a screening technique for identifying vulnerable areas in a given scenario.

Limitations

The limited number of categories of life zone or biome types in the Holdridge model results in a coarse resolution of vegetative description. In reality, vegetative pattern (i.e., physiognomic structure, species composition, and biomass) varies within any one life zone or classification unit. The approach also assumes that the vegetation (life zone) moves as a fixed unit in time and space (i.e., all species within a vegetation type/life zone respond to changes in climate in the same manner). Species are more likely to migrate at different rates, resulting in new assemblages (IPCC, in press).

The Holdridge model, like all models for climate vegetation classification, is correlative and is based on a limited set of variables. Although the bioclimatic indices used in the classification may do a sufficient job of bounding present vegetative patterns, the actual patterns are a function of additional factors not explicitly considered in the model (e.g., soils), which may vary differently (both temporally and spatially) under the changed climatic conditions. The Holdridge model does not consider how changes in water-use efficiency resulting from increased atmospheric concentrations of CO_2 may influence the distribution of vegetation, particularly in the case of decreasing soil moisture.

Perhaps most important, the approach represents an equilibrium solution for both climate (i.e., $2\times CO_2$) and vegetation dynamics. Although changes in the climatic pattern as suggested by the GCM simulations may occur on a time scale of decades to a century, the response of vegetation and soils to those changes may occur at different and varying time scales. In areas where biomass values decrease because of moisture stress (i.e., a higher PET ratio), the changes may occur quickly as the environmental conditions become such that the present vegetation can no longer be supported (e.g., a shift from forest to grassland). In contrast, increases in biomass may require much longer periods of time. Major shifts of forest type (e.g., warm temperate to tropical rain forest, or boreal to cool temperate forest) are dependent on the movement of species across the landscape and the ability of new species to invade existing communities. For many forest species, this time scale may be centuries (Davis, 1989).

In contrast to the Holdridge model, a gap model can predict species composition, the structure and associated productivity of vegetation, and standing biomass through time. Gap models, however, are limited in the spatial extent to which the results can be extrapolated because the models are site-specific. Using these models to provide total coverage over broad regions would be virtually impossible because of computational and data limitations. As an alternative, sampling approaches have been constructed to provide large-scale coverage over broad environmental gradients (Solomon, 1986; Bonan and Hayden, 1990; Bonan et al., 1990; Smith et al., 1992b).

Perhaps the greatest limitation to the gap model approach is in the development of parameters describing the response of tree species (e.g., growth, establishment, and mortality) to environmental conditions. Because of the general lack of data from field and laboratory experiments relating the response of tree species to gradients of temperature, moisture, and mineral nutrients, functions relating the response of species to these environmental features must be derived from data on the species' current geographic distribution. This approach assumes that the species can survive only under the environmental conditions associated with its current geographic distribution. If this assumption is not true, then the predictions of the model under changing climate conditions may be too sensitive.

One environmental factor not generally included in gap models is the direct response of tree species to increasing atmospheric concentrations of CO_2. Higher levels of CO_2 could make trees less sensitive to reductions in soil moisture and could increase biomass (Bazzaz and Fajes, 1992). The failure to include plant response to atmospheric CO_2 concentrations is largely due to the lack of data on this factor. Where data are available, the response of tree species of CO_2 can be directly incorporated into the model framework.

Economic Analyses and Adaptation Strategies

Using the Holdridge model, the development of land cover maps for current climate and climate change scenarios allows for the direct evaluation of impacts for given regions of interest. In addition, land cover maps make it possible to examine overall changes in land use suitability for the region assuming changed environmental conditions (e.g., the identification of areas that are currently not suitable for the production of forestry but may become suitable under a given climate change scenario). The identification of these areas could help in developing land use policies or management plans aimed at offsetting declines in forest production resulting from climate change in other areas.

The forest gap model allows for a more detailed analysis of changes in forest productivity and composition for given forested sites. This second approach provides detailed analyses of impacts on the production of timber, pulp, and other forest products. In addition, the individual-based nature of the model makes it possible to evaluate specific management scenarios (e.g., selective harvesting or planting of species) as well as adaptation and mitigation options (e.g., the introduction of species from other regions).

Water Resources Methods

In many countries, development and population growth are accompanied by increasing demand for improved water supply and sanitation. Massive capital expenditures are needed to develop the required infrastructure. With planning and construction times of 20 to 30 years or more and lifetimes of half a century or more for major water development projects, the primary question asked by many water resources managers in transition and developing countries is how climate change might affect the design of new water resource infrastructures.

The water resources assessment method recommended to participants in the U.S. Country Studies Program has three main components: (1) assessment of climate change impacts on the hydrologic resources of a country, (2) assessment of climate change impacts on the water resource management sector, and (3) assessment of the ability of the water management system to adapt to climate change impacts (see Figure 2.6).

Hydrologic and Water System Assessment Techniques

Hydrologic Resources Assessment

The hydrologic assessment, the first step of the water resources vulnerability and adaptation assessment, is the most developed methodology. To undertake this stage of the assessment, hydrologic models that estimate runoff (e.g., river flow) are used. Four classifications or methodologies for modeling hydrologic processes were identified by Todini (1988). In increasing order of data needs, these approaches can be identified as (1) stochastic models, (2) lumped integral models, (3) distributed integral models, and (4) distributed differential models.

Stochastic Models. This approach develops relationships that describe an output variable such as runoff in terms of input variables like precipitation and temperature without prescribing the physical processes that occur.

Lumped Integral Models. This approach uses the fewest parameters that can describe a basin's response to climatological events. These models are designed to look at medium-to-large watershed areas, and are often called water balance models. These models usually use monthly mean values of temperature and precipitation. The catchment or subcatchment is modeled as a single, homogenous unit subject to uniform events and parameters.

Distributed Integral Model. In this approach, a catchment is divided into subbasins and spatial heterogeneity is taken into account, giving a more realistic representation of the actual catchment. These models attempt to give physical meaning to model parameters. This type of model is really an elaboration of the lumped integral model. This is typically run with daily values. This type of model includes the STANFORD, SACRAMENTO, and National Weather Service models (Todini, 1988).

Distributed Differential Model. This is the most sophisticated of the modeling methods and is generally limited to the laboratory. Here catchment behavior is represented in differential form in both space and time. Mass and momentum equations are developed for each subsystem and are linked by matching boundary conditions at each time step (Todini, 1988). There are no known

Figure 2.6. Flow Diagram for Water Resources Sector Assessment

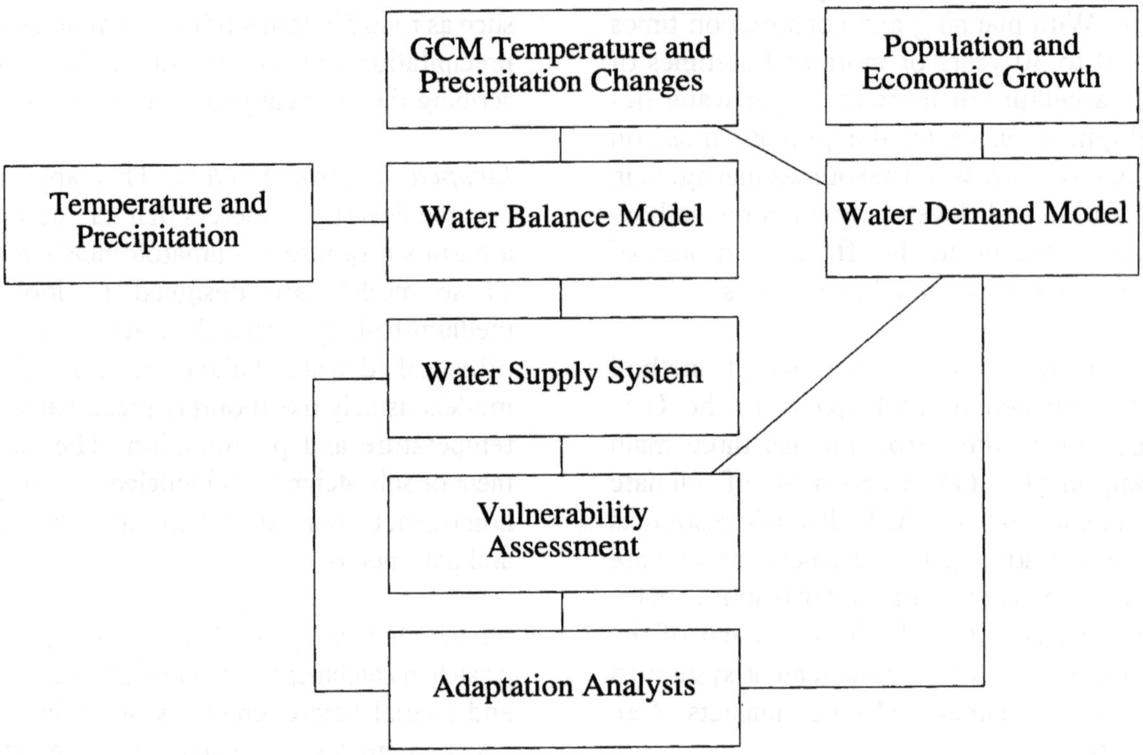

applications of this model type to assess the impact of climate change on basin discharge.

A wide variety of hydrologic modeling experience, spanning all of Todini's classifications as well as a great range of available data, exists among the countries performing water resources vulnerability and adaptation assessments. To accommodate this diversity of analytical capabilities, two approaches are recommended: the stochastic and the lumped integral modeling approach.

For countries with limited streamflow data, an annual stochastic model is recommended. This approach uses only mean annual precipitation and mean annual temperature to predict basin runoff. Although not very accurate for modeling current runoff, it performs adequately in predicting the change in basin runoff in response to changes in temperature and precipitation. Thus it provides a good indication of a river basin's sensitivity to climate change.

For countries with sufficient streamflow and climate data, a monthly lumped integral model or water balance approach is recommended. The models recommended are CLIRUN (Kaczmarek, 1993) which is a stand-alone model, and WATBAL (Yates, in press), which extends the CLIRUN approach. WATBAL has compared favorably to distributed integral models for assessing the impacts of climate change on a river basin's mean monthly runoff.

CLIRUN is a stand-alone monthly water balance model. This hydrologic model requires users to provide data on both effective precipitation and potential evapotranspiration. WATBAL was developed as a framework for the CLIRUN modeling approach. It extended the CLIRUN approach into a climate impact assessment tool for studying river basin response to climate change (Yates, 1994). The uniqueness of this lumped conceptual model to represent the water balance is characterized by the use of continuous functions of relative storage to represent surface outflow, subsurface outflow, and evapotranspiration. In this modeling approach, storage is lumped as a single conceptualized "bucket" (Figure 2.7). The model contains five parameters: (1) direct runoff, (2) surface runoff, (3) subsurface runoff, (4) maximum catchment water-holding capacity, and (5) base flow.

Countries with extensive hydrologic modeling experience and existing distributed integral and distributed differential models are encouraged to use these models, but also are encouraged to use the water balance approach to allow for more meaningful comparisons across countries and regions.

Water Resource Management Systems Assessment

Analyzing the reservoir systems of a river basin is one of the most important components of assessing the vulnerability of basin water resource management systems. There are two main classes of river basin management models: optimization models and simulation models. Both are very useful for assessing climate change impacts. A monthly simulation approach is recommended as the most likely to be successful given the variation in data availability. As in the hydrologic assessment, any country with existing capabilities more advanced than the recommended ones is encouraged to use them in addition to the recommended methodologies.

Countries conducting vulnerability and adaptation assessments are provided with the monthly river basin simulation model IRIS. IRIS was jointly developed by the International Institute for Applied Systems Analysis (IIASA) and Cornell University and has been used on many river basin studies throughout the world, including some assessments of climate change (Salewicz, 1995). This tool models reservoirs and their operation, hydropower generation, and water demand such as agricultural, municipal, industrial, environmental, and recreational demand. A monthly time step is used in this model. In addition, the model can directly use the results of the hydrologic assessment of runoff changes as input.

Limitations

All the hydrologic modeling techniques require the selection of a time step. For most vulnerability and adaptation assessments, data are lumped on a monthly basis. Monthly models might not capture the true response of the basin to precipitation events distributed throughout the month, so it is important to understand what kind of error is introduced when lumping temporally. For example, if data are provided daily and then lumped uniformly over the month, information that provides insight into basin response to storm events such as flooding can be lost. The month's total monthly precipitation could occur during one storm. If precipitation is applied uniformly over the month, true soil moisture dynamics might not be captured using the monthly time step.

Figure 2.7. Conceptual Framework for Linking Predicted Climate Changes to Environmental Conditions

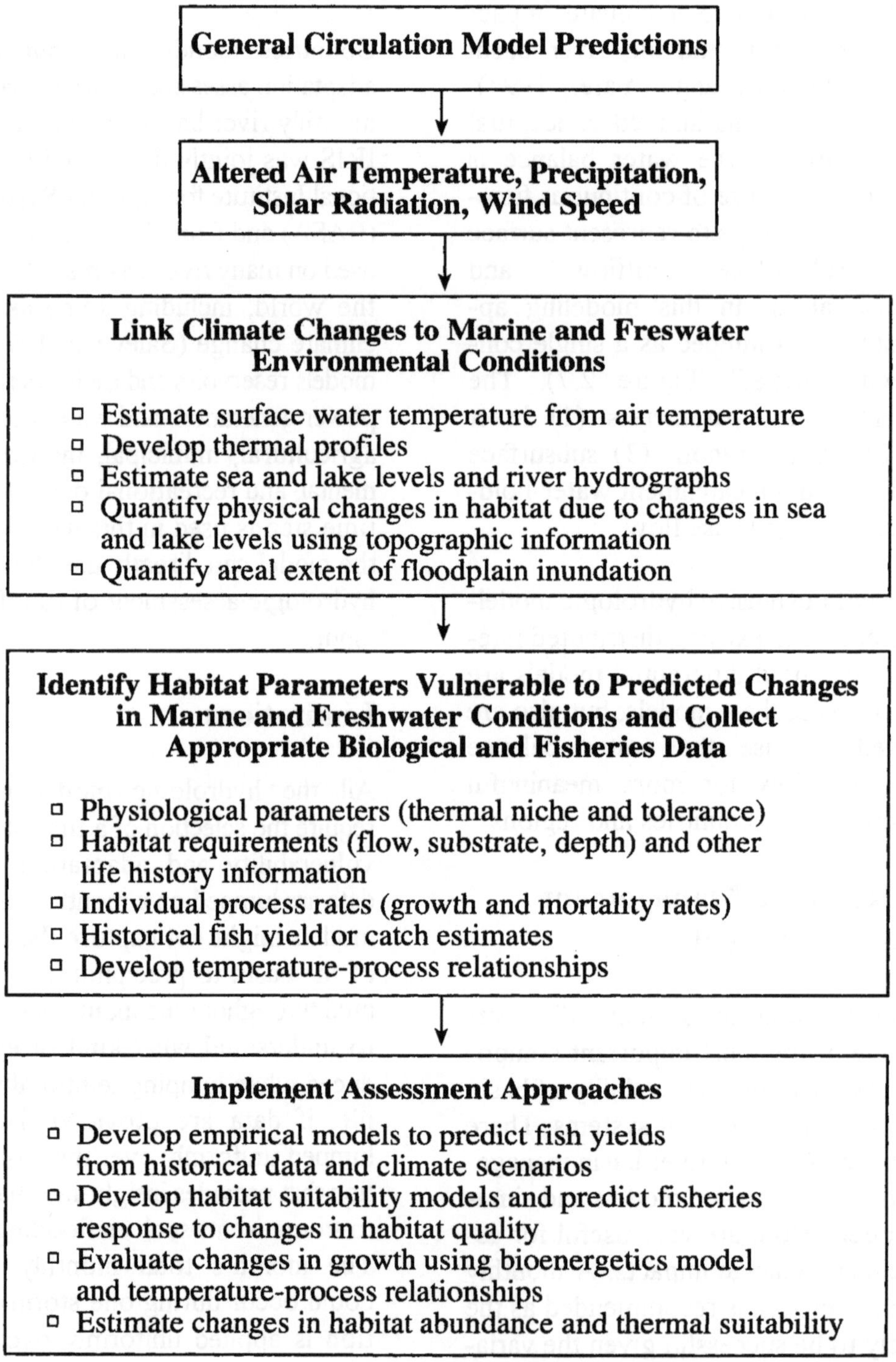

Vulnerability and adaptation assessments that use annual approaches (stochastic models) are also limited by the processes of lumping data. Climate changes may result in significant changes in the seasonal distributions of these parameters occurring during the year. This issue cannot be addressed using the annual approaches.

Adaptation Strategies

The technological, economic, and policy adaptations to climate change available to each country will differ greatly, depending on the hydroclimatic zone, the level of economic development, and the relative sensitivity of the water resources system to potential climatic change. The same tool recommended for assessing the vulnerability of river basin water resources systems, the IRIS model, is also well-suited for assessing adaptations. Structural adaptations such as new reservoirs, canal linings, and groundwater extraction can be analyzed with this simulation model. In addition, operational adaptation, changed allocation priorities, and pricing structures can be evaluated. This tool focuses mainly on the technological aspects of water resources adaptation.

Coastal Resources Methods

The principal impact of sea level rise for most coastal areas is land loss through submergence of lowlands and loss of wetlands, as well as erosion of beaches. Coastal environments are very diverse (e.g., from coral reef atolls to volcanic islands). Consequently, no single model can be used to determine the impacts of global climatic change. However, all approaches to land loss estimates must use detailed, up-to-date information that provides the following types of data:

- geomorphic type (e.g., sandy beach, marshy delta, rocky headland)
- relative elevation (e.g., ability to estimate 1.0 m and 2.0 m contour locations)
- land use (e.g., agricultural areas versus tourist beaches)
- affected populations (e.g., estimated by number of domiciles in impacted areas).

Common Methodology

The IPCC (1990) formalized a seven-step procedure for analyzing coastline vulnerability to accelerated sea level rise. The procedure was termed the "common methodology" (Figure 2.8). The common methodology is a comprehensive stepwise approach for assessing vulnerability to accelerated sea level rise. It also assesses physical, ecological, and socioeconomic impacts in coastal zones. The common methodology uses scenarios for global changes, national development, and local response options.

Aerial Videotape-Assisted Vulnerability Analysis

Maps never contain the richness and diversity of information required by this type of assessment. Only aerial imagery can provide the needed detail, but conventional vertical aerial photography is often too small in scale and too expensive to acquire and analyze.

Figure 2.8. IPCC Common Methodology Steps

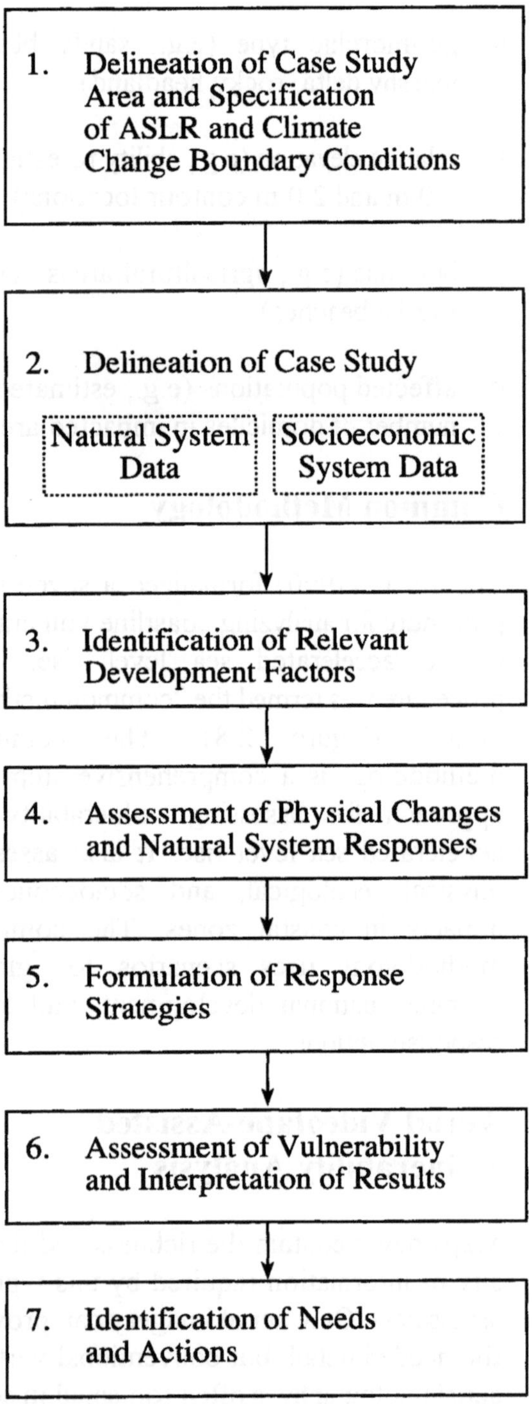

Therefore, other approaches must be used that offer simplicity in acquisition at a much lower cost yet yield reasonable estimates of coastal land loss.

A new technique called aerial videotape-assisted vulnerability analysis (AVVA) circumvents these problems and thus plays an important part in the Country Studies Program approach to vulnerability and adaptation assessment of coastal resources (Leatherman et al., 1995). The technique uses detailed field data to identify land and infrastructures that are at risk and determines protection costs for a range of response options. The AVVA approach involves three data collection activities:

- aerial video recording of the coastline
- limited ground-truth information
- archival research in the country.

The AVVA technique addresses the same problems as and is consistent with the IPCC common methodology (IPCC, 1990). Specifically, AVVA supports step 2 of the common methodology, which is the delineation of the case study area through the collection of natural system data and socioeconomic system data. The physical data, combined with relevant socioeconomic information, help identify relevant development factors and, combined with land loss estimates, provide a quantitative assessment of the expected physical changes and the vulnerability of the natural system (steps 3-6 of the common methodology). Thus, the IPCC common methodology, AVVA, and economic impact and adaptation analyses constitute the full primary approach to coastal impact and adaptation assessment.

The field work procedures for the AVVA method are as follows:

- divide the coast into a working geomorphic classification using published information and expert knowledge and obtain the best maps
- videotape the coastline at low and (sometimes) high altitudes of about 70-100 m and 300 m, respectively, including a recorded commentary and accurate positioning
- view the videotapes and update the geomorphic classification
- visit as many representative coastal types as possible and collect topographic data and information concerning land and property values, agriculture, future plans for development, and so on.

The first stage of the analysis consists of using the video record to develop an inventory of the coastal zone, including coastal geomorphology, coastal land use and development, and estimates of coastal elevation. Thus, the coastline is classified into sections with similar characteristics, which are then considered to be homogeneous units.

The video record is important to the analysis for the following reasons: (1) infrastructure losses are estimated directly from the video record, as well as any available maps, by overlaying the predicted recession and estimating the number of buildings that would be destroyed; and (2) the lengths of coastline requiring protection are measured from the video record.

AVVA has been constructed as a series of modules, so that individual elements of the overall procedure can be improved while the remaining elements of the approach remain robust. Therefore, as understanding of the elements improves, AVVA can easily and rapidly be repeated to generate new and improved estimates of the impacts of a sea level rise.

Estimation of Land Loss

Land loss is estimated using two different methods depending on the shore type.

- *Bruun Rule.* For sandy beaches and erodible bluffs along open coast, the Bruun rule (Hands, 1983; Leatherman, 1991) is applied. The Bruun rule estimates erosion on sand beaches in response to sea level rise based on beach and wave characteristics. The following formula is used:

  $$R = G \times S \times [L/(B + h)]$$

  where:

  R = shoreline recession due to sea level rise S
  G = overfill ratio
  S = sea level rise (projected)
  L = active profile width from dune to depth of closure
  B = dune or erodible bluff height
  h = depth of closure.

- *Drowning Concept.* For coastal wetlands and nonoceanic coastal lowlands, the direct inundation or "drowning" concept is applied. This concept is based on simple inundation and assumes no erosion. A sea level rise of 0.5 m per century results in a loss of wetlands. The area inundated by sea level rise is assumed to be all wetlands. Some wetland migration is assumed where uplands are undeveloped and have a low slope.

Limitations

Two important problems can emerge in preparing these assessments: quantifying land loss and estimating the cost of response measures. The Bruun rule can be used to estimate land loss due to beach erosion on sandy coasts. However, uncertainties require that both high and low erosion estimates be constructed. The cost of response is also difficult to assess because of a lack of experience with the range of options.

Inundation is most significant in deltas and in wetlands around estuaries. For these low-lying areas, the video record is insufficient for estimating elevation and must be combined with available maps plus expert judgment. The video record may be used to check the validity of the maps and help define the present extent of wetlands and mangroves.

Economic Analysis and Adaptation Strategies

The ultimate goal of any economic analysis of future sea level rise is to compute time trajectories of true economic cost in the context of future social, cultural, institutional, and economic development within threatened coastal zones. It is necessary to project time series of the future values of all economic activities that integrally depend on the physical characteristics of coastal zones and to judge their inherent vulnerability to sea level rise along specific scenarios. The time series will have to incorporate any socioeconomic adaptation that might be expected in response to projected vulnerability, and will have to be cast against the costs of more proactive protective strategies to determine which activities should be preserved, and to what extent, and which should be abandoned.

Options for adapting to sea level rise can be divided into four broad categories[5]:

- no protection (also called "retreat")
- present protection maintained
- developed areas protected
- total protection.

Fisheries Resources Methods

The vulnerability of fisheries to climate change depends on the nature of the climate change, the nature of the fishery, and its species and habitats. Changes in climatic conditions such as air temperature and precipitation affect fisheries by altering habitat availability or quality. Specifically, fisheries habitats may be affected by changes in (1) water temperature; (2) the timing and duration of extreme temperature conditions; (3) the magnitude and pattern of annual stream flows; and (4) surface water elevations and shorelines of lakes, reservoirs, and nearshore marine environments (Carpenter et al., 1992).

[5] IPCC (1990) refers to three options: retreat, accommodation (e.g., elevating structures), and protection.

Although climatic changes will affect most species to some degree, some species may be particularly vulnerable. In the event of climate-related reductions in habitat quality or quantity, fishes in isolated habitats or drainage basin habitats would be unable to migrate to potentially more suitable habitats in other basins. In riverine systems, many species rely on inundated floodplains for reproduction, nursery habitats, and successful recruitment. The life cycles of these species are strongly dependent on the temporal cycle of rainy and dry seasons (Welcomme, 1985; Lowe-McConnell, 1987). Thus these species are particularly vulnerable to changes in the timing, magnitude, and duration of rainy or dry seasons.

Available Methods

Because of the wide range of physical habitat responses to climate change and the diversity of the potentially affected fisheries, no one assessment approach is available to adequately evaluate the potential impacts of climate change on fisheries resources. This discussion only briefly mentions the available assessment options. Benioff et al. (in press) should be consulted for more details about these methods. The assessment methods identified for use in the U.S. Country Studies Program follow four approaches:

- estimating changes in the availability of thermal habitats by examining changes in the thermal structure of lakes and streams
- estimating effects of temperature changes on physiological processes, particularly growth and feeding
- estimating impacts of changes in physical habitat features (e.g., flow rates, water depth) on critical life history stages such as migration period and spawning time
- using empirically derived models to examine the relationships between fishery resources and environmental variables, such as temperature and precipitation.

The methods for assessing climate change impacts on fisheries resources include approaches that are readily straightforward to implement, do not require extensive computer equipment or specialized analytical instrumentation, can be applied to a variety of species and habitats, and can be completed in a timely manner.

Linking Climate Change Predictions to Ecological Responses

Although GCMs provide predictions of air temperatures, as well as the magnitude, duration, and distribution of precipitation, they do not provide direct information on water quality or hydrologic parameters that affect fisheries resources. Thus, fisheries vulnerability assessments must link predicted atmospheric climate changes to changes in the environmental conditions of aquatic habitats (Meisner et al., 1987; Christie and Regier, 1988; Kennedy, 1990). Many of the assessment methods require linking GCM predictions of air temperature and precipitation to changes in water temperature, hydrology, and sea level. Yet other methods use the GCM outputs directly. Only when these linkages are made can the ecological and biological response (such as growth rates, reproductive success, mortality, and distribution) of fisheries resources to climate change be identified and evaluated. The linkage of GCM outputs with the aquatic environment requires technical input from research on other vulnerability sectors, particularly those addressing water resources and coastal resources.

Assessing the Effects of Climate Change on Fishes

The assessment of climate change impacts on fisheries resources relies on a weight-of-evidence approach. In this approach, multiple lines of evidence are used to identify the nature of potential impacts and evaluate their significance. The methods also represent a suite of approaches that can be tailored for a particular country, habitat type, climate variable, and fishery resource.

The assessment methods target different aspects of the fisheries resources. Some methods evaluate the effects on habitat availability and suitability, while others evaluate the effects of climate change on growth, feeding, mortality, and timing of life history parameters. Still other methods evaluate the effects of climatic conditions on annual catch or yield. The suggested methods do not address such ecologically important factors as nutrient cycling and primary productivity, eutrophication and other water quality issues, community structure and function, predator/prey interactions, or fishing methods, success, and pressure.

The methods are represented by three categories of predictive models:

- *Bioenergetics models.* These models target fishes inhabiting lacustrine habitats and permits predictions of fish growth and biomass under different temperature scenarios (Hill and Magnuson, 1990).
- *Habitat suitability models.* These models target both freshwater and marine fisheries resources and permit evaluations of changes in the suitability of particular habitats relative to climate-driven changes in environmental conditions.
- *Empirical models.* Empirical models are used to predict a particular response by the fishery to a change in a particular climate-related variable or set of variables. For example, empirical models have been developed that permit predictions of natural mortality in fish stock under different annual water temperatures (Pauly, 1980) and also permit predictions of annual fish catch under different precipitation-driven river discharge conditions (Sagua, 1993).

Specific methods to evaluate freshwater fisheries resources and coastal marine penaeid shrimp resources are available. These methods evaluate the potential effects of a single climate-affected variable, namely water temperature, stream flow, or lake or sea level, and may or may not be species-specific. Other methods employ species-specific bioenergetics, habitat suitability, and natural mortality approaches; can be used in either freshwater or marine habitats; and can simultaneously evaluate multiple climate-affected variables. The choice of methods is dependent on the fishery of concern, the nature of the habitat for that fishery, the availability of environmental and ecological data, and the nature of the predicted change in climate.

Limitations

Environmental, biological, and technical issues, as well as time, effort, and cost constraints, largely rule out laboratory and field studies to generate new data. Consequently, assessments must rely largely on existing available data. Ecological, physiological, hydrological, and fisheries yield data are often limited, particularly in developing countries. In the absence of extensive, long-term, experimental field and laboratory investigations, the methods identified provide the best predictions possible within the limitations of the methods. Since it may be difficult to validate any particular method, it is recommended that as many of the suggested methods as possible be implemented to adequately evaluate impacts to fisheries.

The species-specific approaches will most likely be the most problematic to employ, largely due to the relative absence of species data on life history and physiology. In the absence of species-specific data, professional judgment may be used to estimate life history and bioenergetics variables. Difficulties associated with developing empirical models for predicting catch from environmental conditions will be largely associated with the absence of hydrological and climatological data.

As previously discussed, the overall approach for identifying potential impacts from climate change to fisheries resources follows a "weight-of-evidence" approach. This approach relies on multiple lines of evidence to evaluate the potential for adverse or beneficial impacts to fisheries from climate change. However, the methods identified in this report are not fully integrated and could give conflicting results or, more likely, results in the same direction but of differing magnitudes. For example, estimates of mortality may indicate adverse impacts under a particular climate-temperature scenario, whereas estimates of maximum sustainable yield may indicate a positive impact under the same temperature scenario. The evaluation of multiple results may rely heavily on the professional judgment of the fishery biologists performing the vulnerability assessment. It should be remembered that the results of the fisheries vulnerability assessment are not intended to provide a quantitative, definitive identification of the nature and magnitude of impacts to fisheries resources that would occur for a particular change in climate. Rather, the assessment is meant to provide an indication of the potential of adverse impacts to fisheries resources, and provide a preliminary indication if, or which, fisheries resources may be at risk.

WILDLIFE METHODS

Ecosystems and their wildlife populations are complex entities whose structure and function are intimately determined by climatic conditions. Responses to climate change involve patterns and processes over a wide range of temporal and spatial scales, from the nearly instantaneous physiological responses of individual organisms, to broad-scale geographic shifts in populations and biomes over decades and centuries, to changes in genetic makeup over millennia. Methods to assess the vulnerability of plant and animal populations and the potential for adaptive management must address the diverse range of population and ecosystem responses. Two methods — climatic correlation and habitat suitability — are recommended to U.S. Country Study Program participants.

Assessment Techniques

Climatic Correlation

The climatic correlation technique (CLIMCORR; Markham and Malcolm, in press) capitalizes on the fact that the geographic distribution of many species, communities, ecosystems, and biomes is strongly correlated with climate. In some cases, a geographic limit may actually be caused by climatic thresholds. In others, climate may correlate with other factors that themselves determine the geographic limit (MacArthur, 1972; Root, 1988a, 1988b). CLIMCORR helps identify univariate or multivariate associations between the baseline climate and the geographical wildlife element of interest and subsequently use these associations to predict the distribution of the wildlife element under some future climate scenario.

A series of menu options allow a user to (1) map various climate variables, including monthly rainfall and temperature averaged over the year and over different seasons, as well as various measures of extremes and variability in precipitation and temperature; (2) map bioclimatic indices, including Holdridge life zones (Holdridge, 1947) and Bailey Moisture Provinces (Bailey, 1979); (3) map differences in Holdridge life zones between climate data sets; (4) examine associations between current climate and the geographic distribution of a species or other naturally occurring element; and (5) plot "climate profiles" that describe the predicted distributions (see Bennet et al., 1991) of species or other elements under baseline and future climate conditions.

Habitat Suitability Indices

Habitat Suitability Indices (HSI) is a software package developed by the U.S. Fish and Wildlife Service that allows the user to create simple models to quantify the capacity of an area to support a species (U.S. Fish and Wildlife Service, 1981). The suitability of an area is a function of the species' requirements for food, shelter, reproduction, etc., and other limiting factors such as climatic conditions, the presence or absence of interacting species, and human impacts. In a series of single-parameter models, the relationship between a habitat feature and its suitability is defined; a completely unsuitable habitat is assigned a value of 0 and an optimally suitable habitat is assigned a value of 1. The single-parameter models are aggregated to assign an overall suitability (HSI) value, again ranging between 0 and 1. A major appeal of the methodology is its flexibility: the single-parameter models can be of virtually any form and can be aggregated in several different ways.

Limitations

For the climate correlation technique, a key assumption is that the geographic distribution is in equilibrium with climate conditions. The method does not indicate how far or fast the landscape element can move in response to climate change, only the amount and geographic distribution of any expected change. It is a static, linear, and noninteractive description of a very dynamic process.

The HSI provides flexibility; however, the models are hypotheses of species-habitat relationships, not proven statements of cause and effect. The ultimate utility of the model will depend on the extent of current knowledge and research.

Many management and intervention techniques can be used as part of an overall adaptation strategy or in response to particular impacts. These include ecosystem restoration, prescribed fire and fire exclusion, species relocation and colonization, assisted migration or reintroduction, control of alien or invasive species, control of disease, irrigation or drainage, and food and water provision (Markham and Malcolm, in press).

HUMAN HEALTH METHODS

Scientists anticipate that global climate change will affect a variety of causes of human mortality and morbidity (McMichael, 1993). Estimating the vulnerability of human health will therefore involve a number of different approaches, each targeted at a specific human disease or group of diseases (Table 2.4). The methods for assessing the vulnerability of human health range from extrapolations of the association between hot humid air masses and mortality to the integrated modeling of vector-borne diseases in the context of climate, ecosystem, and societal change.

Since human health is ultimately dependent on the integrity of ecosystems and human social systems, changes in global climate are likely to affect human health not only in terms of human diseases but also because of effects in other sectors. As these other sectors collectively represent much of the foundation of public health, their potential impact is large. Human health consequences of changes in these sectors should therefore be considered in comprehensive assessments.

Initial Approaches

For each country, the first step in assessing regional human health vulnerability to the effects of global climate change is to identify sensitive populations and pertinent diseases within the country's borders. For example, populations living at low elevations that are vulnerable to the health effects of a sea level rise or populations susceptible to certain vector-borne diseases can be identified and mapped. By mapping these populations and their location with respect to geoclimatic boundaries of vector species, qualitative inferences can be made about regional vulnerability to health effects from specific climate predictions. For example, the *Anopheles* mosquito, which carries malaria, is limited by the 16° C winter isotherm. If the climate prediction for an area adjacent to a region where malaria is endemic indicates an elevation in mean monthly winter temperatures to above the 16°C isotherm, it can be inferred that the area is vulnerable to an influx of malaria. Whether malaria outbreaks will actually happen depends on many other factors. Simple mapping can also help identify populations at risk from extreme events based on geographic location, quality of shelter, socioeconomic status, or other factors.

Infectious Disease Modeling

For many countries, the threat of increased incidence of vector-borne diseases and changes to the areas they thrive in is of great concern. The IPCC (in press) identified malaria, schistosomiasis, and dengue/dengue hemorrhagic fever as among the vector-borne diseases most likely to be affected by climate change. Studies relating climatic factors to the incidence of vector-borne diseases have developed predictive equations

Table 2.4. Summary of Methods for the Assessment of Human Health Vulnerability to Global Climate Change and Stratospheric Ozone Depletion

Method	Special Tools	Appropriate Diseases	Data Requirements
Mapping	GIS Remote sensing	Infectious diseases	Incidence by geographic area, vector ranges, and habitats
Extrapolation		Heat-related mortality	Baseline incidence and prevalence data; baseline climate data
Mathematical Models	TARGETS Matsuoka and Kia, 1995	Infectious diseases Malnutrition	Substantial demographic data; Incidence, prevalence, vector population data; climate parameters

to model the association. They indicate that the incidence of mosquito-borne diseases is related to the number vectors per host, the proportion of bites that result in human infection, the proportion of bites by susceptible mosquitos on infectious humans that produce an infection, the daily number of blood feedings per vector, the vector's daily survival probability, and the number of days between the vector's ingestion of an infected blood meal and the point at which it is capable of infecting a new host (also known as the extrinsic incubation period). All of these variables can be modeled by equations that contain climate variables. The number of feedings and the extrinsic incubation period are temperature-dependent, whereas the survival probability and vector density are dependent on both temperature and humidity (Martens et al., 1994).

The relationships between incidence of disease and climate-affected factors can serve as the basis for assessments of vulnerability to vector-borne diseases. Data on the dependence of the equation's parameters on climate factors are available for malaria, dengue, and other arthropod-based diseases, and climate prediction models can provide estimates of future changes in temperature and precipitation. An important limitation of using such an approach, however, is that it neglects many other important factors in disease transmission and behavior that are also likely to change over time, such as the percentage of susceptible persons in the population, degree of malnutrition in the population, pesticide use and resistance, availability of drugs, and drug resistance. Increases in temperature are not going to have significance for a human disease, for example, if they lead to migration of the population away from the area. An integrated systems approach is used to try to account for these other factors in the changes in disease incidence.

Integrated Systems-Based Modeling

The model chosen for use by U.S. Country Studies Program participants was the TARGETS (Tool to Assess Regional and Global Environmental and Health Targets for Sustainability) model developed by the National Institute of Public Health and Environmental Protection of the Netherlands (Rotmans et al., 1994). TARGETS uses a systems approach to interrelate social, economic, and environmental forces in predicting the behavior of human diseases or other human behaviors. Climate data are generated by GCMs linked to the Integrated Model to Assess the Greenhouse Effect (IMAGE) (Rotmans, 1990). Discrete but interrelated components of the model include modules for land and water use, energy resource development, population growth, and toxic chemical release. The use of a systems approach, which includes inputs of societal factors, makes it possible to explore the relative impacts of different levels of population growth, resource allocation, and energy use among other human factors. These factors can be dynamically modeled or input as fixed projections based on available data sources such as the U.N. Food and Agricultural Organization. For U.S. Country Studies Program participants, scaled-down, disease-specific modules for malaria and schistosomiasis were created that dynamically model population growth, nutrition, and other diseases (i.e., measles, diarrhea) but use fixed projections for other areas like land use and water resources. The main data inputs required by this version of TARGETS are population, food intake, GNP, governmental health investment levels, and climate and baseline disease data. TARGETS' main outputs include disease incidence and prevalence, as well as years of healthy life lost (DALYs) as an indicator of socioeconomic impact.

Limitations

Infectious disease modeling serves as a basis for vulnerability assessments, but it does not account for many factors in disease transmission and behavior that are also likely to change over time. An integrated systems approach accounts for many of these factors, but there are a number of limitations of the TARGETS and other integrated models. The first is the aggregation of assumptions and variability inherent in integrated statistical models. Although it is useful to integrate the many determinants of human health, the quantitative relationships between these factors are not known with great precision, and iterative repetitions of predictive equations will result in large degrees of uncertainty. This is one reason that fixed projections are used for certain socioeconomic data instead of dynamic models. A second problem is that the resolution of the GCMs is still too crude to allow for accurate local climate predictions. TARGETS uses a grid of 0.5° latitude by 0.5° longitude. This scale is more refined than the 2°×4° scale of most GCMs, but it still may be too large to take into account some local climate-altering geographical features such as mountain ranges or lakes. Thus, there is a need to combine the outputs of an integrated predictive model such as TARGETS with an evaluation of local geoclimates that could influence potential disease spread. Alternatively, a range of different climate projections for a specific region could be used to explore the range of possible outcomes. For these reasons, integrated models (TARGETS specifically) should be viewed as providing insight into the interactions between various societal and natural systems and suggesting trends in the behavior of human disease systems, rather than providing quantitative predictions of actual disease incidence or impact.

Adaptation Assessments

Nations assessing their vulnerability to climate change should also assess the need to adapt to the potential effects of climate change. Since there is a lag between emissions of greenhouse gases and climate change and because emissions are likely to rise in the future (Houghton et al., 1992), it is likely that climate will change, and that nations will have to adapt to this consequence.

Adaptation assessments should distinguish between adaptations taken as a reaction to climate change and those taken in anticipation of climate change. The former measures are implemented after climate change effects are felt, whereas the latter are implemented before effects are felt in order to mitigate these effects (Smith, in press). The effectiveness of options for adapting to climate change should be evaluated based on two criteria:

- *Flexibility*. Policies should meet their objectives under a variety of climate change situations.
- *Efficiency*. The benefits of policies should be greater than costs, particularly when accounting for the present value of benefits and costs.

Some adaptation measures need to be implemented in anticipation of climate change because their effectiveness would be eliminated or significantly reduced if taken as a reaction to climate change. Such measures are high priority adaptation measures and can be grouped by the following categories:

- *Irreversible or catastrophic impacts*. Adaptation policies here would address irreversible effects of climate change such as loss of life or extinction of species or catastrophic effects such as extensive loss of property or destruction of resources. Reactive measures cannot reverse climate change impacts here.
- *Long-term decisions*. Building infrastructure in climate-sensitive resources such as water resources or developing coastal areas involves long-term decisions that could be affected by climate change. It may be far less costly to adjust these decisions before they are made (e.g., marginally change the infrastructure) than after climate changes (NAS, 1992).
- *Unfavorable trends*. Continuation of some trends such as fragmentation of habitat or development of low-lying coastal areas may make it more difficult for society or nature to adapt to climate change. Adjusting these trends to account for climate change may make adaptation more successful.

Adaptation Assessment Process

A four-step process is proposed for countries to use in identifying and assessing adaptation options. The process is displayed in Figure 2.9.

Assess Vulnerability

Countries should begin by assessing the vulnerability of climate-sensitive sectors to climate change using methods such as those described in this chapter. It is important that these assessments consider how society and nature may adapt on their own. So-called autonomous adaptations include farmers switching crops or changing practices (Carter et al., 1994). These assessments should be used to identify sectors that may have adverse or positive effects of climate change that may justify anticipatory adaptation measures.

Define Scope of Policy Assessment

The next step is to define the scope of the adaptation policy assessment. Within climate-sensitive sectors, specific regions, groups of individuals, types of species, etc. may be especially at risk from climate change

Figure 2.9. Adaptation Assessment Steps

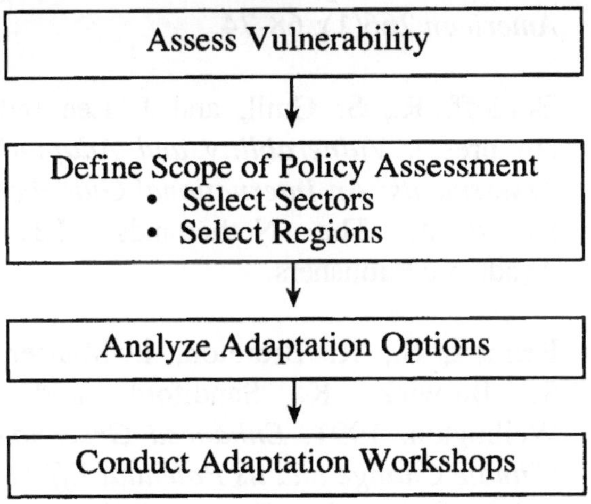

Source: Benioff et al., in press.

or may have a particular opportunity to benefit from climate change. The results of the vulnerability assessments can be used to identify vulnerable regions and populations. The assessment should focus on these subsectors.

Analyze Adaptation Options

Once sensitive subsectors are identified and the scope narrowed, analysis of the effectiveness of adaptation measures can begin. Countries should assess how well current policies and alternative policies perform under current (i.e., known) climate variation. Benioff et al. (in press) describes a number of adaptation policies. Countries may also consult OTA (1993) and Smith and Lenhart (in press) as well as using their own knowledge and experience. The adaptation assessment should also examine how well the current and alternative policies perform under climate change scenarios. In addition, the assessment should consider the feasibility of alternative policies; that is, whether there are significant barriers to implementing these policies and whether these barriers can be overcome.

Two methods for assessing adaptation options have been made available to U.S. Country Studies Program participants and are described in more detail in Benioff et al. (in press). The first is a decision matrix. To evaluate the policies, their objectives need to be defined. Common metrics are used to estimate the level of performance of adaptation options. This approach makes it possible for decision makers to directly compare the performance of adaptation options across objectives. Results can be compared with costs to determine cost-effectiveness. The requirement that all attributes be quantified in a common metric may be difficult to meet when nonquantifiable objectives are included. Results of the evaluation can be displayed and analyzed in a decision matrix.

A second option is to use the Adaptation Strategy Evaluator (Smith et al., 1995). The ASE uses multicriteria objective analysis, which enables users to examine quantitative and nonquantitative objectives. Software has been developed for analyzing climate change adaptation on agriculture, coastal resources, and water resources and has been used in Venezuela and Egypt and distributed to other participants in the U.S. Country Studies Program.

Both of these methods are more likely to produce results acceptable to decision makers if they are closely involved in the analysis.

Conduct Adaptation Workshops

The assessment of adaptation options should culminate in a one- to two-day workshop with policy makers to discuss vulnerability to climate change and adaptation options. The workshop should review the results of the vulnerability assessments and analysis on the performance of current and alternative policies under current climate and conditions of climate change. After that, policy makers should analyze and discuss adaptation options. The policy makers could fill in a decision matrix or use ASE to evaluate the alternatives. When the analysis is completed, the policy makers should discuss next steps such as implementation or future research. Researchers can act as rapporteurs to write up the results of the workshops.

REFERENCES

Bailey, H.P. 1979. "Semi-Arid Climates: Their Definition and Distribution." In A.E. Hall, G.H. Cannell, and H.W. Lawton (eds.), *Agriculture in Semi-arid Environments.* Berlin, Germany: Springer-Verlag.

Baker, B.B. 1991. *A National Analysis of the Potential Effects of Climate Change on Rangeland Ecosystems.* PhD Dissertation. Fort Collins, CO: Colorado State University.

Baker, B.B., R.M. Bourdon, and J.D. Hanson. 1992. "Forage: a Simulation Model of Grazing Behavior for Beef Cattle." *Ecological Modelling* 60: 257-279.

Baker, B.B., J.D. Hanson, R.M. Bourdon, and J.B. Eckert. 1993. "The Potential Effects of Climate Change on Ecosystem Processes and Cattle Production on U.S. Rangelands." *Climatic Change* 25: 97-117.

Bazzaz, E.A. and E.D. Fajes. 1992. "Plant Life in a CO_2-Rich World." *Scientific American* 266(1): 68-74.

Benioff, R., S. Guill, and J. Lee (eds.) (In press). *Vulnerability and Adaptation Assessments: An International Guidebook.* Dordrecht, The Netherlands: Kluwer Academic Publishers.

Bennett, S., R. Brereton, I. Mansergh, S. Berwick, K. Sandiford, and C. Wellington. 1991. *Enhanced Greenhouse Climate Change and its Potential Effect on Selected Victorian Fauna.* Arthur Rylah Institute Tech. Rep. No. 132, Department of Conservation & Environment, Victoria.

Boer, G.J., N.A. McFarlane, and M. Lazare. 1992. "Greenhouse Gas-Induced Climate Change Simulated with the CCC Second-Generation General Circulation Model." *Journal of Climate* 5: 1045-1077.

Bonan, G.B. and B.P. Hayden. 1990. "Using a Forest Stand Simulation Model to Examine the Ecological and Climatic Significance of the Late-Quaternary Pine-Spruce Pollen Zone in Eastern Virginia, U.S.A." *Quaternary Research* 33: 204-218.

Bonan, G.B., H.H. Shugart, and D.L. Urban. 1990. "The Sensitivity of Some High-Latitude Boreal Forests to Climatic Parameters." *Climatic Change* 16: 9-29.

Bos, E., M.T. Vu, A. Levin, and R.A. Bulatao. 1992. *World Population Projections 1994-95 Edition.* Published for the World Bank. Baltimore: Johns Hopkins University Press.

Bos, E., M.T. Vu, E. Massiah, and R.A. Bulatao. 1994. *World Population Projections 1994-95 Edition*. Published for the World Bank. Baltimore: Johns Hopkins University Press.

Burnash, R.J., R.L. Ferral, and R.A. McGuire. 1973. *A Generalized Streamflow Simulation System, Conceptual Modeling for Digital Computers*. Sacramento, California: National Weather Service and California Department of Water Resources.

Carpenter, S.R., S.G. Fisher, N.B. Grimm, J.F. Kitchell. 1992. "Global Change and Freshwater Ecosystems." *Annual Review of Ecological Systems* 23: 119-139.

Carter, T.R., M.L. Parry, H. Harasawa, and S. Nishioka. 1994. *IPCC Technical Guidelines for Assessing Climate Change Impacts and Adaptations*. London: Department of Geography, University College.

Christie, G.C. and H.A. Regier. 1988. "Measures of Optimal Thermal Habitat and Their Relationship to Yields of Four Commercial Fish Species." *Canadian Journal of Fisheries and Aquatic Sciences* 45: 301-314.

Crutcher, H.L., and J.M. Meserve. 1970. *Selected Level Heights, Temperatures and Dew Points for the Northern Hemisphere*, NAVAIR 50-1C-52 (revised), Chief of Naval Operations, Washington, DC.

Cure, J.D. and B. Acock. 1986. "Crop Responses to Carbon Dioxide Doubling: A Literature Survey." *Agricultural and Forest Meteorology* 38: 127-145.

Davis, M.B. 1989. "Lags in Vegetation Response to Greenhouse Warming." *Climatic Change* 15: 75-82.

Gates, W.L., P.R. Rowntree, Q.C. Zeng. 1990. "Validation of Climate Models." In J.T. Houghton, G.J. Jenkins, and J.J. Ephraums (eds.). 1990. *Climate Change: The IPCC Scientific Assessment*. New York: Cambridge University Press.

Gates, W.L., J.F.B. Mitchell, G.J. Boer, U. Cubasch, and V.P. Meleshko. 1992. "Climate Modeling, Climate Prediction and Model Validation." In J.T. Houghton, B.A. Callander, and S.K. Varney (eds.). *Climate Change 1992: the Supplementary Report on the IPCC Scientific Assessment*. Cambridge: Cambridge University Press.

Hands, E.B. 1983. The Great Lakes as a Test Model for Profile Responses to Sea Level Changes, in P.D. Komar (ed.), *Handbook of Coastal Processes and Erosion*, Boca Raton, FL: CRC Press.

Hansen, J., G. Russell, D. Rind, P. Stone, A. Lacis, S. Lebedeff, R. Ruedy, and L. Travis. 1983. "Efficient Three-Dimensional Global Models for Climate Studies: Models I and II." *Monthly Weather Review* 3: 609-622.

Hanson, C.L., E.L. Neff, and A.D. Nicks. 1980. "Estimating SCS Runoff Curve Numbers On Native Grazing Lands." In W.G. Knisel (ed.), *CREAMS: A Field Scale Model for Chemicals, Runoff, and Erosion from Agricultural Management Systems*. U.S. Dept. of Agriculture Conservation Research Report 26: 398-404.

Hanson, J.D., et al. 1992. *SPUR2 Documentation and User's Guide.* U.S. Department of Agriculture, ARS, Great Plains Systems Research Technical Report-1, Fort Collins, CO.

Hanson, J.D., et al. 1993. "Comparison of the Effects of Different Climate Change Scenarios on Rangeland Livestock Production." *Agricultural Systems* 41: 487-502.

Hill, D.K. and J.J. Magnuson. 1990. "Effects of Global Climate Warming on the Growth and Prey Consumption of Great Lakes Fish." *Transactions of American Fishery Society* 119: 265-275.

Holdridge, L.R. 1947. "Determination of World Plant Formations from Simple Climate Data." *Science* 105: 367-368.

Holdridge, L.R. 1967. *Life Zone Ecology.* San Jose, CA: Tropical Science Center.

Houghton, J.T., B.A. Callander, and S.K. Varney (eds.). 1992. *Climate Change 1992 — The Supplementary Report to the IPCC Scientific Assessment.* WMO/UNEP Intergovernmental Panel on Climate Change. Cambridge: Cambridge University Press.

(IPCC) Intergovernmental Panel on Climate Change. 1990. *Strategies for Adaption to Sea-Level Rise.* Report of the Coastal Zone Management Subgroup, Intergovernmental Panel on Climate Change, Response Strategies Working Group, The Hague, the Netherlands.

(IPCC) Intergovernmental Panel on Climate Change. 1995. "WG1 Contribution to the IPCC Second Assessment Report: Summary for Policy Makers." November 30. Mimeograph. Rome: IPCC.

(IPCC) Intergovernmental Panel on Climate Change. (In press). *Scientific-Technical Analyses of Impacts, Adaptations, and Mitigation of Climate Change.* New York: Cambridge University Press.

Kaczmarek, Z. 1993. "Water Balance Model for Climate Impact Analysis" *ACTA Geophysica Polonica* 41(4): 1-16.

Kennedy, V.S. 1990. "Anticipated Effects of Climate Change on Estuarine and Coastal Fisheries." *Fisheries* 15(6): 16-24.

Leatherman, S.P. 1991. "Modeling Shore Response to Sea-Level Rise in Sedimentary Coasts." *Progress in Physical Geography* 14: 447-464.

Leatherman, S.P., R.J. Nicholls, and K.C. Dennis. 1995. "Aerial Videotape-Assisted Vulnerability Analysis: A Cost-Effective Approach to Assess Sea-Level Rise Impacts." *Journal of Coastal Research*, Special Issue 14: 15-25.

Leggett, J., W.J. Pepper, and R.J. Swart. 1992. "Emissions Scenarios for IPCC: An Update." In J.T. Houghton, B.A. Callander, and S.K. Varney, (eds.). *Climate Change 1992 — The Supplementary Report to the IPCC Scientific Assessment.* WMO/UNEP Intergovernmental Panel on Climate Change. Cambridge: Cambridge University Press.

Lowe-McConnell, R.H. 1987. *Ecological Studies in Tropical Fish Communities.* Cambridge, Great Britain: Cambridge University Press.

MacArthur R.H. 1972. *Geographical Ecology: Patterns in the Distribution of Species*. Princeton, NJ: Princeton University Press.

Markham, A. and J. Malcolm. (In press). "Biodiversity and Wildlife Conservation: Adaptation to Climate Change." In J. Smith, N. Bhatti, G. Menzhulin, R. Benioff, M. Campos, B. Jallow, and F. Rijsberman (eds.). *Adaptation to Climate Change: Assessment and Issues*. New York: Springer-Verlag.

Martens, W.J.M., J. Rotmans, and L.W. Niessen. 1994. "Climate Change and Malaria Risk — An Integrated Modeling Approach." Rijksinstituut voor volksgezondheid en milieuhygiene (RIVM) Report No. 461502003. Bilthoven, the Netherlands. March.

Matsuoka, Y., Kai K., 1995. "An Estimation of Climatic Change Effects on Malaria." *Journal of Global Environmental Engineering* 1: 43-57.

McMichael, A.J. 1993. "Global Environmental Change and Human Population Health: a Conceptual and Scientific Challenge for Epidemiology." *International Journal of Epidemiology* 22: 1-8.

Mearns, L.O., C. Rosenzweig, and R. Goldberg. 1992. "Effect of Changes in Interannual Climatic Variability of CERES-Wheat Yields: Sensitivity and $2\times CO_2$ General Circulation Model Studies." *Agricultural and Forest Meteorology* 62: 159-189.

Meisner J.D., J.L. Goodier, H.A. Regier, B.J. Shuter, and W.J. Christie. 1987. "An Assessment of the Effects of Climate Warming on Great Lakes Basin Fishes." *Journal of Great Lakes Research* 13: 340-352.

Mitchell, J.F.B., S. Manabe, T. Tokioka, and V. Meleshko. "Equilibrium Change." in Houghton, J.T., G.J. Jenkins and J.J. Ephraums (eds.). 1990. *Climate Change: The IPCC Scientific Assessment*. New York: Cambridge University Press.

Möller, F. 1951. "Vierteljahreskarten des Niederschlags für die ganze Erde." *Petermanns Geographische Mitteilungen*. Justus Perthes, Gotha. pp. 1-7.

(NAS) National Academy of Sciences. 1992. *Policy Implications of Greenhouse Warming*. Washington, DC: National Academy Press.

(OTA) Office of Technology Assessment. 1993. *Preparing for an Uncertain Climate*, vols. 1 and 2, OTA-0-567 and OTA-0-568, Office of Technology Assessment, Washington, DC.

Otter-Nacke, S., D.C. Godwin, and J.T. Ritchie. 1986. *Testing and Validating the CERES-Wheat Model in Diverse Environments*. AgGRISTARS YM-15-00407. No. 20244. Houston: Johnson Space Science Center.

Pauly, D. 1980. On the Interrelationships between Natural Mortality, Growth Parameters, and Mean Environmental Temperature in 175 Fish Stocks." *Journal du Conseil International pour Exploration de la Mer* 39(2): 175-192.

Peart, R.M., J.W. Jones, R.B. Curry, K. Boote, and L.H. Allen, Jr. 1989. "Impact of Climate Change on Crop Yield in the Southeastern USA." In J.B. Smith and D.A. Tirpak (eds.), *The Potential Effects of Global Climate Change on the United States*. Washington, DC: U.S. Environmental Protection Agency.

Pepper, W., J. Leggett, R. Swart, J. Wasson, J. Edmonds, and I. Mintzer. 1992. *Emission Scenarios for the IPCC: An Update*. Prepared for the Intergovernmental Panel on Climate Change Working Group I. Washington, DC: ICF, Incorporated.

Root, T. 1988a. "Energy Constraints and Avian Distributions and Abundances." *Ecology* 69: 330-339.

Root, T. 1988b. "Environmental Factors Associated with Avian Distributional Boundaries." *Journal of Biogeography* 15: 489-505.

Rosenzweig, C. and A. Iglesias (eds.). 1994. *Implications of Climate Change for International Agriculture: Crop Modeling Study*. EPA 230-B-94-003. Washington, DC: U.S. Environmental Protection Agency.

Rosenzweig C. and M.L. Parry. 1994. "Potential Impact of Climate Change on World Food Supply." *Nature* 367: 133-138.

Rotmans, J. 1990. *IMAGE: an Integrated Model to Assess the Greenhouse Effect*. Dordrecht/Boston/London, The Netherlands: Kluwer Academic Publishers.

Rotmans et al. 1994. Global Change and Sustainable Development — A Modeling Perspective for the Next Decade." Rijksinstituut voor volksgezondheid en milieuhygiene (RIVM) Report No. 461502004. Bilthoven, the Netherlands. June.

Sagua, V.O. 1993. *The Effects of Climate Change on the Fisheries of the Sahel*. Food and Agriculture Organization of the United Nations, Rome.

Salewicz, A. 1995. Impact of Climate Change on the Operation of Lake Kariba Hydropower Scheme on the Zambezi River. In *Water Resources Management in the Face of Climatic and Hydrologic Uncertainties*. Z. Kaczmarek et al. (eds.).

Schlesinger, M.E. and Z.C. Zhao. 1988. "Seasonal Climate Changes Induced by Doubled CO_2 as Simulated by the OSU Atmospheric GCM/Mixed Layer Ocean Model." *Journal of Climate* 2(5): 459-95.

Schutz, C. and W.L. Gates. 1971. *Global Climate Data for Surface, 800 millibars, 400 millibars*. R-915-ARPA. Rand Corporation. January.

Shugart, H.H. 1984. *A Theory of Forest Dynamics*. New York: Springer-Verlag.

Shugart, H.H. and D.C. West. 1980. "Forest Succession Models." *BioScience* 30: 308-313.

Smith, A., N. Chan, H.Q. Chu, C. Helman, and B. Kim. 1995. *Documentation of Adaptation Evaluator Systems*. Report prepared for the U.S. Environmental Protection Agency. Washington, DC: Decision Focus, Inc.

Smith, J.B. (In press). "Setting Priorities for Adapting to Climate Change." *The Environment Professional.*
Smith, J.B. and S.S. Lenhart. (In press). "Climate Change Adaptation Policy Options." *Climate Research.*

Smith, T.M., R. Leemans, and H.H. Shugart. 1992a. "Sensitivity of Terrestrial Carbon Storage to CO_2-Induced Climate Change: Comparison of Four Scenarios Based on General Circulation Models." *Climatic Change* 21: 367-384.

Smith, T.M., G.B. Bonan, H.H. Shugart, and J.B. Smith. 1992b. "Modeling the Potential Response of Vegetation to Global Climate Change." *Advances in Ecological Research* 22: 93-113.

Solomon, A.M. 1986. "Transient Responses of Forests to CO_2-Induced Climate Change: Simulation Modeling Experiments in Eastern North America." *Oecologia* 68: 567-569.

Strzepek, K.M. and J.B. Smith (eds.). 1995. *As Climate Changes: International Impacts and Implications.* New York: Cambridge University Press, 213 pp.

Sulzman, E.W., K.A. Poiani, T.G.F. Kittel. 1995. "Modeling Human-Induced Climatic Change: A Summary for Environmental Managers." *Environmental Management* 19: 197-224.

Taljaard, J.J., H. van Loon, H.L. Crutcher, and R.L. Jenne. 1969. *Climate of the Upper Air: Southern Hemisphere. 1: Temperatures, Dewpoints, and Heights at Selected Pressure Levels.* NAVAIR 50-1C-55. Chief of Naval Operations. Washington, DC. 135 pp.

Tegart, W.J. McG., Sheldon, G.W., and Griffiths, D.C. 1990. *Climate Change — The IPCC Impacts Assessment.* WMO/UNEP Intergovernmental Panel on Climate Change. Canberra: Australian Government Publishing Service.

Todini, E. 1988. "Rainfall-Runoff Modeling — Past, Present, and Future." *Journal of Hydrology* 100: 341-352.

U.S. Fish and Wildlife Service. 1981. *Standards for the Development of Habitat Suitability Index Models.* 103 ESM. Washington, DC: U.S. Fish and Wildlife Service, Division of Ecological Services.

Walter, H. 1985. *Vegetation of the Earth and Ecological Systems of the Geo-Biosphere.* 3rd ed. Berlin, Germany: Springer-Verlag.

Welcomme, R.L. 1985. *River Fisheries.* FAO Fisheries Technical Paper No. 262, Food and Agriculture Organization of the United Nations, Rome.

Wight, J.R. and J.W. Skiles, (eds.). 1987. *SPUR: Simulation of Production and Utilization of Rangelands. Documentation and User Guide.* ARS 63. U.S. Department of Agriculture, Agricultural Research Service, 372 pp.

Wigley, T.M.L. and S.C.B. Raper. 1992. "Implications for Climate and Sea Level of Revised Press. IPCC Emissions Scenarios." *Nature* 357: 293-324.

Wilson, C.A. and J.F.B. Mitchell. 1987. "A Doubled CO_2 Climate Sensitivity Experiment with a Global Climate Model Including a

Simple Ocean." *Journal of Geophysical Research* 92: 13315-43.

Woodward, F.I. 1987. *Climate and Plant Distribution.* Cambridge: Cambridge University Press.

Yates, D. (In press). WATBAL: A Monthly Water Balance Model for Climate Change Impact Assessment. *International Journal of Water Resources Development.*

Yates, D. and K.M. Strzepek. 1994. "Comparison of Water Balance Models for Climate Change Assessment Runoff." Draft working paper, IIASA, Laxenburg, Austria.

Chapter 3

Summary of Results by Vulnerability Sector

INTRODUCTION

The U.S. Country Studies Program was designed to develop and enhance the capacity of developing countries and countries with economies in transition to study climate change. There are 55 countries participating in the program, which provides technical and financial assistance for conducting greenhouse gas inventories, mitigation assessments, and vulnerability and adaptation assessments, and preparing national climate change plans.

One of the benefits of building capacity in climate change research is that more developing and transition countries will be able to assess their vulnerability to climate change and their adaptation options. This assessment capability is important because many people in these countries live in arid or semiarid regions that are drought prone, in basins or deltas that are flood prone, or in coastal areas that are exposed to sea level rise risks, which are indicative of particularly vulnerable regions (IPCC, in press). Furthermore, reactive adaptation options at the local level may be limited by lack of capital or access to technology. Consequently, advance planning at the national level will play an important role in improving adaptation capabilities.

This chapter puts the preliminary vulnerability and adaptation results from 13 participating countries in perspective by reviewing them across four sectors — agriculture, forests, water resources, and coastal resources — and comparing them with the existing literature on climate change impacts. Readers are encouraged to read the individual country chapters for more information about methodologies, results, and conclusions. In comparing the results from the U.S. Country Studies Program with the IPCC (in press) summary and other sources of impact information, we note that in almost every instance, the results in this report do not repeat previous national-level research efforts. However, they tend to confirm past estimates of regional impacts.

Table 3.1 shows the sectors that were analyzed by the 13 countries contributing to this report.[1] In this chapter, we summarize biophysical impacts, economic impacts, and adaptation for each sector. The summaries focus on the preliminary biophysical impacts predicted by the research efforts. Ongoing research in these and other countries is analyzing the effects of the biophysical impacts on economic and social systems and potential feedback mechanisms between socioeconomic conditions and biophysical impacts. For instance, changes in agricultural productivity will affect national and international food product markets by altering

[1] The grasslands/livestock sector was examined by only Kazakhstan, and thus is not summarized in this chapter.

J. B. Smith et al. (eds.), Vulnerability and Adaptation to Climate Change, 49–68.

Table 3.1. Summary of Vulnerability and Adaptation Assessments by Sector

Country	Agriculture	Grasslands[1]	Forest	Water Resources	Coastal Resources
The Gambia	x				x
Zimbabwe	x				
Bangladesh	o		o	o	o
Kazakhstan	x	x		x	
Mongolia	x		x	x	
Sri Lanka					x
Czech Republic	x		x	x	
Slovakia	o		o	o	
Estonia	o		x		x
Poland					x
Ukraine			x	x	x
Peru					x
Venezuela			x		x

[1] Not summarized in this chapter.
Note: x = have completed preliminary assessments.
o = assessments still in progress.

prices. These price changes may, in turn, encourage changes in farm-level operations such as shifting to different crops. An assessment of the actual impact of climate change on agriculture ideally incorporates these kinds of feedback effects. Additional ongoing research efforts are assessing adaptation measures for vulnerable sectors. Some preliminary adaptation results are provided below, including the effects of farm-level adaptation on agricultural yields and the benefits and costs of sea level rise adaptation.

For example, the agricultural yield impacts summarized by region in IPCC (in press) show wide ranges of yield changes that generally include yield increases as well as yield decreases. Rosenzweig and Iglesias (1994) found that lower latitude regions tended to be more negatively affected than higher latitude regions, and that the primary causes of crop losses included shortened growing periods and decreased water availability. Furthermore, the beneficial physiological effects of CO_2 and the simulated adaptation measures tended to offset potential yield losses. They also found that maize yields tended to be more negatively affected than other crops, which may be indicative of relatively low direct CO_2 benefits. The discussion in the agriculture section concurs with these findings.

Forest impact research can study how the distribution of vegetation zones shifts under different equilibrium climate conditions using one of several global vegetation models (Neilson, 1993; Smith and Shugart, 1993). Based on climate change simulations of potential distributions, the IPCC (in press) reports that about one-third of global forested area will undergo major vegetation changes. Greater changes are expected in the higher latitudes where boreal forests decline due to changes in climate. Smaller changes are estimated for temperate and tropical forests, and the direction of change differs across the vegetation models. For example, Smith et al. (1992, 1995) show that, according to the Holdridge Life Zone Classification Model, cool temperate forest area increases because gains from boreal forest and nonforest biome conversion exceed losses to warm temperate forests and nonforest biomes. However, warm temperate forest area declines primarily because of conversion to subtropical forest.

Although vegetation modeling can identify potential changes in vegetation zones, additional research of the transient responses of forest ecosystems to climate perturbations is required to study whether the transitions between zones is likely to occur with or without management. This research can only be studied at the forest stand or region level using forest gap models, which show how climate change might alter species composition and growth rates over time (Solomon, 1986; Smith et al., 1995).

Climate change is expected to have major impacts on regional water resources such as runoff increases in high latitude regions where precipitation may increase, and runoff decreases in lower latitudes due to combined temperature increases and precipitation decreases (IPCC, in press). Changes in water availability may have significant impacts in developing countries where current water usage is low. Hydrological changes will also affect water quality, e.g., lower dissolved oxygen levels due to warmer temperatures.

Coastal populations and infrastructure are vulnerable to sea level rise impacts. The IPCC (in press) estimates that currently 46 million people are at risk each year from storm surges, and that a sea level rise of 0.5 m (1.0 m) may increase the risk to 92 million (118 million), based on current population levels. There are no estimates of potential global land and infrastructure losses. Delta regions such as those in Egypt and Bangladesh, and small island nations, are particularly vulnerable.

AGRICULTURE

Agriculture is an important economic sector in many developing countries and countries in transition, contributing a relatively large share of GNP and employing a large proportion of the workforce. For example, Zimbabwe reports that agriculture accounts for about 40% of annual exports, 11% to 14% of GDP, and directly or indirectly employs 80% of the workforce. Furthermore, developing countries often have a relatively large population that is dependent on subsistence agriculture. Consequently, climate change impacts on agriculture can have significant effects on national and local economic and social systems.

A summary of crop yield studies across all global regions in IPCC (in press) showed that most regions had wide ranges of yield changes that included yield increases and decreases. Yields in lower latitude regions tended to be more negatively affected than yields in higher latitude regions, and direct CO_2 physiological benefits and farm-level adaptation measures were important factors for offsetting potential yield reductions. (Rosenzweig and Iglesias, 1994).

Eight of the many countries in the Country Studies Program that are evaluating the sensitivity of agricultural to climate change present study summaries in this report: The Gambia, Zimbabwe, Kazakhstan, Mongolia, Czech Republic, Estonia, Slovakia, and Bangladesh. Of these, the first five listed have reported preliminary impact results in their chapters. Estonia, Slovakia, and Bangladesh are still working to generate agricultural impact assessments.

Biophysical Impacts

The five countries with results used the Decision Support System for Agrotechnology Transfer version 3 (DSSAT 3) crop yield models to estimate how yields at the farm level (e.g., tons per hectare) could be affected by changes in climate variables such as precipitation, temperature, and evapotranspiration. Their analyses typically focused on the sensitivity of important cash or staple crops such as maize. Climate change data comprised various GCM predictions for the $2\times CO_2$ scenario or selected incremental scenarios that combine changes in temperature and precipitation. Changes in crop yields were derived by contrasting productivity under baseline climate conditions with productivity under altered conditions. Table 3.2 summarizes the percentage changes in yields by country and crop.

These results indicate that climate change tends to decrease crop productivity in low latitude countries such as The Gambia and Zimbabwe. For example, in The Gambia, all maize and millet simulations show yield declines. In Zimbabwe, results for four study sites showed that maize yields could increase or decrease. The range reported in Table 3.2, -13.6% to -11.5%, represents

Table 3.2. Summary of Changes in Agricultural Yields

Country	Scenario	Crop	Percent Change in Farm Level Yields	Direct CO_2 Effect Included	Adaptation Included
The Gambia	CCCM, GFD3, GISS	Maize	-26% to -15%	Yes	No
		Millet (early)	-44% to -29%		
		Millet (late)	-21% to -14%		
		Groundnuts	+40% to +52%		
Zimbabwe	CCCM, GFD3	Maize (all sites)	-13.6% to -11.5%	Yes	No
		(Masvingo site)	-98%		No
		(Masvingo site)	+3% to +16%		Yes[1]
Kazakhstan	CCCM, GFD3[2]	Wheat (spring)	-70% to -25%	Only for	No
	Incremental Scenarios	Wheat (winter)	-35% to +17%	GCMs	
Mongolia	GFD3, GISS[3]	Wheat (spring)	-74.3% to +32.0%	No	Yes[4]
Czech Republic	Incremental Scenarios	Wheat (winter)	-3% to +16%	Yes	No

[1] Adaptation measures for Masvingo include earlier planting date and short season varieties.

[2] Also estimated impacts for 2030 using GF01 climate data. Spring wheat yield declined by 27% and winter wheat yield increased by 21%.

[3] Also considered impacts if historical warming trend continues. In contrast to the GCMs, this trend predicts a cooler and wetter climate in June and July, which may lead to yield increases compared to baseline.

[4] Adaptation measure shifts planting date from May 15 to May 1 or April 25. Yield loss range is smaller for earlier planting dates (-5% to -67%).

average yield changes across the sites, which were calculated using national production shares as weights. (Table 3.2 also shows yield increases due to adaptation for the Masvingo study site.) However, notable increases in groundnut yields were predicted for The Gambia. Groundnuts are historically the country's primary cash crop, so this result has important economic consequences. In contrast to the low latitude results, wheat crop yields in the higher latitude countries (Kazakhstan for winter wheat only, Mongolia, and the Czech Republic) include potential yield increases as well as decreases.

As Table 3.2 shows, most of the simulations included the direct beneficial effect of CO_2 on plant growth. Increased CO_2 levels tend to enhance the photosynthesis process and reduce a plant's demand for water. These benefits are expected to help offset detrimental agricultural impacts or augment advantageous impacts caused by climate change (Bazzaz and Fajer, 1992). Plants such as wheat that are classified as C_3 plants are more likely to benefit from CO_2 increases than C_4 plants such as maize. If CO_2 benefits had been excluded from the analysis, the estimated yields under climate change would be lower than those reported.

Earlier studies have simulated agricultural impacts in developing countries and countries in transition (e.g., Parry et al., 1988a,b; Parry et al, 1992; Downing, 1992; Qureshi and Hobbie, 1994; Rosenzweig and Iglesias, 1994). Table 3.3 is taken from the IPCC's (in press) summary of these studies. About

Table 3.3. Selected Crop Study Results from IPCC Assessment Report

Region	Crop	Yield Impact (%) ($2\times CO_2$)	Comments
Latin America	Maize	-61 to increase	Data from Argentina, Brazil, Chile, and Mexico; range is across GCM scenarios, with and without CO_2 effect.
	Wheat	-50 to -5	Data from Argentina, Uruguay, and Brazil; range is across GCM scenarios, with and without CO_2 effect.
	Soybean	-10 to +40	Data from Brazil; range is across GCM scenarios, with CO_2 effect.
Former Soviet Union	Wheat	-19 to +41	Range is across GCM scenarios and region, with CO_2 effect.
	Grain	-14 to +13	
Africa	Maize	-65 to +6	Data from Egypt, Kenya, South Africa, and Zimbabwe; range is over studies and climate scenarios, with CO_2 effect.
	Millet	-79 to -63	Data from Senegal; carrying capacity fell 11-38%.
	Biomass	Decrease	Data from South Africa; agrozone shifts.
South Asia	Rice	-22 to +28	Data from Bangladesh, India, Philippines, Thailand, Indonesia, Malaysia, and Myanmar; range is over GCM scenarios, with CO_2 effect; some studies also consider adaptation.
	Maize	-65 to -10	
	Wheat	-61 to +67	
China	Rice	-78 to +28	Includes rainfed and irrigated rice; range is across sites and GCM scenarios; genetic variation provides scope for adaptation.

Source: IPCC (in press).

two-thirds of the studies in Table 3.3 indicate that yields may increase or decrease; the remaining studies show declines in yields. Yield declines appear to be most frequent in Africa. The U.S. Country Studies Program results are very similar. They show a strong trend toward maize and millet yield losses in Africa, but mixed results in other regions. The lack of much overlap by region and crop between the two tables suggests that the U.S. Country Studies Program adds a number of new study results to the existing inventory of studies.

We can compare the results in Table 3.2 for Zimbabwe and The Gambia with other studies of agricultural impacts in or near those countries. Downing (1992) reported maize yield impacts for 2°C to 4°C warming for three regions in Zimbabwe (not including those discussed in Chapter 6). Simulated yield reductions ranged from 5% to 27%, excluding beneficial effects of CO_2. Rosenzweig and Iglesias (1994) also showed maize decreases in Zimbabwe of 8% to 14%. These findings are consistent with the results in this report, which show reductions near the low end of this range, which include direct CO_2 effects. Downing also reported that regional millet yields in Senegal may decline by 33% to 45% throughout most of the country with a 4°C warming and a 20% reduction in annual precipitation. These are consistent with the 14% to 43% reductions in early and late millet yields reported here by The Gambia under the Goddard Institute for Space Studies (GISS) scenario, which has a mean temperature increase of 4.1°C and a precipitation decrease of 21%.

Adaptation

Two of the assessments in Table 3.2 simulate the effects of farm-level adaptation measures on crop yields. Zimbabwe considered the joint effect of an earlier planting date and a switch to a short-season variety of maize. Combined, these two measures dramatically altered the predicted climate change yield for the Masvingo region (one of four regions studied) from a 98% decline from baseline yield to a 3% to 16% increase over baseline conditions. Mongolia also considered the effects of earlier planting dates on productivity for several regions that initially showed yield decreases under climate changes. The adaptation measure partially offsets the yield reductions, but the impact was not as dramatic as the one in Zimbabwe's study.

Rosenzweig and Parry (1994) show that adaptation measures such as these can offset a large proportion of global yield reductions. Their more extensive adaptation scenario, which exceeds the adaptation measures assumed in the assessments in this report, almost eliminates reductions in global cereal productivity under the Geophysical Fluid Dynamics Laboratory equilibrium (GFD3) and GISS climate assumptions. Darwin et al. (1995) show even greater potential benefits from adaptation. The results reported here reemphasize the potential benefits of adaptation measures, and the importance of considering them in the analysis.

Given the relative importance of adaptation in offsetting potential yield declines, future research efforts should consider how likely the farm-level adaptation efforts such as planting different cultivars or crops are to occur. Actual adaptation will depend on the availability of information, materials, and funding, and on the market conditions that are expected to prevail. If any of these factors are unfavorable, then adaptation may not occur.

Economic Impacts

Whether the yield impacts are positive or negative, they are potentially large. Given the relative importance of agriculture in some of these countries (e.g., 27% of GDP and 75% of employment in The Gambia), such impacts — if they are applicable for all areas of crop production — are likely to have significant effects on regional and national economies. Thus far, economic impacts have been qualitatively addressed in the vulnerability assessments. Future research efforts will integrate results such as these with national macroeconomic models to evaluate economy-wide impacts. Further integration with international trade models can determine how climate-induced productivity changes in other countries affect domestic agriculture, and how domestic productivity shifts affect international trade flows. Previous studies of global agricultural impacts with trade flows (Kane et al., 1992; Rosenzweig and Parry, 1994; Reilly et al., 1994; Darwin et al., 1995) suggest that markets can help ameliorate negative effects by shifting agriculture to more productive areas. However, they also show that agriculture in low-latitude developing countries, in general, is worse off under climate change.

FORESTS

Changes in climate parameters such as temperature and precipitation will cause some areas to become unsuitable for existing forest vegetation patterns (IPCC, in press). These regions may become too hot or too dry for the types of ecosystems they currently sustain. Smith et al. (1995) used the Holdridge model to estimate global changes in vegetation distribution for several GCMs climate change scenarios. They found relatively large shifts to warmer forest types, e.g., boreal forests converting to cool temperate, cool temperate converting to warm temperate. Cool temperate and tropical forest areas tended to increase across the scenarios, and boreal, warm temperate, and subtropical forest areas declined. Such shifts in vegetation zones will affect forest and forest product uses as well as ecosystems. Neilson (1993) also found significant shifts in forest cover.

Assessments of forest vulnerability and adaptation have been prepared by seven countries for this report: Bangladesh, Mongolia, Czech Republic, Estonia, Slovakia, Ukraine, and Venezuela. Five of these report preliminary vulnerability results; Bangladesh and Slovakia do not yet have preliminary results. The five reporting countries have regions with temperate forests or biomes and subtropical/tropical (dry and wet) forests; no boreal regions were assessed.

The climate change scenarios used differed across the countries. In most instances, $2\times CO_2$ scenarios were analyzed. However, the Czech Republic used climate parameters for 2010 and 2030 instead of 2075. These were estimated by linear interpolation assuming that the $2\times CO_2$ climate changes occur in 2075.

Biophysical Impacts

Mongolia, Ukraine, and Venezuela use the Holdridge model to identify areas where climate change would lead to life zone shifts. Classifications of existing vegetation zones are compared to those predicted by the GCMs' $1\times CO_2$ scenarios to determine which GCM best approximates current vegetation conditions. Equilibrium $2\times CO_2$ climate conditions for the selected GCM or GCMs were used to generate new life zone maps. Regions that fall into different zones between the $1\times CO_2$ and $2\times CO_2$ maps are likely to undergo significant change in vegetation type under climate change. Figure 3.1 illustrates the shift in vegetation zones for Venezuela between current climate conditions and predicted climate conditions for the United Kingdom Meteorological Office (UKMO) $2\times CO_2$ scenario. Although these Holdridge model impacts indicate potential vegetation shifts, only transient analyses can show whether forest ecosystems can successfully migrate during the next 70 to 100 years. The Czech Republic uses its forest vegetation zone classification, which is similar to the Holdridge model, to study the effects of climate change on forests. Finally, Estonia used a forest gap model to predict how climate change may affect two forest sites. This approach shows how forest stand composition gradually changes over time as climate conditions change.

Table 3.4 summarizes the reported impacts. Only Ukraine reports a decline in forested area; the other countries predict vegetation zone shifts that suggest changes in forest composition rather than changes from a forest to a nonforest biome. The estimated decline in Ukraine is between 39% and 47%, with large regions of cool temperate moist forest converting to steppe or desert scrub regions under drier conditions.[2] Furthermore, small boreal forest regions in Ukraine are converted to temperate forests. Both Mongolia and Venezuela show large shifts from one type of vegetation zone to another. Mongolia estimates that about 66 million ha of land switches from cool temperate desert and steppe to warm temperate desert and steppe. In Venezuela, about 35 million ha change from subtropical forest to tropical forest, and between 40 million and 50 million ha shift from moist to dry or very dry forest. Finally, the results for both the Czech Republic and Estonia suggest that forest areas currently dominated by Norway spruce will become forests dominated by pine or deciduous trees. In Estonia, this change represents a shift from boreal forests to temperate forests.

Global vegetation modeling by Smith et al. (1995) showed that tropical forest areas could increase primarily through conversions from subtropical forest areas, which were expected to decline. The climate change results in Venezuela show this same trend from subtropical to tropical forests. In the next two to five decades, however, land use activities are expected to have larger impacts on tropical forests than climate change (IPCC, in press). The baseline vegetation distribution in Venezuela's analysis was derived from $1\times CO2$ climate conditions; it did not include potential future land use effects on forest distribution. Consequently, the analysis may overestimate the climate-

[2] Since this analysis was completed, work was undertaken to revise the analysis using data from 66 meteorological stations. It is possible that these results may change significantly.

Figure 3.1. Map of Venezuela Showing Holdridge Life Zones

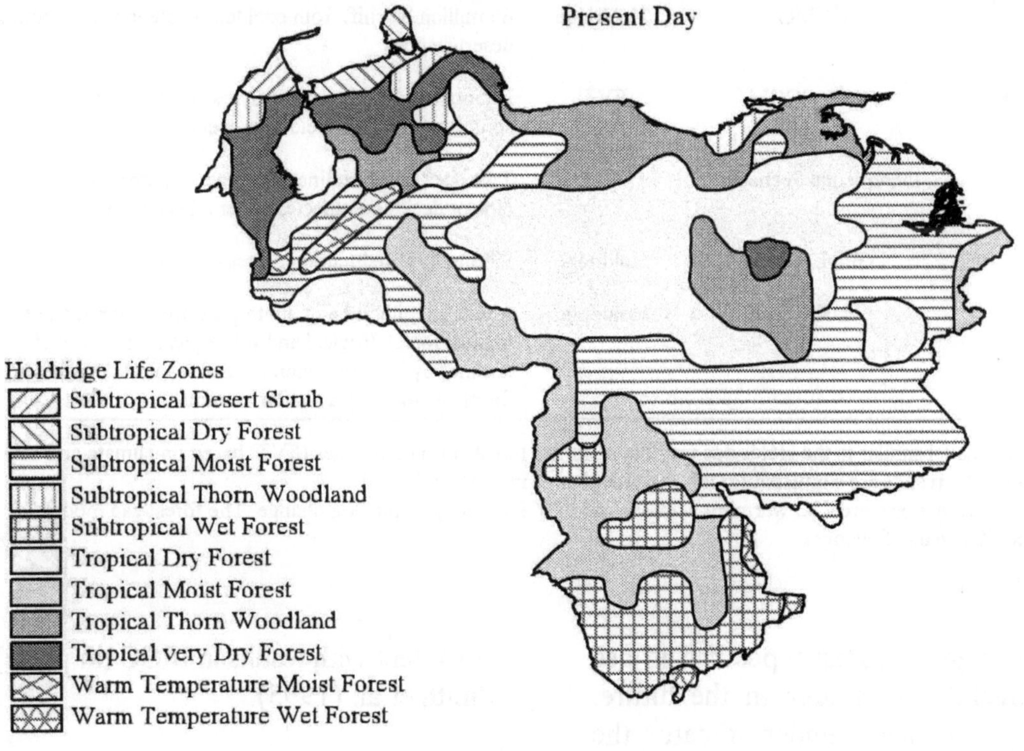

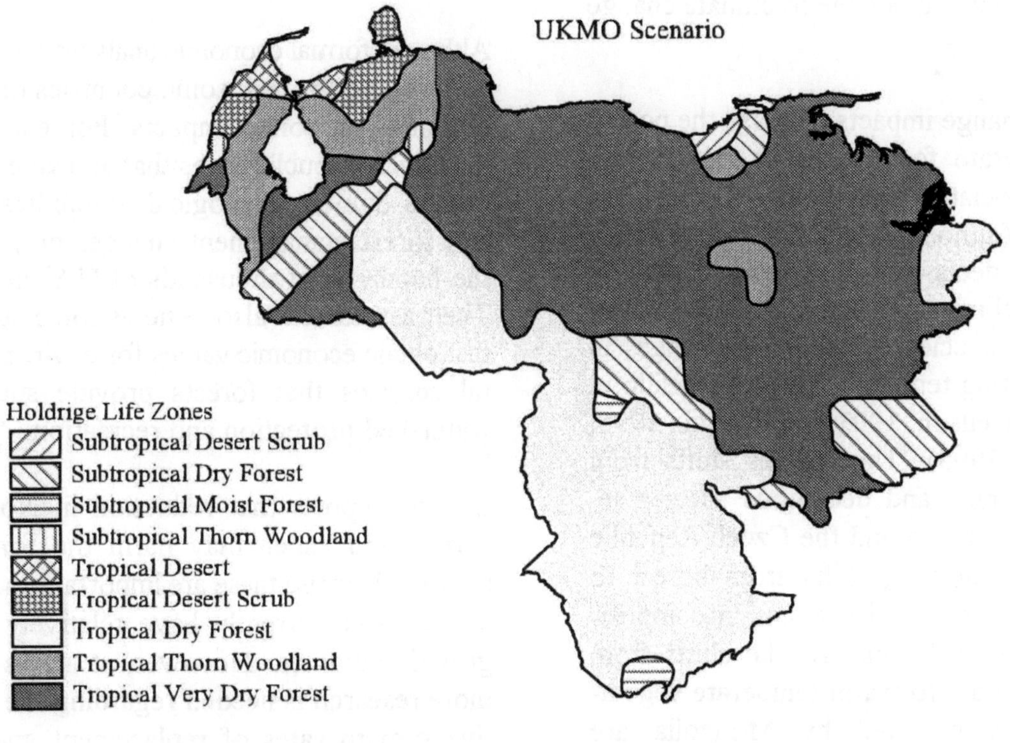

Table 3.4. Summary of Forest Impacts

Country	Scenario	Model	Impact
Mongolia	UKMO	Holdridge	66 million ha shift from cool temperate to warm temperate desert/steppe
Czech Republic	GISS, CCCM	FVZ[1]	National and study area analyses show shift from Norway spruce to pine and deciduous species
Estonia	Incremental Scenarios[2]	Gap	Two sites show decline of spruce and aspen and increase in pine or deciduous species under all scenarios
Ukraine	GFDL, GISS	Holdridge	39% to 47% reduction in forested area
Venezuela	UKMO	Holdridge	About 35 million ha of subtropical forest converted to tropical forest; tropical and subtropical moist or wet forest declines by 40-50 million ha, and tropical dry and very dry forests increase by a similar amount

[1] Based on a model similar to the Holdridge one that identifies forest vegetation zones (FVZ) based on climate and soil conditions. The results are simulated for temperature change in 2010 and 2030.

[2] Incremental scenarios were 4°C warming with 0% and ±10% annual precipitation change. The forest gap model simulated 500 years of changes.

induced changes in forest types if land use reduces overall forest area in the future. Conversely, it may underestimate the changes if future use degrades forest quality, making it more vulnerable to climate change impacts.

Climate change impacts increase the potential temperate forest area as the climate zones associated with boreal forests shift poleward (although the actual forested area may either decline or increase depending on transient climate impacts), and may cause significant species composition changes in many existing temperate forests (Melillo et al., 1993; Neilson, 1993; Smith et al., 1995; VEMAP, 1995). The species shifts from spruce to pine and deciduous species reported by Estonia and the Czech Republic are consistent with shifts from boreal to temperate forest and changes in composition, respectively. Finally, the shifts from cool temperate to warm temperate vegetation zones reported by Mongolia are consistent with results in IPCC (in press) and Smith et al. (1995).

Economic Impacts and Adaptation

Although formal economic analyses have not yet been performed, some countries discuss potential economic impacts. For example, the Czech Republic notes that annual expenditures due to "ecological destabilization" and forest management changes may be in the hundreds of thousands of U.S. dollars. Their assessment also reviews some studies that obtain economic values for environmental services that forests provide such as watershed protection and recreation.

Estonia reports that reduction in Norway spruce and aspen may harm the forestry industry, because these are important harvest species and currently have relatively high growth volumes. However, it notes that more research is needed regarding the relative growth rates of replacement species, including the magnitude of any direct CO_2 effects, before concluding that harvest volumes will necessarily decline.

The Czech Republic suggests that anticipatory adaptation measures such as interplanting deciduous trees in the spruce forests will improve biological diversity in the stands and provide trees that are suited to the altered climate. Furthermore, its assessment concludes that adaptation options are comparable with measures that "ensure stability and well-managed forest use" despite changes in climate. Adaptation measures for the other countries are being formulated, as are compatibility checks with forest sink mitigation policies.

WATER RESOURCES

Changes in temperature and precipitation levels will alter the hydrologic cycle and affect water supplies. The impacts depend on the direction and timing of changes. For example, higher precipitation levels may be beneficial in drier climates, particularly if they result in higher runoff during low flow periods. However, if they occur during already high runoff seasons, they may induce flooding. In general, the IPCC (in press) notes that estimated precipitation increases in high latitude regions may lead to runoff increases, and runoff will tend to decrease in lower latitudes due to combined temperature increases and precipitation decreases. Riebsame et al. (1995) studied five basins in developing countries and found that basins such as the Nile, which have low runoff-to-precipitation ratios and high potential evapotranspiration-to-precipitation ratios, will be more sensitive to climate change than basins at the opposite end of the spectrum such as the Uruguay.

Six of the countries that are studying the effects of climate change on water basins report assessment summaries in this document: Bangladesh, Kazakhstan, Mongolia, Czech Republic, Slovakia, and Ukraine. Of these, Kazakhstan, Mongolia, Czech Republic, and Ukraine also report preliminary impacts, which are discussed below. Bangladesh and Slovakia are still conducting research. All reporting countries use GCM results for $2\times CO_2$, and some include incremental scenarios.

Physical Impacts

The four reporting countries provide basin runoff impacts estimated from four different, national water balance models. Table 3.5 summarizes the effects of climate change on annual runoff. Each country simulated runoff impacts of $2\times CO_2$ for at least two GCMs. Mongolia and the Czech Republic also simulated runoff for several incremental scenarios to derive basin sensitivity to a wide range of conditions.

In these results, only Kazakhstan's two study basins show consistently lower annual runoffs. The other countries' basins show potential runoff increases as well as decreases. The ranges for some basins are very broad, suggesting that the basins are very sensitive to climate change. Mongolia reports that its basins tended to be more sensitive to precipitation changes than to temperature changes. This can be seen in their incremental scenario results, which are shown in Table 3.6. The Czech Republic found its low-to-medium runoff basins to be

Table 3.5. Summary of Water Resource Impacts

Country	Scenario	Basin	Runoff Change (% annual change)
Kazakhstan	CCCM, GFDL	Uba/Ulba	-23% to -29%
		Tobol	-23% to -26%
Mongolia	Incremental Scenarios,[1] GISS, GFDL, UKMO, CCCM	Bogd	-24% to +32%
		Bulgan	-53% to +79%
		Khalkhgol	-39% to +25%
Czech Republic	Incremental Scenarios,[2] GISS, GFDL, CCCM	Elbe, Zelivka, Upa, Metuje	-45% to + 15%[3]
Ukraine	GFDL, UKMO	Dnieper	-20% to +128%

[1] 2°C and 4°C warming with 0%, ±10%, and ±20% annual precipitation change.
[2] 2°C and 4°C warming with 0%, ±5%, and +10% annual precipitation change.
[3] Results not reported for individual basins.

Table 3.6. Annual Percent Change in Runoff for the Selected Basins in Mongolia under Incremental Scenarios[1]

Bogd	P + 0%	P + 10%	P + 20%	P - 10%	P - 20%
T = 0	0%	+18%	+32%	-8%	-21%
T + 2	+3%	+16%	+30%	-10%	-23%
T + 4	+2%	+13%	+29%	-11%	-24%
Bulgan					
T = 0	0%	+42%	+79%	-20%	-42%
T + 2	-2%	+29%	+63%	-28%	-47%
T + 4	-13%	+15%	+47%	-36%	-53%
Khalkhgol					
T = 0	0%	+3%	+19%	-23%	-33%
T + 2	-16%	-3%	+12%	-27%	-36%
T + 4	-20%	-8%	+6%	-31%	-39%

[1] P = precipitation, T = temperature.

more sensitive to climate change than its larger basins.

Ukraine's study of the Dnieper basin is particularly important because that basin provides about 80% of Ukraine's water supply. The analysis predicts a range of runoff changes from more than double the current runoff to a decline of 20%. Finally, although Slovakia has not completed its analysis, it noted that preliminary simulations through 2030 showed 20% decline in annual runoff.

Thus far, groundwater impacts are qualitatively assessed. The Czech Republic notes that groundwater runoff may decline even with a 5% increase in precipitation, if temperature increases by 2°C to 4°C. Bangladesh plans to include water salinity, inundation, and groundwater supply impacts in its study.

Regarding potential seasonal variations such as changes in the timing or volume of spring runoff, Kazakhstan notes that although the water balance model cannot predict runoff variability on a daily basis, the distribution of monthly mean precipitation under climate change will probably reduce the risk of flooding events.

The IPCC (in press) summarizes regional studies of climate change impacts on surface water and groundwater supplies, as well as water quality. Table 3.7 shows these summaries for selected countries. Among the results for northern China, it notes that water basins located in the plains tend to be more sensitive to precipitation changes than basins located in mountain regions. This is similar to Mongolia's finding that the Bulgan basin, the lower portion of which is located in a lowland area with sandy soil, was more sensitive to climate change than the Bogd basin, which is located in a mountainous region.

Economic Impacts and Adaptation

Both the Czech Republic and Ukraine provide preliminary demand information. The Czech Republic reports preliminary estimates of baseline demand in 2075, which are shown in Table 3.8. Aggregate demand would range from 3.2 billion to 3.7 billion m^3, which is less than total current runoff of 10.1 billion m^3. Although the baseline demand estimate does not include climate change impacts, the study comments that agricultural irrigation demand may increase by as much as 40% due to climate change impacts.

Table 3.7. Selected River Basin Summaries from IPCC

Country/River Basin	GCMs/Climate Scenarios	Hydrologic Impact
China: Haihe, Huaihe, and Yellow River Basins	Seven GCM scenarios	Decrease in annual runoff of 2-12%
Hungary: Danube Basin	Incremental scenarios	Winter streamflow increased, summer stream flow decreased, spring flood peak shifted in timing
India: Kolar and Sher Basins	Incremental scenarios	Sher Basin is more sensitive; impacts large enough to influence storage design and operations
Nepal: Lantang Kola Catchment	Not known	River flow increase of 100% with stable glacier area; +30% if glacier area decreases by 30%
South Asia	ECHAM3-Hamburg	Increase in runoff over northeastern India, south China, and Indonesia; decline in runoff possible over north China and Thailand
Ukraine: Poles'e River Basin	IPCC scenarios	Increase in annual runoff of 6%

Source: IPCC (in press).

Table 3.8. Projected Demand for Water in 2075 in Czech Republic

Water Use	Demand (cubic meters)
Drinking Water	1.0-1.3 billion
Irrigation	575 million[1]
Hydropower	924-1,201 million
Industry	652 million
Total	3.2-3.7 billion

[1] Demand could increase by 40% to offset agricultural impacts of drier climate.

Ukraine notes that its economic transition status makes it difficult to project water demand in 2075. As an approximation to demand and supply analysis, it compares simulated annual runoff with current demand levels for three stretches of the Dneiper River. In almost all instances, runoff exceeds demand by a large margin; the ratio of demand to supply ranges from 0.18 to 0.66. Only the Geophysical Fluid Dynamics Laboratory 1% transient (GF01) scenario projects a supply deficiency along one river stretch (demand to supply ratio of 1.08). Ukraine also estimates annual hydropower losses of US$10 million to US$17 million.

Additional demand and supply analysis is needed to determine the socioeconomic effects of the predicted changes in runoff. Estimates of future baseline demand that include foreseeable adaptations to climate change in other sectors (e.g., energy, agriculture) are necessary to approximate the total expected shortfall or surplus.

Adaptation measures are qualitatively assessed by Kazakhstan and the Czech Republic. Kazakhstan briefly discusses surface water and groundwater development strategies. In doing so, it raises the important consideration that surface water development projects will affect water supply in neighboring countries, and consequently involves an additional institutional barrier. The Czech Republic lists several adaptation options, including implementing conservation policies, implementing drought management programs, and improving real-time operation of reservoirs.

COASTAL RESOURCES

One of the most likely effects of climate change is a rise in sea level. Expected impacts on coastal resources include erosion of beaches and cliffs, and inundation of low-lying coastal areas. Urban areas such as The Gambia's capital city of Banjul could suffer from land and infrastructure losses. Rural areas could lose agricultural land, wetlands, and mangrove forests. Coastal recreation areas or protected reserves such as the Laguna de Tacarigua National Park in Venezuela could be dramatically altered. The IPCC (in press) estimates that the coastal population at risk from storm surges alone may more than double if sea level rises by 0.5 m or more. This does not include those affected by erosion or permanent inundation. Nicholls and Leatherman (1995) summarized case study assessments for 11 developing countries. Their conclusions included the following: the most of the at-risk land was in deltas; other at-risk land included high value land near sandy beaches, and coastal wetlands would be significantly reduced worldwide.

Eight of the countries that are studying the potential effects of sea level rise are providing assessment summaries in this report: The

Gambia, Bangladesh, Sri Lanka, Estonia, Poland, Ukraine, Peru, and Venezuela. All except Bangladesh have also provided results (in some cases these results are preliminary). These countries estimate sea level rise impacts based on several different predictions of sea level rise for 2100. The two most common scenarios are a 0.5 m rise and a 1.0 m rise.

Physical Impacts

Erosion and inundation impacts were estimated using a combination of geomorphological information and topographical maps. Coastline shifts for selected sea level rise scenarios were generated and land losses estimated as the difference between present and predicted coastlines. Preliminary impacts were estimated for the entire coastline in The Gambia and Poland, and for selected vulnerable study sites for Sri Lanka, Estonia, Ukraine, Peru, and Venezuela. Table 3.9 summarizes land loss by scenario for each country.

The Gambia estimates that a sea level rise of 1.0 m will inundate coastal mangroves and could threaten the loss of the capital city of Banjul unless protective measures are undertaken. Other urban areas at risk include Gdańsk, Poland; Lima, Peru; and the Lake

Table 3.9. Summary of Coastal Resources Impacts

Country	Scenario (m)	Land Loss (km^2)	Other Impacts	Economic Damages	Adaptation Costs
The Gambia	0.2	5	N/A		US$3.9 million (7 km revetment)
	0.5	46			US$3.1 million (7 km seawall)
	1.0	92		US$217 million (land)	
	2.0	182			
Sri Lanka[1]	0.3	47	N/A	N/A	N/A
	1.0	106			
	2.0	197			
Estonia[2]	1.0	60			
Poland	0.1 to 2.5	845 to 2,203	40,860 to 234,840 population at risk	US$10 billion to US$75 billion capital value at risk	US$2.3 billion to US$9.8 billion for full protection, US$0.01 billion to US$0.50 billion for annual maintenance
Ukraine[3]	0.5	32	N/A	N/A	N/A
	2.0	370			
Peru[4]	1.0	78.32	N/A	N/A	N/A
Venezuela[5]	0.5	85.54	N/A	US$153 million (land + buildings)	N/A
	1.0	124.74			

1 Estimate for southwest coast; combines erosion and inundation losses.
2 Estimate for two study sites only.
3 Estimate for wetland losses along the Black Sea coastline.
4 Estimate for selected vulnerable areas. Sea level rise of 0.3 m was also analyzed, but no results were reported.
5 Estimate for five vulnerable areas. Economic damages represent lost land values and building values for four study sites, excluding potential oil and gas industry losses.

Maracaibo region of Venezuela (a key petroleum industry location). Recreational beaches or shoreside reserves and infrastructure would be lost in regions of most of the countries, including four of Venezuela's five study sites, both of Estonia's study sites, and Sri Lanka's southwest coast.

In addition to loss of land, Poland reports that between 40,860 and 234,840 people are living in areas classified as at risk, depending on the sea level rise scenario. Poland's impact estimates are based on projections of future coastal development. This approach provides a more accurate impact estimate than an approach that estimates impacts based on current coastal development levels such as in The Gambia or Venezuela.

For comparison purposes, Table 3.10 summarizes the sea level rise impacts reported in Nicholls and Leatherman (1995) and IPCC (in press), which generally assume a 1.0 m sea level rise and existing coastal development levels. These include results for Poland and Venezuela that we can compare with results in Table 3.9. For Poland, the land loss and at-risk population impacts of a 1.0 m rise shown in Table 3.10 fall within the impact ranges shown in Table 3.9 for sea level rise scenarios of 0.1 m to 2.5 m. The national estimate of land loss for Venezuela in Table 3.10 is about 45 times greater than the land loss estimate for the case studies that is reported in Table 3.9. However, the national damage values are of the same order of magnitude as the case study damages. We will discuss this point further in the next section.

Table 3.10. Sea Level Rise Impacts Reported in Nicholls and Leatherman (1995) and IPCC (in press)

Country	Scenario (m)	Land Losses (km^2)	Population at Risk	Economic Damages	Adaptation Costs[1]
Bangladesh	1.0	25,000	13 million	N/A	N/A
Malaysia	1.0	7,000	N/A	N/A	N/A
China	1.0	125,000	72 million	N/A	N/A
Egypt	1.0	4,200 to 5,250	6 million	N/A	N/A
Nigeria	1.0	18,398 to 18,803	3.2 million	US$18.1 billion buildings, land, and oil infrastructure	US$1.4 to US$1.8 billion
Senegal	1.0	6,042 to 6,073	0.1 to 0.2 million	US$1.2 billion buildings and land	US$0.9 to US$2.2 billion
Poland	1.0	1,700	235,000	US$24 billion capital value at risk	US$1.5 billion
Argentina	1.0	>3,430 to 3,492	N/A	US$10.7 billion buildings and land	US$1.8 to US$3.3 billion
Uruguay	1.0	94	10,000	US$1.8 billion	US$3.1 to US$3.8 billion
Venezuela	1.0	5,686 to 5,730	60,000	US$349 million[1]	US$1.7 to US$2.6 billion

[1] Adaptation costs shown for total protection; adaptation costs for protecting important areas only are about 30% to 60% of total protection costs.

[2] Minimum value, excludes ports.

Source: Adapted from Nicholls and Leatherman (1995) and IPCC (in press).

Economic Impacts and Adaptation

Table 3.9 includes estimates of economic damages for some of the countries. The Gambia estimates that the value of lost land in the Banjul region from a 1.0 m sea level rise is US$217 million based on land loss estimates and market land values. Poland estimates that between US$10 billion and US$75 billion in capital value is at risk under sea level rise scenarios from 0.1 m to 2.5 m.

Venezuela estimates land and building value losses of about US$153 million for four of its five study sites; the damage value does not include economic impacts to the oil industry in the Lake Maracaibo region, which are expected to be large. If the study site approach in Table 3.9 provides better damage estimates than the national economic damage estimate reported in Table 3.10, then the national estimate probably understates damages by a significant amount. As noted above, the land loss reported in Table 3.10 is 45 times greater than the land loss for the case study sites. However, the national economic damages estimate is only about twice as large as the case study damage estimate. This seems unlikely, particularly because the US$153 million in Table 3.9 excludes oil industry damages. Consequently, actual damages must be greater than the value reported in Table 3.10.

Table 3.9 also shows adaptation costs that were provided for two studies. The Gambia estimated costs for protecting its capital of Banjul. Two options included a 7 km revetment, costing US$3.9 million, and a 7 km seawall, costing US$3.1 million. The Gambia also estimated costs for protecting mangrove regions with dikes. Approximately 16 km of dikes about 1.5 m to 2 m tall would be needed, and the construction cost would be US$542,000.

Poland provides an extensive analysis of three adaptation options by coastal region: retreat, partial protection, and full protection. Depending on the sea level rise scenario, the capital costs for full protection could range from US$2.3 billion to US$9.8 billion, with annual maintenance costs ranging from US$0.1 billion to US$0.5 billion. Comparing the incremental adaptation cost of going from partial to full protection with the avoided damages, the report concludes that the incremental net benefits of full protection are positive; i.e., incremental net benefits are about two to three times greater than incremental capital costs.

Sri Lanka discusses four adaptation options: retreat, selective protection, full protection, and reclamation/expansion seaward. The retreat strategy has been used to adapt to constant coastal erosion during the past six decades. However, that strategy is considered inappropriate for densely populated areas and main transportation routes that may be at risk under climate change conditions. Selective protection would let wetlands migrate landward. Sri Lanka currently has a preservation policy that discourages development near coastlines and protects erodible areas (e.g., revetments currently protect several kilometers of coastline), and the country could expand the policy to provide full protection. The final adaptation measure, reclamation, would use artificial beach nourishment to protect areas such as river mouths that cannot be protected by barriers.

CONCLUSION

The vulnerability and adaptation assessment results presented in this report show potentially significant climate change impacts on four major economic sectors for many of the reporting countries. Furthermore, many of these findings are consistent with earlier studies of agricultural yield changes, vegetation shifts, and sea level rise impacts.

Most of the results reported thus far pertain to biophysical impacts, and some of these are still in the preliminary stages of analysis. Additional research will finalize the biophysical impacts and subsequent socioeconomic impacts of the vulnerability assessments. The results from the vulnerability studies will provide countries with the information necessary to initiate adaptation assessments, which will help countries identify, screen, and evaluate adaptation measures, and prepare their adaptation plans. Several of the countries summarizing their progress in this report have begun to identify potential adaptation options for their vulnerable sectors. In some instances, preliminary cost estimates have been prepared. For example, Poland's sea level rise study indicated that total protection benefits appear to exceed protection costs, and that the incremental benefits of moving from partial to total protection also exceed the incremental costs. Additional analysis is needed to identify potential implementation barriers such as lack of funding, and prepare an implementation plan that overcomes those barriers.

REFERENCES

Bazzaz, F.A. and E.D. Fajer. 1992. "Plant Life in a CO_2-Rich World." *Scientific American* 26(1): 68-74.

Darwin, R., M. Tsigas, J. Lewandrowski, and A. Raneses. 1995. *World Agriculture and Climate Change: Economic Adaptations*. Agricultural Economic Report Number 703. Washington, DC: U.S. Department of Agriculture.

Downing, T.E. 1992. *Climate Change and Vulnerable Places: Global Food Security and Country Studies in Zimbabwe, Kenya, Senegal, and Chile*. Research Report No. 1, Oxford, UK: Environmental Change Unit, University of Oxford.

IPCC (in press). *Climate Change 1995: The IPCC Second Assessment Report, Volume 2: Scientific-Technical Analyses of Impacts, Adaptations, and Mitigation of Climate Change* [Watson, R.T., M.C. Zinyowera, and R.H. Moss (eds.)]. Cambridge: Cambridge University Press.

Kane, S., J. Reilly, and J. Tobey. 1992. "An Empirical Study of the Economic Effects of Climate Change on World Agriculture." *Climatic Change* 21: 17-35.

Melillo, J.M., A.D. McGuire, D.W. Kicklighter, B. Moore, C.J. Vorosmarty, and A.L. Schloss. 1993. "Global Climate Change and Terrestrial Net Primary Productivity." *Nature* 363: 234-240.

Neilson, R.P. 1993. "Vegetation Redistribution: A Possible Biosphere Source of CO_2 During Climate Change." *Water, Air, and Soil Pollution* 70: 659-673.

Nicholls, R.J. and S.P. Leatherman. 1995. "Global Sea Level Rise." In *As Climate Changes: International Impacts and Implications*. K.M. Strzepek and J.B. Smith (eds.) Cambridge: Cambridge University Press.

Parry, M.L., T.R. Carter, and N.T. Konijn (eds.). 1988a. *The Impacts of Climate Variations on Agriculture: Volume 1: Assessments in Cool Temperate and Cold Regions*. Dordrecht: Kluwer Academic Publishers.

Parry, M.L., T.R. Carter, and N.T. Konijn (eds.). 1988b. *The Impacts of Climate Variations on Agriculture: Volume 2: Assessments in Semi-Arid Regions*. Dordrecht: Kluwer Academic Publishers.

Parry, M.L., M.B. de Rozari, A.L. Chong, and S. Panich. 1992. *The Potential Socio-economic Effects of Climate Change in Southeast Asia*. Nairobi: United Nations Environment Programme.

Qureshi, A. and D. Hobbie (eds.) 1994. *Climate Change in Asia: Thematic Overview*. Manila: Asian Development Bank.

Reilly, J., N. Hohmann, and S. Kane. 1994. "Climate Change and Agricultural Trade: Who Benefits and Who Loses?" *Global Environmental Change* 4(1): 24-36.

Riebsame, W.E., K.M. Strzepek, J.L. Wescoat, Jr., R. Perritt, G.L. Gaile, J. Jacobs, R. Leichenko, C. Magadza, H. Phien, B.J. Urbiztondo, P. Restrepo, W.R. Rose, M. Saleh, L.H. Ti, C. Tucci, and D. Yates. 1995. "Complex River Basins." In *As Climate Changes: International Impacts and Implications*. K.M. Strzepek and J.B. Smith (eds.) Cambridge: Cambridge University Press.

Rosenzweig, C. and A. Iglesias (eds.). 1994. *Implications of Climate Change for International Agriculture: Crop Modeling Study*. EPA 230-B-94-003. Washington, DC: U.S. Environmental Protection Agency.

Rosenzweig, C. and M.L. Parry. 1994. "Potential Impact of Climate Change on World Food Supply." *Nature* 367: 133-138.

Smith, T.M. and H.H. Shugart. 1993. "The Transient Response of Terrestrial Carbon Storage to a Perturbed Climate." *Nature* 361: 523-526.

Smith, T.M., R. Leemans, and H.H. Shugart. 1992. "Sensitivity of Terrestrial Carbon Storage to CO_2-Induced Climate Change: Comparison of Four Scenarios Based on General Circulation Models." *Climate Change* 21: 367-384.

Smith, T.M., P.N. Halpin, H.H. Shugart, and C.M. Secrett. 1995. "Global Forests." In *As Climate Changes: International Impacts and Implications*. K.M. Strzepek and J.B. Smith (eds.) Cambridge: Cambridge University Press.

Solomon, A.M. 1986. "Transient Response of Forests to CO_2-Induced Climate Change: Simulation Modeling Experiments in Eastern North America." *Oecologia* 68: 567-579.

VEMAP Participants. 1995. "Vegetation/Ecosystem Modeling and Analysis Project (VEMAP): Comparing Biogeography and Biogeochemistry Models in a Continental-Scale Study of Terrestrial Ecosystem Responses to Climate Change and CO_2 Doubling." *Global Biogeochemical Cycles* 9: 407-437.

Chapter 4
Regional Summary

Saleemul Huq,[1] Luis Jose Mata,[2] Ivana Nemesova,[3] Sekou Toure[4]

1 **Bangladesh Centre for Advanced Studies, Bangladesh**
2 **Institute of Fluid Mechanics, Universidade Central de Venezuela, Venezuela**
3 **Institute of Atmospheric Physics, Czech Academy of Sciences, Czech Republic**
4 **Ecole Nationale Supérieure des Travaux Publics, Cote d'Ivoire**

INTRODUCTION

The countries participating in the U.S. Country Studies Program are found throughout the world and represent four distinct regions: Africa, Asia-Pacific, Central and Eastern Europe, and Latin America. These regions contain mainly developing countries or countries in transition to market economies. This chapter provides an interim regional summary of research by 13 countries at various stages of completion in their vulnerability and adaptation assessments. This chapter summarizes by region the vulnerability and adaptation results that have been reported in this document. This chapter also identifies some of the common geographic, climatic, demographic, and/or socioeconomic characteristics that define the regions and influence the vulnerability of the countries within those regions to climate change. These results contribute to the literature on potential vulnerability and adaptation on climate change and are presented in a manner that will facilitate comparisons among countries in a region or between various regions. While it may be possible to identify general positive or negative trends within a region, the results are in almost all cases preliminary, caution should be used in comparing results that were developed using different methodologies, and the number of countries reporting results for each sector does not provide adequate information to assess the region as a whole.

SUMMARY FOR THE AFRICAN REGION

Region-Specific Issues

Geography and Climate

The African continent encompasses a rich mosaic of ecological settings. Africa's vast expanse of 30,244,050 km^2 stretches from the Mediterranean in the north to the meeting point of the Atlantic and the Indian oceans in the south (Figure 4.1).

The continent is characterized by several climatic regimes and ecological zones. All parts of Africa, except the Republic of South Africa and the Mediterranean countries north of the Sahara, have tropical climates. These tropical climates may be divided into three distinct climatic zones: (1) wet tropical climates, (2) dry tropical climate, and (3) alternating wet and dry climates.

J. B. Smith et al. (eds.), Vulnerability and Adaptation to Climate Change, 69–108.

Figure 4.1. Map of the African Region

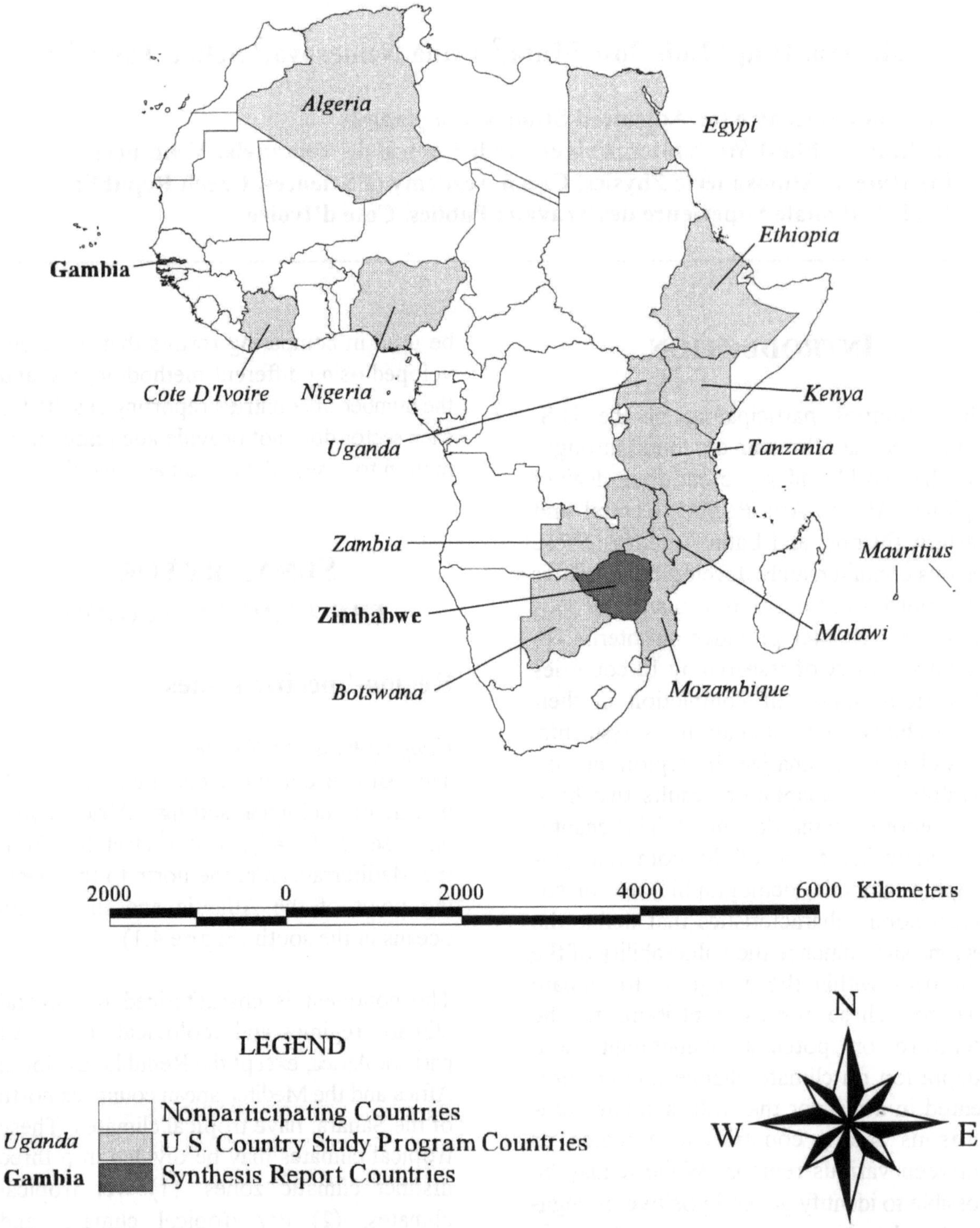

Water resources are unevenly distributed on the continent, and include rivers such as the Nile, Senegal, and Kagera with large drainage basins that may stretch across several countries. Coastal and tropical areas are blessed with large quantities of water resources; in contrast, the landlocked countries suffer from severe water shortages.

The region is also known for its shrinking tropical dense forest, which has been severely degraded. The continent is losing 3.6 million ha of its tropical forests annually, representing an average annual loss rate of 0.6%, and the continent is rehabilitating only a few thousand ha during the same period. (African Development Bank, 1991).

Africa includes some of the world's most unique and diverse coastal and marine ecosystems. The region includes five oceanic realms, four coastal realms, and five faunal provinces. Landward of the marine realms, the terrestrial biogeography of these coastal zones is even more varied, ranging from submerged coastal forest to desert scrub. Together these systems harbor a wealth of economically and biologically important resources, from individual species to productive habitats.

Demographic Conditions and Trends

The African region typically has high population density on a small percentage of the land. In addition, estimates of the population in Africa show rapid growth rates relative to other parts of the world. Population is estimated to grow from 627 million in 1990 to 2.4 billion in 2075. The region's population is estimated to account for 13.4% of the world population in 2000, and this figure is estimated to reach 23.1% by 2075 (World Bank, 1994). In general, rapidly increasing populations have put enormous pressure on the scarce productive lands available in many countries, particularly those inland countries where extreme climatic events have occurred recently. Land degradation has been attributed for the most part to poor cultivation methods, deforestation, and overgrazing. Population pressure also significantly affects many areas with fragile ecosystems. This is the case along the West African coast, where coastal zone ecosystem productivity is already threatened by high population densities, urbanization, and destructive patterns of natural resource exploitation.

Socioeconomic Conditions and Trends

Economic indicators project relatively poor economic growth in Africa in the future. The economies are characterized by low GNP; some of the poorest countries of the world are located here. With the exception of a few countries, the region is in economic turmoil. Economic conditions have generally been worsening for over a decade now, and poverty is widespread. Development priorities are thus oriented toward meeting the most basic needs, often at the expense of sound environmental policies. The deterioration of economic conditions is also leading to the deterioration of educational capacity building.

Climate Change Vulnerability

Vulnerability to climate change is highly integrated between various natural resource sectors and highly dependent on local or regional resources, management systems, and socioeconomic conditions. Thus, the factors described below are intended to characterize the region, rather than identify the most critical factors contributing to vulnerability in a particular region.

- **Desertification and Drought.** Deserts such as the Sahel and the Kalahari cover a significant portion the total land area of the continent. The Sahara consists of a strip of marginal land that is 500 to 1,000 km wide and stretches 7,000 km from west to east, from the Atlantic beaches of Senegal, The Gambia, and Mauritania to the barren mountains of Ethiopia and Somalia. In the past 50 years, about 650,000 km^2 have been transformed into deserts; currently between 50,000 to 70,000 km^2 a year are turning barren (WRI, 1994). In past years, Africa has seen unimaginable human suffering because of extreme climatic events such as drought and famine. Several inland countries have been affected, including Ethiopia. A dramatic decline in rainfall in the Sahel has contributed to two devastating droughts since 1968. If climate changes toward more frequent extreme events in these region, the results could be catastrophic for the continent. In studies on climate variability, IPCC (in press) and Mearns et al. (in press) found that periodic droughts have significant negative affects on agriculture production and higher-order social impacts.

- **Dependence on Agriculture.** Africa is also known for its high dependence on agriculture. Among the developing regions of the world, only South Asia shares Africa's level of dependence on agriculture resources relative to industry. Agriculture employs approximately two-thirds of the labor force in both Africa and South Asia (Biodiversity Support Program, 1993). The agricultural labor force represented 74% of the total labor force of Africa in 1970 (WRI, 1994). This figure was down to 63% in 1990, but remained far above the world average, which was 47% in 1990 (WRI, 1994). With little prospect for economic progress in the near future, this sector will remain the primary employer. Thus any major disruption in this sector due to climate variability, or any other cause, may be detrimental to the whole continent.

- **Sensitive Biological Resources.** Several species and ecosystems are specific to the African region, and because of the threat of their extinction or destruction, have been protected. For example, over 9 million ha of mangroves are found within the coastal wetlands of West Africa (a small portion of which is protected). Threats to these species or ecosystems may come from variabilities in global as well as local climate conditions, and the impacts could be more acute for some of the already fragile species such as the mangroves.

Participating Countries

The countries in the African region that are participating in the U.S. Country Studies Program are shown in Figure 4.1. Of the participating countries, two have submitted chapters to this report: The Gambia and Zimbabwe. Both are Round 1 countries, and the results presented are fairly complete. Table 4.1 lists the sectors analyzed by the countries, as well as the scenarios used, the assessment methods used, a summary of results, and the progress of the work.

Table 4.1. African Countries Contributing to the Interim Synthesis Report

Country	Sectors	Scenarios	Vulnerability Assessment Methods	Summary of Results	Progress
The Gambia (Round 1)	Agriculture	CCCM, GFD3, GF01,[1] GISS, UKMO	DSSAT 3	Change in annual yields: -15% to -26% (Maize) -29% to -44% (Early Millet) -14% to -21% (Late Millet) +40% to +52% (Groundnuts)	Biophysical results
	Coastal	SLR of 0.2 m, 0.5 m, 1.0 m, 2.0 m	IPCC	Land loss for entire country: 46 km^2 under 0.5 m SLR 92 km^2 under 1.0 m SLR	Biophysical results
Zimbabwe (Round 1)	Agriculture	CCCM, GFD3	CERES-Maize	Decline in annual yields: -13.6% to -11.5%[2]	Biophysical results

[1] Results of the transient scenarios are not included.
[2] Weighted average of all study sites.

Agriculture

Agriculture has been, is, and will remain the main contributor to the economies of most African nations. However, nowhere in the world is the unsustainability of agricultural practices as profound as in most of Africa. These conditions make the agriculture sector in Africa more vulnerable to climate change. Pressure on marginal lands is intensifying the process of the already alarming desertification, and periodic droughts exacerbate this process. In other areas, population pressures and government policies have forced peasants to abandon their traditional patterns of cultivation and adopt unsustainable practices, such as using shorter fallow periods in areas where shifting cultivation and slash and burn has been practiced. Despite the importance of agriculture in the economies of most African countries, the region cannot feed itself and is dependent to a large extent on imports of basic goods.

Although it is difficult to reliably predict regional impacts to agriculture because of a lack of reliable geographic resolution in future climate estimates and in estimates of autonomous adaptations, some trends for the region have been identified. Recent studies (IPCC, in press; Downing, 1992; Magadza, 1994; Rosenzweig and Iglesias, 1994) indicate that the agriculture sector in most of Africa will be sensitive to climate change, although some regions may benefit from warmer and wetter conditions. IPCC (in press) found that Africa could experience relatively large reductions in yields due to climate change. A number of factors contributing to vulnerability have been identified, including crop yields, the viability of the farming sector, the sensitivity of the regional or national economy to farm sector impacts, and factors that contribute to exposure to hunger (IPCC, in press). Applying these factors to various geographic regions, many of the populations in sub-Saharan Africa appear to be the most vulnerable to climate change. Large areas of the region are arid or semi-arid; the average per capita income is among the lowest in the world; and the economy of the region is highly dependent on agriculture. In addition, in recent decades droughts and political and civil instability have increased the vulnerability of the region. North Africa appears to be relatively less sensitive, with a smaller share of the population (32%) directly dependent on agriculture and a diversity of agricultural crops (FAO, 1992 as cited in IPCC, in press). However, North Africa is generally arid and the current climate already limits the land suitable for agriculture.

Both The Gambia and Zimbabwe reported on the agriculture sector. Both countries used the Decision Support System for Agrotechnology Transfer version 3 (DSSAT 3) model (Benioff et al., in press) with general circulation model (GCM) scenarios to assess the vulnerability of important crops in the region. Both assessments included the direct effects of CO_2.

Agriculture in The Gambia

The Gambia selected several crops (maize, early millet, late millet, and groundnuts, or peanuts) for the vulnerability assessment. The assessment was conducted at four study sites, each of which represented an area where one of the selected crops is cultivated. The DSSAT 3 was used with $2\times CO_2$ scenarios from the Canadian Climate Centre Model (CCCM), Geophysical Fluid Dynamics Laboratory equilibrium (GFD3), and Goddard Institute for Space Studies (GISS) models (Benioff et al., in press) to generate estimated changes within the sector. The assessment found that the predicted climate change associated with a doubling of CO_2 levels for 2075 could lead to a decrease in dry matter production and nitrogen utilization in all the cereal crop studied.

A decrease in maize biomass production (total dry matter production) of 15% to 35% was estimated for The Gambia. Total dry matter production of early millet is estimated to decrease by 4% to 43%. For late millet the estimated decrease in dry matter production was 14% to 21%. Groundnuts, on the other hand, would do better under the $2\times CO_2$ conditions: dry matter production would increase from the current 7,950 kg dm/ha by 40% to 52%. (The range of results is due to use of different GCM scenarios.)

Agriculture in Zimbabwe

Zimbabwe assessed the vulnerability of maize at four representative study sites. The assessment used the CERES-Maize model (Benioff et al., in press) and the DSSAT 3 with climate change scenarios from the GFD3 and CCCM models. The results using both GCMs suggest that maize yields may increase by up to 37% or decrease by as much as 98%, depending on the study site. However, the average yield change across the sites (calculated using national production shares as weights) is -13.6% to -11.5%. The models also suggest that yields at the site with the largest estimated decrease could be increased over baseline conditions by 3% to 16% for the CCCM and GFD3 models, respectively, when an earlier planting date and short season variety of maize is used. However, the preliminary nature of this assessment must be borne in mind. Adaptation options seem to be an indispensable tool in this sector to offset climate change impacts.

Coastal Resources

Entire countries in the African region would be affected by a 1.0 m sea level rise, including Cape Verde, São Tomé and Príncipe, Comoros, Mauritius, Réunion, and

Seychelles, where coastal areas represent 100% of the country (WRI, 1994). Analysis of climate variability in Africa is illustrated by the three main ecological coastal zones: the west coast, the east coast, and the Mediterranean coast.

The west coast is 8,000 km long, from Mauritania to Namibia, and stretches between the Tropic of Cancer and 30°S latitude. This zone spans a broad range of habitats and biota and includes the pristine islands of the Bijagós Archipelago; the offshore island nations of Cape Verde, São Tomé, and Príncipe; and the remote central Atlantic islands of San Helena and Ascension. The west coast is often buffeted by storm surges and is currently at risk from erosion, inundation, and extreme storm events. Coastal erosion has already been reported to reach 23 to 30 m per year (Ibe and Quelennac, 1989). Inundation could be of significant concern. As reported by The Gambia, under a 1.0 m sea level rise the capital city of Banjul would be at risk of inundation in the next 50 to 60 years. Other major cities such as Abidjan, Lagos, and Port Harcourt, all situated at sea level, would also be very vulnerable (Awosika et al., 1993). Finally, tide waves, storm surges, and hazards may also increase and may modify littoral transport (Allersman and Tilmans, 1993).

The coastal zone of East Africa will also be affected, although, unlike West Africa's Atlantic coast, this area experiences calm conditions through much of the year. This area includes the coastal lowlands extending from Sudan to South Africa, the near-shore islands off the coast of Tanzania and Mozambique, and the oceanic islands of Madagascar, the Seychelles, Comoros, Mauritius, and Réunion. Along the east coast of Africa, sea level rise and climatic variation may decrease the attenuation of the coral and patch reefs that have evolved along major sections of the continental shelf. The desert margins of the Red Sea feature some of the richest coral reefs in the world. The coral reefs farther south, extending from Kenya to the Tropic of Capricorn, are well distributed around most of the oceanic islands. They buffer the coastline against the impact of breakers and the full force of storms and cyclones. The lessening of this buffer effect as a result of climate change would increase the potential for erosion of the east coast. Many principal East African cities are located inland. However, despite their low densities, coastal cities in this region are experiencing high population growth rates (Dar-es-Salaam, 6.7%, Mombassa, 5%) (World Bank, 1994). This increase combined with the likelihood of a 1.0 m sea level rise could create conditions for significant negative impacts on the tourism-oriented economies, ecology, and natural habitats of this area.

The Gambia has assessed land loss on the coast and sheltered coast of the country due to inundation under 0.2 m, 0.5 m, 1.0 m, and 2.0 m sea level rise scenarios. The Gambia estimated that potential land loss would be 45.89 km² and 92.32 km² under a 0.5 m and a 1.0 m sea level rise, respectively, and that land retreat under a 1.0 m sea level rise would be approximately 2,348 km². A 1.0 m sea level rise would threaten many ecosystems and human developments, including the mangrove systems of the country and Banjul, the capital city. The assessment suggests that damages from a 1.0 m sea level rise would be approximately US$217 million, based on estimates of land loss and market values.

Sand management and repair of the groin system were identified as important short-term adaptation policies, and planning and regulations were described as important anticipatory adaption policies.

Summary and Discussion

Of the four regions being studied in the report, Africa may be on the whole the poorest and have the highest proportion of the GNP sensitive to climate. Africa is also projected to have a very high population growth rate. Thus it may be one of the regions most sensitive to climate change.

Because Africa is highly dependent on agriculture and subject to periodic drought and famine, the potential impact of climate change on agriculture is very important for the region. The results of the vulnerability assessments for The Gambia and Zimbabwe indicate that some crops and some locations are quite vulnerable to climate change while others are much less vulnerable. With appropriate adaptation policies, many vulnerabilities could be reduced. These studies are consistent with many recent regional studies in that they demonstrate the variability in estimated yield impacts among countries, scenarios, methods of analysis, and crops (IPCC, in press). Thus, the ability to extrapolate these finding to assess the region as a whole is limited. Other comparisons to recent research on the vulnerability of agriculture to climate change can also be made.

- The Gambia estimated a decrease in dry matter production in all the cereal crops studied, while the production of groundnuts, a major cash crop, was estimated to increase. These findings, like many recent assessments (IPCC, in press) suggest the importance of variations in agriculture systems, climates, resources, and economic characteristics across and within countries in determining the extent of vulnerability and the potential for adaptation. This is the first study to examine climate change impacts in The Gambia.

- Zimbabwe reported an average yield decrease (calculated using national production shares as weights) of -13.6% to -11.5% in maize production. This is consistent with recent studies of maize yields in Zimbabwe by Muchena (1994) as cited in IPCC (in press) that estimated decreases of -40% to -10% under GISS, Geophysical Fluid Dynamics Laboratory (GFDL), and United Kingdom Meteorological Office (UKMO) scenarios (Benioff et al., in press) and -17% to -5% (under +2/+4°C; ±20% precipitation), respectively.

The coastal resources in the region, particularly in West Africa, appear to be at risk from sea level rise. However, adaptation policies may be relatively easier to implement in Africa, where infrastructures are still being built.

- The Gambia estimated that under a 1.0 m land loss would be 92.32 km^2 and land retreat would be approximately 2,348 km^2. The associated damages would be approximately US$217 million. A global assessment of vulnerability to sea level rise ranked The Gambia as the sixth most vulnerable country according to the population at risk per 1 million inhabitants (Delft Hydraulics, 1993). Thus, the vulnerability of The Gambia may be higher than other countries of the region.

It should be stressed that this summary is drawn from the results of only two countries and a few sectors. It is important that the preliminary data presented here be supplemented by work being done in other countries and by more complete assessments as they become available. Nonetheless, the results presented here are consistent with other studies that show a relatively high sensitivity to climate change in Africa.

SUMMARY FOR THE ASIAN-PACIFIC REGION

Region-Specific Issues

Geography and Climate

The Asian-Pacific region includes both the major land mass of Asia as well as thousands of islands in the Indian and Pacific oceans (Figure 4.2). Thus it is probably the most heterogenous region of the world, not only in terms of geography but also in terms of its climate and socioeconomic conditions (for the purposes of this report Kazakhstan is included in the Asian-Pacific region).

One of the main climatic features of significance in this region is the typhoons generated in the Pacific Ocean that are a threat to the island countries in the Pacific, including Japan and the Philippines, as well as to the mainland, principally China. Another major ocean-related climatic feature is the South Asian Monsoon, which originates in the Indian Ocean in the early part of summer and travels northward. It brings with it the majority of the precipitation that falls in the South Asian region, and is the most critical factor in the rainfed agriculture of the region. The Bay of Bengal also generates a number of cyclonic storms every year. An important impact of climate variability in the Asian-Pacific region is drought due to low precipitation and high temperatures. This can cause catastrophic impacts on agriculture and populations, as has happened in both China and India in recent decades.

Demographic Conditions and Trends

The Asian-Pacific region, with a population of more than 3 billion in 1990, has the highest population of any region of the world, and it is also growing at a relatively high rate of 1.54%; it is projected to grow to more than 6 billion by 2075 (World Bank, 1994). It includes the most populous countries of the world, such as China and India, and also the most densely populated, such as Singapore and Bangladesh. The average population density of the region in 1993 was 1,229 per km^2, compared to the global average of 427 per km^2 (WRI, 1994).

A particular feature of the population in this region is its concentration in some of the areas that are more vulnerable to climate change, such as the coasts of Bangladesh, Indonesia, and the Philippines, as well as the river deltas of the Ganges in India and Bangladesh and Hwang Ho in China. These populations will most likely be subjected to relatively higher risks due to flooding and sea level rise.

Figure 4.2. Map of the Asian-Pacific Region

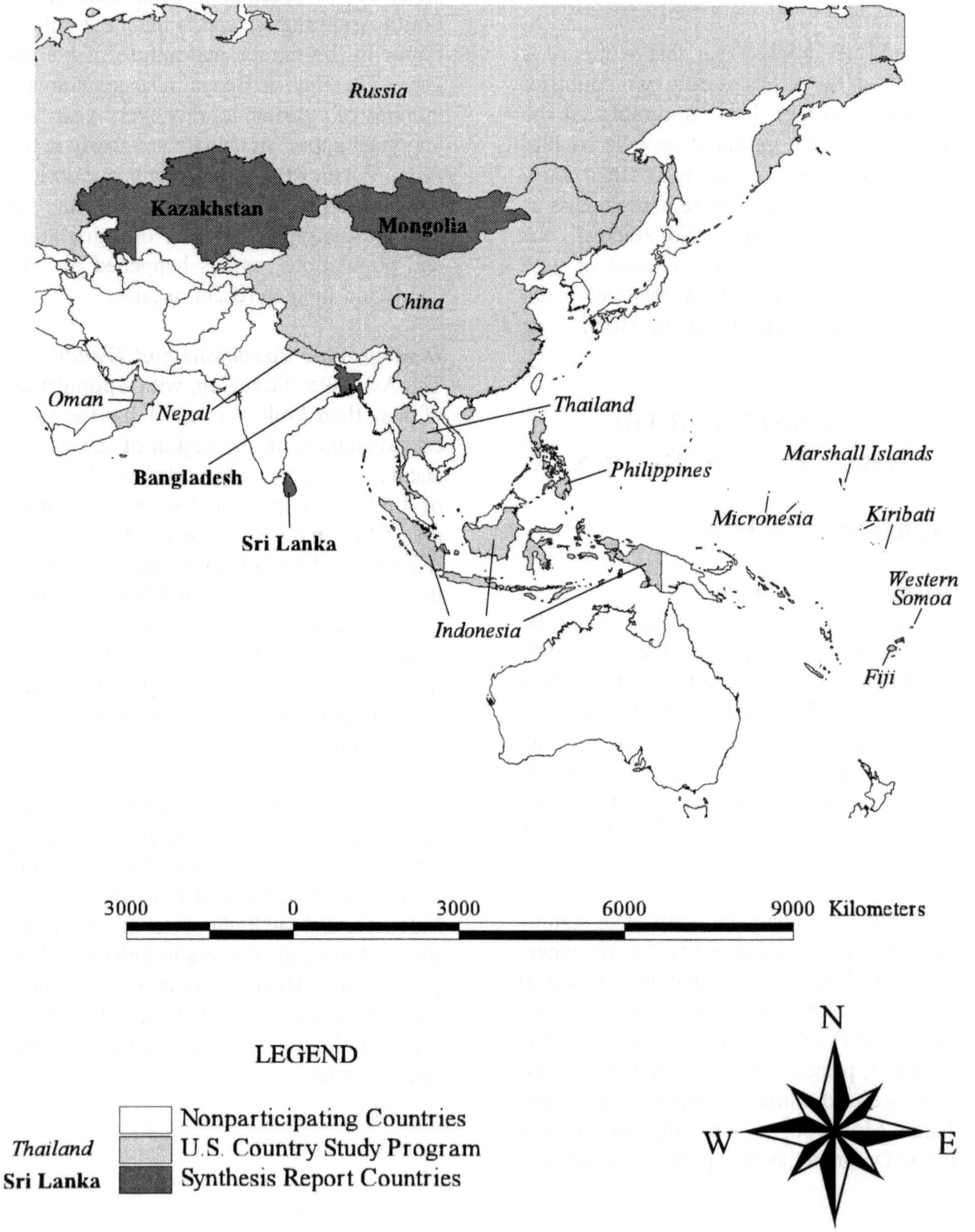

Socioeconomic Conditions and Trends

The region has a large variety of socioeconomic conditions, both among and within countries. For example, Japan, with a per capita GNP of over US$26,000, lies within the region, as do the populous and less developed countries of South Asia with per capita GNPs of US$200 to 300 (WRI, 1994). Thus these countries present a spectrum from some of the richest to some of the poorest in the world.

The populations of the region are still largely rural: nearly 66% of the population was living in rural areas in 1995. The rural population is largely dependent on agriculture, which to a large extent is still rainfed and hence is relatively more vulnerable to climate change impacts (WRI, 1994). For example, agriculture contributes between one-quarter and more than one-half the GDP in some of the most populous countries, including China, India, Indonesia, Bangladesh, and Nepal (WRI, 1994).

Climate Change Vulnerability

Vulnerability to climate change is highly integrated between various natural resource sectors and highly dependent on local or regional resources, management systems, and socioeconomic conditions. Thus, the factors described below are intended to characterize the region, rather than identify the most critical factors contributing to vulnerability in a particular region.

- **Extensive Coastal Areas.** Sea level rise would endanger the very existence of some small island countries such as the Maldives in the Indian Ocean and Kiribati in the Pacific Ocean. It would also affect very sizeable populations living in the low-lying deltas of the major rivers, such as those along the coasts of Bangladesh, India, and China. Sea level rise would also threaten significant biodiversity reserves such as the mangrove forests of Bangladesh, India, and Indonesia, and many of the island countries as well as other important coastal wetlands. A recent global assessment of vulnerability to sea level rise found that almost half of the world's population at risk from sea level rise is located along the coasts of the Indian Ocean, and a weighted ranking of the most vulnerable coastal regions of the world (incorporating risk to population, GNP, and wetlands) indicated that the Indian Ocean coast of the Asian region is the most vulnerable coastal region of the world (Delft Hydraulics, 1993).

- **Estimated High Temperatures and Low Precipitation**. Higher annual average temperatures, if not accompanied by significantly higher annual precipitation, would make the more drought-prone parts of South Asia, China, and Central Asia much more vulnerable and could lead to significant shifts in vegetation types. For example, a recent study found that the more arid parts of China are particularly vulnerable to the possibility of mid-latitude continental drying (Lin, 1994, as cited in IPCC, in press). Much of the rainfed agriculture and pasture would be adversely affected. The significance of this estimated impact is heightened by the large populations that are potentially affected.

- **Vulnerability to Extreme Events**. The potential for increases in the severity and frequency of extreme events could have the most devastating impact of all since those events could significantly damage agriculture and infrastructure (from floods and droughts) and human lives (from cyclones and typhoons). The entire South Asian monsoon-dependent agriculture sector is very vulnerable, not to mean temperatures and precipitation, but to the timing of the onset of the monsoon and its length. In addition, the high concentration of populations in coastal areas may be vulnerable to storms and storm surges. While climate scientists have not yet resolved how large-scale atmospheric patterns such as the El Niño/Southern Oscillation and tropical monsoons may be affected by long-term climate change (IPCC, in press), and there is conflicting evidence concerning estimated changes in the frequency and intensity of tropical cyclones (IPCC, in press), the potential for damages from changes in these events is quite high.

Participating Countries

The countries in the Asian-Pacific region that are participating in the U.S. Country Studies Program are shown in Figure 4.2. Of the participating countries, four have submitted chapters to this report: Bangladesh, Kazakhstan, Mongolia, and Sri Lanka. Mongolia and Kazakhstan are Round 1 countries and thus the results presented are more complete; Sri Lanka and Bangladesh being Round 2 countries have not presented complete results in this interim report. Sri Lanka has presented only coastal sector results and Bangladesh has presented only methods and no results. Table 4.2 lists the sectors analyzed by the countries, as well as the scenarios used, the assessment methods used, a summary of results, and the status of the work.

Agriculture

Agriculture is the most important economic activity for many countries in the region, although its significance varies from country to country. The total cropland is nearly 0.5 million km^2, representing about 40% of the world's total. The Asian-Pacific region produces nearly half of the world's cereals output and the major portion of the world's rice (WRI, 1994). The vulnerability of the agriculture sector in the region is further increased because the region generally supports a large population per hectare of cropland: 5.4 people per hectare in South Asia, 5.7 people per hectare in Southeast Asia, and 12.6 people per hectare in East Asia (FAO, 1992, as cited in IPCC, in press). For all other regions of the world, the ratio of population per hectare of crop land is less than 3.7 (FAO, 1992, as cited in IPCC, in press). The contribution of agriculture to GDP varies across countries, from 3% in Japan to 37% in Bangladesh, 31% in India, 28% in China, and as high as 59% in Nepal (WRI, 1994). A significant portion of agriculture in the region is still dependent on rainfed conditions, particularly in South and Central Asia, which makes it relatively more vulnerable to climate change.

Agriculture in South and Southeast Asia is dependent on tropical storms, and changes in the severity or frequency of these storms would have significant impacts. However, it is currently difficult to estimate how tropical storms may change and effect specific regions as a result of climate change. Recent studies estimating the impact of climate change on rice yields in South and Southeast

Table 4.2. Asian and Pacific Countries Contributing to the Interim Synthesis Report

Country	Sectors	Scenarios	Vulnerability Assessment Method	Summary of Results	Progress
Bangladesh (Round 2)	Agriculture	GFDL	CERES-Rice	N/A	Methods only
	Water	GFDL	MIKE11	N/A	Methods only
	Coastal	SLR	Modified IPCC	N/A	Methods only
	Forest	To be selected	Water salinity projections	N/A	Methods only
Kazakhstan (Round 1)	Agriculture	GFD3, GF01,[1] CCCM	CERES-Wheat	Change in annual yields: -70% to -25% (spring wheat) -35% to +17% (winter wheat)	Biophysical results
	Grassland	GFD3, GF01,[1] CCCM	Dynamic Productivity Model	Decline in wool productivity: -11% to -19% Decline in lambing: -2% to -26%	Biophysical results
	Water	GFD3, GF01,[1] CCCM	KazNIGMI Model	Change in annual runoff: -23% to -29% (Uba/Ulba Basin) -23% to -26% (Tobol Basin)	Biophysical results
Mongolia (Round 1)	Agriculture	GISS, GFD3, Historical	CERES-Wheat (spring)	Change in annual yields: -74.3% to +32.0%	Biophysical results
	Water	GISS, GFD3, CCCM Incremental	Basin Conceptual Model	Change in annual runoff: -24% to +32% (Bogd Basin) -53% to +79% (Bulgan Basin) -39% to +25% (Khalkhgol Basin)	Biophysical results
	Forest	UK89	Holdridge model	66 million ha shift from cool temperate to warm temperate desert/steppe	Biophysical results
Sri Lanka (Round 2)	Coastal	SLR 0.3 m, 1.0 m, 2.0 m	Modified IPCC	Land loss for the southwest coast: 106 km^2 under a 1.0 m SLR	Biophysical and preliminary socioeconomic results

[1] Results of the transient scenarios are not included.

Asia have varied results across the region and among the climate scenarios (IPCC, in press). Estimates of the impacts of climate change for East Asia, including China (mainland and Taiwan), North and South Korea, and Japan, range broadly (Tao, 1993; Zhang, 1993; Sugihara, 1991; Seino, 1994; Horie, 1993; and Matthews et al., 1994 as cited in IPCC, in press). For example the IPCC (in press) concluded that rice yields in many South Asia countries could change by -22% to +28%. For China, climate change would most likely cause a general northward movement of agroclimatic regions (Lin, 1994, as cited in IPCC, in press), and yield results show generally negative trends (Zhang, 1993, and Lin, 1994, as cited in IPCC, in press).

The countries in the Asian-Pacific region that have reported on the agriculture sector include Bangladesh, Kazakhstan, and Mongolia. Bangladesh has not yet completed its assessment. All three countries used (or are using) the DSSAT 3 to assess vulnerability to climate change, although Kazakhstan and Bangladesh both used (or are using) the model in combination with their own country-specific models. The IPCC (in press), which provides a comprehensive summary of the estimated impacts of climate change on crop yields, does not include crop yield data from Mongolia or Kazakhstan.

Agriculture in Mongolia

Mongolia used the CERES-Wheat model (Benioff et al., in press) and the DSSAT 3 to looked at spring wheat in the central region of the country. This region is the major crop production center and wheat is the staple crop. The team tested the GISS and GFD3 scenarios. The overall impact of the different climate change scenarios is a decrease in spring wheat yield that ranges from 19% to 74% at five of seven study areas. The Baruunharaa study site, with an estimated increase in yields under both scenarios, was the notable exception to the trend toward decreased yields. Mongolia also examined a number of adaptation measures and their influence on climate change impacts. One adaptation was a change in the planting dates for the spring wheat from May 11 to April 25, which resulted in yields 8% to 80% higher than estimated yields without adaptation (measuring yields as share of baseline yields). However, yields would still be lower than without climate change.

Agriculture in Kazakhstan

Kazakhstan analyzed both spring and winter wheat yield in northern Kazakhstan using a combination of the DSSAT 3 model with their own (KazNIGMI) model, which was developed and tested for grain yield in Kazakhstan. Climate change scenarios from the GFD3, CCCM, and Geophysical Fluid Dynamics Laboratory 1% transient (GF01) (tenth decade) (Benioff et al., in press) models were used in the assessment. Spring wheat yields were estimated to decrease from 7% to 62% under all scenarios. At the same time, the results for winter wheat indicate that crop yields would increase from 1% to 32%. When aggregating results across all study sites, the assessment estimated decreases in spring wheat yields under all scenarios ranging from 27% to 56% and increases in winter wheat yields under all scenarios ranging from 17% to 21%. Given that arable land for winter wheat is less than 1% of the total arable land, this increase is not significant; however, switching from spring wheat to winter wheat could be an adaptation alternative. Several adaptation measures were qualitatively addressed, including improved irrigation, which may be too expensive; changing planting dates, which is quite feasible; switching from spring to winter wheat, which may be possible in a third of the area; increasing fallow area to allow better yields of winter wheat; implementing snow storage, which was a traditional but discontinued practice; changing wheat varieties to more productive, drought-tolerant ones; and applying fertilizers, pesticides, and weed control to increase yields.

Agriculture in Bangladesh

Bangladesh is using the GFD3 scenario in conjunction with a combination of the CERES-Rice model and their own Agro-Ecological Zone model for rice as the main crop. They have reported only preliminary findings because the work is still in progress.

From the phenological response assessment, Bangladesh found that high yielding varieties of rice cultivars are more susceptible to drought conditions and to inundation than local varieties. The team has also identified some possible adaptations, including changing the rice genotype to sustain yields and changing agricultural management practices.

Water Resources

Fresh water resources are very important in the Asian-Pacific region, which has approximately one-quarter of the world's internal annual renewable water resources. The annual river runoff at over 14,000 km^3 is the largest of any region in the world. However, because of its relatively high population, the per capita availability is less than half the world average. A number of the world's largest rivers flow through countries of the region such as Bangladesh with more than 1,000 km^3 in runoff annually. The water resources are of particular importance to the agriculture sector, which accounts for 86% of withdrawals in the region (WRI, 1994). Hence any adverse impact on water resources will have a severe cross-sectoral impact on agriculture as well.

Although there are large uncertainties associated with water resources assessments, some studies (IPCC, in press) show that water resources in developing countries are often small-scale, isolated, and under considerable stress, and thus may have a difficult time adapting to climate change effectively. This description of the water resources applies to areas within the Asian-Pacific region, and the water resources in the region are estimated to be under increasing pressures due to the high rate of population growth, one of the highest in the world (1.54%; World Bank, 1994). However, a recent study (Kaczmarek et al., 1995, as cited in IPCC, in press) suggested that climate change could have a net positive impact on water resources of the Asian continent.

The countries that reported results on the water sector are Kazakhstan and Mongolia; Bangladesh has presented only methods. The IPCC (in press) does not report results for Mongolia or Kazakhstan. Each of the countries used (or is using) a different national water balance model to estimate basin runoff impacts.

Water Resources in Mongolia

Mongolia used their own hydrological model for water resources, the Basin Conceptual Model (which had been tested and calibrated in the country), for the Khalkhgol, Bogd, and Bulgan river basins. These study sites represent a range of geographic and climatic conditions. The assessment used incremental scenarios and the GISS, GFD3, United Kingdom Meteorological Office (UK89) (Benioff et al., in press), and CCCM scenarios to develop estimates of vulnerability. The results showed that all three rivers were more sensitive to changes in precipitation than to changes in temperature. For example, the incremental scenario analysis showed that a 20% increase in precipitation alone produced an increase in runoff from 32% to 79%, while a 2°C temperature rise reduced flows by between 2% and 16% in the Bulgan and Khalkhgol rivers. However, for the Bogd study site, which is dominated by snowmelt, a 2°C temperature increase raised annual runoff by 3%. Most of the GCM scenarios estimated increases in runoff in the three river basins, ranging from 6.5% for the Khalkhgol river basin under the UK89 scenario to 26.3% for the Bulgan river basin under the CCCM scenario, with the exception of an estimate of a small decrease in runoff for the Bulgan basin under the UK89 scenario.

Water Resources in Kazakhstan

Kazakhstan used their own hydrological and water balance model called KazNIGMI in conjunction with the 2×CO_2 climate change scenarios of the GFD3 and CCCM models. The models were applied to two watersheds, one in the plains (Tobol River) and one in the mountains feeding off glaciers (Uba and Ulba rivers). The results showed reductions of 30% to 35% in runoff in years of excess water and 20% to 25% in years of water shortages. Kazakhstan also examined some adaptation strategies, which include greater use of underground water and diversion of the Volga and Siberian rivers to compensate for any water shortages.

Water Resources in Bangladesh

Bangladesh is using the GFDL scenario coupled with a hydrological model called MIKE 11. This hydrological model has been tested and calibrated for the country. They will be looking at changes in water level, salt water intrusion, drought, groundwater availability, and flooding. Although no adaptations are discussed, they recognize the importance of the water sector to other sectors, including agriculture, forests, and coasts, which will also be studied.

Coastal Resources

The Asian-Pacific region contains a wide variety of coastal morphologies, types, and activities. Some of the island countries in the Indian and Pacific oceans such as the Maldives and Kiribati consist almost entirely of coastal areas, and hence what happens to the coast affects the whole country. The very existence of these countries is threatened by sea level rise. Other large continental countries such as China and India also have large proportions of their populations living on the coasts. The total length of coastline in the region is 163,609 km, which represents over one-quarter of the world's total. The total population living in coastal agglomerations was nearly 300 million in 1980 and is expected to reach nearly 500 million by 2000 (WRI, 1994). The Indian Ocean coast of Asia and the coast of Southeast Asia were found to be the most vulnerable coastal regions of the world based on a weighted ranking incorporating risk to population, GNP, and wetlands (Delft Hydraulics, 1993). This same study found that the coasts of the Asian-Pacific region are vulnerable with respect to both their socioeconomic systems and their coastal wetlands.

Sri Lanka and Bangladesh reported on their coastal resources, but Bangladesh has presented only its methods. While Bangladesh has been the topic of sea level rise studies (Huq et al., 1995), recent comprehensive reviews of sea level rise impacts (i.e., Nicholls and Leatherman, 1995) do not include results from Sri Lanka.

Coastal Resources in Sri Lanka

Sri Lanka analyzed impacts of 0.3 m, 1.0 m, and 2.0 m sea level rise on the southwestern coast of the country, which is very vulnerable to sea level rise (from Devundara Head to Negombo). The potential land loss along the southwestern coast due to erosion under a 1.0 m sea level rise would be 11.5 km^2 and land loss due to inundation was estimated to be 94.8 km^2 under a 1.0 m sea level rise. The area is one of the most densely populated, and it is also economically important for agriculture and tourism; therefore, the impacts are expected to be quite severe. A quantitative survey of socioeconomic characteristics of the coast (type of land use, value, important human activities, and structures) has not been completed. Therefore, only preliminary estimates of the socioeconomic activities within the study site are presented. The study team also reviewed a number of possible adaptations, including protection of coastal areas by physical structures, which has been practiced in the past with limited success, retreat with conservation, and adjustments to salinity intrusion.

Coastal Resources in Bangladesh
Bangladesh is examining the effects of a 0.3 m and 1.0 m sea level rise, and has selected the southeastern part of the country, which contains the main port cities, tourism, fisheries, and agriculture, as the site for their assessment.

Forestry

The Asian-Pacific region contains some of the world's most important forest areas, which are important in terms of both resources and biodiversity. The total forest area in the region is 275 million ha, which represents 16% of the world's total (WRI, 1994). These areas are being depleted at a rate of 1.2% per annum, compared to the global average of 0.8%. The forests of the region include both tropical rain forests and the temperate coniferous forests of eastern Asia. Tropical forests are threatened more by land use practices than by climate change, and deforestation is expected to continue in response to economic development, population growth, and the associated need for agricultural land (IPCC, in press). The tropical rainforests in Indonesia and Malaysia were depleted by 22.6% and 8.1%, respectively, between 1981 and 1990 (WRI, 1994). In the absence of land use pressure, the global vegetation models provide conflicting results on whether climatic change will increase or decrease the total area of tropical forests; however, large gross decreases and increases in tropical dry forests are expected (IPCC, in press). In contrast, temperate forests are under less pressure from land use changes; in some regions the temperate forest area is likely to be increased by afforestation (potential increase of 44 million ha in Europe and 100 million ha in the United States; Heath et al., 1993 as cited in IPCC, in press). Climate change is estimated to allow temperate forests to displace boreal forests (IPCC, in press; VEMAP, 1995). Temperate forest cover could be lost in areas that become drier (Smith and Shugart, 1993). The two countries that have presented forestry sector assessments in this report are Mongolia and Bangladesh, and Bangladesh did not present results.

Forestry in Mongolia
Mongolia examined the biophysical impact of the UK89 $2\times CO_2$ scenario with the Holdridge Life Zone Classification Model (Benioff et al., in press) and found that there would be a shift of about 66 million ha from cool temperate desert and steppe biomes to warm temperate desert and steppe biomes. The chapter does not discuss any adaptation possibilities. A preliminary conclusion from this analysis is that the characteristics of the Mongolian Gobi desert and steppe zones would be altered in the region of the Khangai Mountains under a $2\times CO_2$ scenario.

Forestry in Bangladesh

Bangladesh is analyzing the impacts to the forestry sector only as they relate to the impacts of sea level rise on the Sundarban mangrove forest along the southwestern coast of the country. Because the Holdridge model is predominately for temperate boreal forests, a methodology using projections of future water salinity was developed to assess the vulnerability of the Sundarbans. This methodology will be applied in their future assessment.

Grasslands/Livestock

Some of the world's most important grasslands are situated in Central Asia, and they support considerable numbers of livestock. The total pasture land in Asia in 1990 was about 759 million ha, which was about one-fifth of the world's total (WRI, 1994). Biogeographical models suggest significant changes in the distribution of grasslands and arid lands worldwide in response to climate change (IPCC, in press). Water availability and water balance appear to be the most influential climate variables in controlling the geographic distributions of rangeland vegetation and production (IPCC, in press). Additionally, extreme events drive rangeland systems and land use alters the composition and productivity of rangelands (IPCC, in press).

Kazakhstan assessed the vulnerability of grasslands and livestock under the GFD3, GF01, and CCCM scenarios. The study team used their own model of grassland productivity (KazNIGMI) to look at impacts and found that wool productivity decreased by 11% to 19% and lambing decreased by 2% to 26%. Although no quantitative assessment of adaptation options was completed, the assessment includes a number of suggested adaptation measures such as changing the period of spring lambing, sheep driving, and shearing to earlier dates; improving the natural vegetation cover; breeding only coarse grained sheep better able to withstand higher temperatures; and developing higher mountain pastures to take advantage of the potential shift of vertical mountain zones with climate change.

Summary and Discussion

In the Asian-Pacific region, the climate change scenarios indicate that whether agricultural yields increase or decrease will depend on the specific location and the crops that are grown. However, the available results suggest a decline in spring wheat, which accounts for a large proportion of total wheat production. Thus far, the results are not sufficiently robust to allow general conclusions to be drawn. However, the high population in the region places pressure on the natural resources and may complicate the process of adaptation.

- Mongolia estimated decreases in spring wheat yields from 19% to 74%. This result may be compared to a previous assessment for China that estimated an 8% decrease in wheat yields (Tao, 1993, as cited in IPCC, in press). Similarly, Kazakhstan estimated a decrease in spring wheat yields, but also estimated an increase in winter wheat yields (-56% to +21% for both spring and winter wheat). This trend coincides with an assessment for Russia and former Soviet Republics that estimated a change in spring and winter of -19% to +41% (Menzhulin and Koval, 1994, as cited in IPCC, in press).

The limited water resources on many islands and in drought-prone areas of China and Central Asia combined with salinity intrusion problems associated with sea level rise make water resources another significantly vulnerable sector. However, as shown by the results from Mongolia's assessment, water availability may increase in some countries of the region, and recent studies suggest that the net impact of climate change on water resources could be positive for the Asian continent (Kaczmarek et al., 1995, as cited in IPCC, in press). Further research is needed before any regional trend may be identified.

- Mongolia and Kazakhstan both used national water balance models to estimate changes in runoff. Under GCM scenarios, Mongolia estimated that there would be increases in runoff ranging from 7% to 26% and Kazakhstan estimated that there would be decreases ranging from 20% to 35%. These assessments are preliminary and do not account for changes in water demand or potential changes in the management of the supply system. Because nonclimatic factors will be the determining factors for water resource management in many regions of the world (IPCC, in press), further research is needed before regional trends can be identified.

Other vulnerability sectors in the region for which results were presented in this report include coastal resources, forestry, and grasslands. The results presented are not inconsistent with findings of previous studies; however, the limited number of results for each sector do not allow conclusions to be drawn for the region as a whole.

- Sri Lanka assessed the vulnerability of its southwestern coast and found that a 1.0 m sea level rise would be associated with a land loss of 11.5 km^2 due to erosion and 94.8 km^2 due to inundation. The study area is densely populated and used for agriculture and tourism; therefore, the impacts are expected to be quite severe.

- Both Mongolia in its assessment of the forest sector and Kazakhstan in its assessment of the grasslands sector estimated a shift to warmer and drier conditions that would be consistent with mid-continental drying. Mongolia found that there would be a shift of about 66 million ha from cool temperate desert and steppe biomes to warm temperate desert and steppe biomes. Such large changes in forests are consistent with results from previous studies (Neilson, 1993; Smith et al., 1995). Kazakhstan found that warmer and drier conditions would have a negative impact on wool productivity (decrease of 11% to 19%) and lambing (decrease of 2% to 26%).

The results in this report reflect the work being conducted in four countries within the Asian-Pacific region and provide an opportunity to compare the results across countries in the region and with previous assessments. However, these results cannot be considered entirely representative of the region; in particular the impacts of climate change on the Pacific Island countries may differ substantially from the majority of the results presented here.

Summary for the Central and Eastern European Region

Region-Specific Issues

Geography and Climate

The Central and Eastern European region contains most countries of the former Eastern Bloc, nine of which are participating in the U.S. Country Studies Program. The geographic locations of the countries are depicted in Figure 4.3.

These countries belong to the northern temperate zone, with a typical rotation of four seasons and variable weather. In this climate, as one moves east, the temperature difference between winter and summer seasons becomes larger, and precipitation generally declines. Recent evaluations of precipitation trends indicate a statistically significant decline, which becomes more pronounced with an eastward shift. A large part of the region has a complicated orography, which affects observed weather patterns and climate impacts, and consequently makes GCM scenario construction a difficult task.

Demographic Conditions and Trends

The Central and Eastern European region is characterized by relatively low projected growth in population. The population growth rate for countries in the region is generally projected to be less than 1% for each five-year interval between 1990 and 2030. In addition, small decreases in population are projected for several of these countries (World Bank, 1994). These projections can be ascribed a high degree of confidence, assuming no significant future immigration or emigration.

The population density of Central and Eastern European countries is moderate to low. The population density ranges from 807 people per 1,000 ha in Bulgaria to 1,265 people per 1,000 ha in Poland, as compared to a worldwide density of 427 people per hectare (WRI, 1994). The region is also highly urbanized, with an urban population exceeding 50% of total population in most countries (WRI, 1994).

Socioeconomic Trends and Conditions

Most of the countries in the region belong to the low- and middle-income group (WRI, 1992). Agriculture accounted for approximately 30% to 45% of the total area in Central and Eastern Europe before 1990. Forest coverage was between 25% to 35% depending on the country, with the exception of Hungary, which had 18% forest coverage (WRI, 1992). During the 1980s there was a slight decrease in agricultural land use and a small increase in the forested land (both less than 1%). Much of Europe's heavy industry is located in this region.

Between the end of World War II and 1990, all of the countries in the region had quite similar state-governed economic structures, and an emphasis on heavy industry accompanied by inefficient use of energy. Since 1990 the transformation of their economies has resulted in significant changes. The economic transformation can be seen not only in the new structure of industry, but also in the transformation of legislation, institutions, and taxation policies. Large-scale privatization of property has played a key role in the transformation. Forestry and agriculture are the main target

Figure 4.3. Map of the Central and Eastern European Region

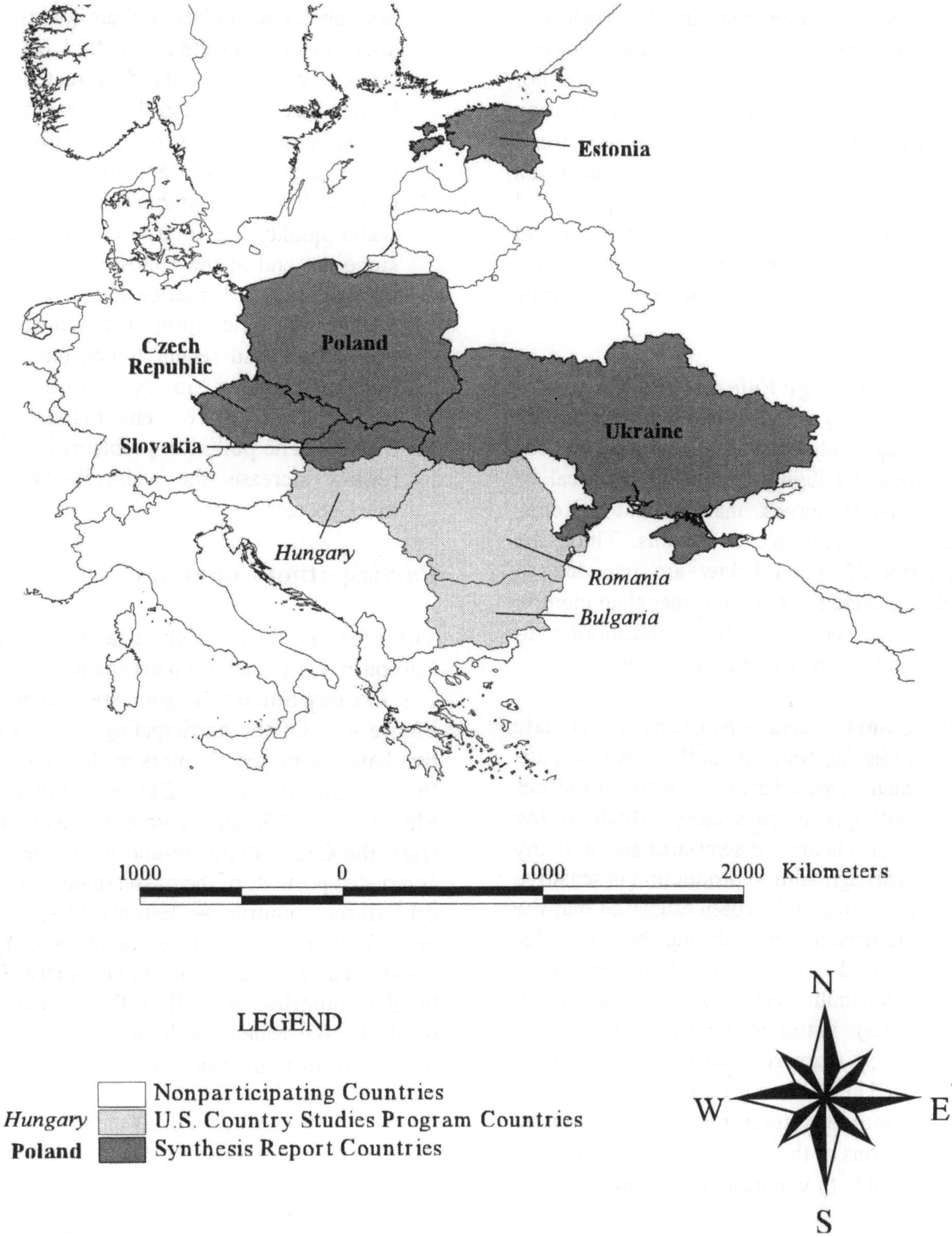

areas for privatization, and these changes are affecting the amount of land used in these sectors. Because of the transition, the socioeconomic development in this region is highly uncertain. Economic models developed for fairly stable economies are not applicable to the current economic situation in Central and Eastern Europe, and trend analyses are not possible because all countries in the region have experienced a sharp decline in overall production and GDP. It is therefore difficult to make socioeconomic projections with a high degree of confidence.

Climate Change Vulnerability

Vulnerability to climate change is highly integrated between various natural resource sectors and highly dependent on local or regional resources, management systems, and socioeconomic conditions. Thus, the factors described below are intended to characterize the region, rather than identify the most critical factors contributing to vulnerability in a particular region.

- **Climate Variability.** Changes in climate variability could affect the region significantly, particularly in the water resources and agriculture sectors. Much of the region is arid or semi-arid and in many areas agricultural production is sensitive to drought. It has been estimated that the increase in aridity during the 1930s decreased yields by 25% to 39% (Menzhulin, 1992, as cited in IPCC, in press). Water resources are also sensitive; Kaczmarek and Kindler (1995, as cited in IPCC, in press) found that water resource systems in Poland, as in many regions of the world, could be very vulnerable to variations in climate.
- **Degradation of Water Resources and Forests.** Pollution problems in the Central and Eastern European region involve not only the degradation of air quality in areas of concentrated industrial enterprises (damaging forests and threatening human health) but also the degradation of water quality due to both industrial and agricultural activities (e.g., overfertilization of arable land). Air and water quality has improved somewhat since the end of the 1980s because the economic transformation has caused a decline in agricultural and industrial production, and recent trends indicate that management in agriculture and industry is more environmentally friendly. The pollution problems in this region increase the vulnerability of these sectors.

Participating Countries

The countries in the Central and Eastern European region that are participating in the U.S. Country Studies Program are shown in Figure 4.3. Of the participating countries, five have submitted chapters to this report: the Czech Republic, Estonia, Poland, Slovakia, and Ukraine. Two of these countries, the Czech Republic and Poland have completed portions of their assessments. The other three countries — Estonia, Slovakia, and Ukraine — have presented preliminary results. Table 4.3 lists the sectors analyzed by the countries, as well as the scenarios used, the assessment methods used, a summary of results, and the status of the work.

Table 4.3. Central and Eastern Europe Countries Contributing to the Interim Synthesis Report

Country	Sectors	Scenarios	Vulnerability Assessment Method	Summary of Results	Progress
Czech Republic (Round 1)	Agriculture	CCCM, GISS, Incremental	CERES-Wheat	Change in annual yields: -3% to +16% (winter wheat)	Biophysical results
	Forestry	CCCM, GISS, Incremental	Forest Vegetation Zones	National and study area analyses show shift from Norway spruce to pine and deciduous species	Biophysical results
	Water	CCCM, GISS, Incremental	BILAN, CLIRUN, SAC-SMA	Change in annual runoff: -45% to +15%[2]	Biophysical results
Estonia (Round 2)	Agriculture	GCMs	CERES-Barley, SPUR2	N/A	Preliminary biophysical results
	Forestry	Incremental	Gap models	Decline of spruce and aspen and increase in pine or deciduous species at two sites	Preliminary biophysical results
	Coastal	SLR 1.0 m	IPCC	Land loss at two study sites: 60 km^2	
Poland (Round 1)	Coastal	SLR 0.1 m to 2.5 m	IPCC	Land loss for entire country: 845 to 2,203 km^2	Biophysical and socioeconomic results
Slovakia (Round 2)	Agriculture	CCCM, GISS, GFD3	DSSAT 3	N/A	Methods only
	Forestry	CCCM, GISS, GFD3	Holdridge model	N/A	Methods only
	Water	CCCM, GISS, GFD3	WATBAL	N/A	Methods only
Ukraine (Round 2)	Forestry	GFD3, GISS	Holdridge model	39% to 47% reduction in forest area[3]	Preliminary biophysical results
	Water	GFD3, GF01,[1] UKMO	National models	Change in annual runoff: -20% to +128% (Dnieper Basin)	Preliminary biophysical results
	Coastal	SLR 0.5 m, 1.0 m	IPCC	Land loss along the Black Sea: 32 km^2 under a 0.5 m SLR 370 km^2 under a 1.0 m SLR	Preliminary biophysical results

1 Results of the transient scenarios are not included.
2 Results not reported for individual basins.
3 Results are preliminary; more recent analysis may result in significantly different estimates of impacts.

Agriculture

In 1990, agricultural activities accounted for a relatively small portion of the GDP of countries in Central and Eastern Europe. For example, only 8% and 14% of the GDP was produced by agricultural activities in Czechoslovakia and Poland, respectively (WRI, 1992). The ownership of much of the region's agricultural land has changed recently as a result of economic transformation, and improvements in agricultural productivity in the region may be possible. In some of the countries of the region, such as the Czech Republic, the contribution of agriculture to GDP and the area of arable soil for cultivation declined significantly due to the transformation of property ownership. The population per hectare of cropland in the region is relatively low (1.5 for the former Soviet Union; FAO, 1992, as cited in IPCC, in press), and agriculture, which is currently limited by cool temperatures, could benefit from increased temperatures. Estimates of potential additional cropland for the world suggest that this region could be one source of substantial additional cropland (Crosson, 1995, as cited in IPCC, in press).

The countries in the Central and Eastern European region that have reported on their agriculture assessments are the Czech Republic, Estonia, and Slovakia. The IPCC (in press) does not report results from these countries. The Czech Republic used the DSSAT 3 to estimate yields of winter wheat. Estonia and Slovakia are Round 2 countries, and therefore, they are still verifying models [DSSAT 3 crop models and SPUR2 (Benioff et al., in press)] and exploring the sensitivity of crop yields to different kinds of fertilization and soil conditions.

Agriculture in the Czech Republic

The Czech Republic has completed its assessment of the agriculture sector. For the assessment of winter wheat yields, two study sites, south Moravia and central Bohemia, were selected. The CERES-Wheat model and the DSSAT 3 were used in conjunction with modifications of historical climate trends to estimate the impact of future climate change. The assessment found that in comparison with current conditions, there would be a reduction in the flowering period of 7 to 17 days, and a reduction in the period of physiological maturity of 8 to 17 days. Taking into account the direct effects of increased CO_2 concentrations, winter wheat yields could change by -3% to +16%, depending on the climate scenario. Results of an assessment of different climatic conditions on pests, diseases, and soil are also discussed in qualitative terms. In addition, a number of adaptation strategies are suggested, including practices to maintain soil fertility, control erosion, implement irrigation, and manage pests.

Agriculture in Estonia

Estonia is currently calibrating the models and verifying the data that will be used in their assessment; therefore, no specific yield results are available at this time. The assessment will examine four study sites — Eerika, Antsla, Olustvere, and Kuusiku — using the CERES-Barley and SPUR2 models. Earlier experiments using biophysical models indicate that a temperature rise of 4°C would decrease barley yields by 20% to 30% (Roostalu et al., 1980). At the same time clover-timothy mixture yields would increase by 10%.

Agriculture in Slovakia

Slovakia is currently using the DSSAT 3 to assess the vulnerability of the agriculture sector. Simple model calculations indicate that an estimated warming of 1° to 2°C and a small change in precipitation would be accompanied by an increase in the evapotranspiration deficit by 13% to 26%, a prolongation of vegetative periods by 6% to 8%, and a decrease in soil moisture of approximately 20% in the main agricultural fields.

Forestry

In response to forest damage from industrial pollution and in recognition of the economic importance of forests in Central and Eastern Europe, efforts have been under way for many years to protect the forest stands and improve their quality. As an example of the effect of these efforts, Slovakia has increased the acreage of forested land by more than 20% since 1920, and the timber resources in the country increased from 193.5 million m^3 in 1950 to 352.2 million m^3 in 1991. Despite the afforestation that has occurred, damage from pollution continues to be a problem. Forest damage surveys conducted in 1989 found that roughly one third of the forests of Czechoslovakia and Poland were moderately to severely damaged (WRI, 1992).

Many of the forests in the region are monocultures and are either temperate or boreal forests. Recent studies have estimated that climate change will enable temperate forest to advance poleward, and in this region temperate forest may displace boreal forest and expand into maritime regions (IPCC, in press). The boreal forest is likely to decrease in area, biomass, and carbon stock (Neilson et al., 1994).

Four countries reported on the progress of their forestry assessments: the Czech Republic, Estonia, Slovakia, and Ukraine. Slovakia is still developing their assessment and, therefore, only discusses possible impacts identified in existing literature. Each of the three remaining countries used a different method of analysis — forest vegetation zones, forest gap models (Benioff et al., in press), and the Holdridge model, respectively, for the Czech Republic, Estonia, and Ukraine.

Forestry in the Czech Republic

The Czech Republic used forest vegetation zones and a nondynamic correlative model based on a relationship between contemporary climate conditions and vegetation types. The impacts on the forestry sector were assessed in terms of a shift of the zones for two time horizons (2010 and 2030) under the GISS and CCCM scenarios. Under the GISS scenario the Czech Republic estimated that the oak, oak-beech, and beech-oak vegetative zones would increase from the present share of 21% of total forest area to 47% by 2030, while vegetation zones with beech, fir, and spruce would be reduced. Under a regional climate scenario that was developed using spatial temperature, precipitation patterns, and statistical methods, forest vegetation zones were predicted to shift toward higher elevations and Norway spruce would decline, which would encourage the establishment of European beech. The adaptation options described were to reduce the high proportion of the Norway spruce and Scots pine outside their natural areas, and to improve forest management. Forest management costs for activities such as road maintenance and possible anticipatory adaptation measures are included in the assessment.

Forestry in Estonia
Estonia has conducted a preliminary assessment of forest vulnerability at two study sites using a modified version of a forest gap model. The preliminary simulations of long-term forest development indicate significant changes would occur in the structure of forest communities. The results of the analysis indicate that an increase in temperature would cause Scots pine to become the predominant species in central Estonia, and the total mass of tree stands may increase up to 15%. With increased temperature and decreased annual precipitation, drought-tolerant species such as Scots pine and oak would grow in central Estonia and the total biomass level would be lower than it is today. Under a scenario of increased temperature and precipitation, Norway spruce would be predominant in central Estonia, with Scots pine and few deciduous species found in the coastal areas.

Forestry in Slovakia
Slovakia is currently using the Holdridge model to assess the vulnerability of its forestry sector. Based on previous analyses and literature, the following preliminary conclusions have been drawn: European fir and Norway spruce could be the species most affected by climate change, beech could expand into higher altitudes, and production of forest stands could increase by 10% to 30%.

Forestry in Ukraine
Ukraine used the Holdridge model in conjunction with the GFD3 and GISS scenarios to assess the vulnerability of the forested area of Ukraine (approximately 10 million ha of forested area). It is estimated that Polissya and the northern part of the forest-steppe zone would remain stable if the climate changed. However, total forest area could decline by 39% to 47%. Preliminary results show that Scots pine will tend to replace existing spruce and aspen forests, assuming a 4°C warming and a ±10% annual precipitation.

Water Resources

In Central and Eastern Europe, water resources are significantly affected by water pollution. Industry, mining, and agriculture all have generated pollution that threatens the quality of both groundwater and surface water supplies. Thus, even in areas with adequate water resources, the quality of the water may result in water shortages or harm to aquatic ecosystems. For a majority of these countries, water withdrawals for industrial purposes exceed withdrawals for either agriculture or domestic uses. For example, industrial water withdrawals in Poland account for approximately 60% of all withdrawals (WRI, 1992). Climate change is estimated to have little impact on industrial water use.

One recent study suggests that climate change could have a net positive impact on global water resources (Strzepek et al., 1995). Another study suggests that in Europe climate change would be associated with some decrease of per capita water availability, while in Asia climate change could have a net positive impact (Kaczmarek et al., 1995, as cited in IPCC, in press). These results are based on particular scenarios and assumptions and should be interpreted with caution; socioeconomic conditions in the region, which are relatively difficult to estimate for the future, are necessary for determining water demand.

The countries that reported results on the water resources sector are the Czech Republic, Slovakia, and Ukraine.

Water Resources in the Czech Republic
The Czech Republic investigated the vulnerability of their water regime using three models [BILAN, CLIRUN (Benioff et al., in press), and SAC-SMA]. In this assessment, four river basins were examined under various incremental climate change scenarios as well as the GISS, CCCM, and GFD3 climate scenarios. The assessment estimated that there would be a tendency toward decreases in runoff, seasonal shifting in the distribution of water yield toward winter, and a difference in the sensitivity of various river basins. It was estimated that annual runoff could decrease by as much as 45% or increase by as much as 15%, depending on the river basin and the climate scenario. The analysis shows that while some sites may be highly vulnerable to changes in climates, others may be able to withstand even relatively large changes. A very preliminary analysis of water resources demand in 2075 was performed, taking into account socioeconomic development. This analysis indicates that if annual average temperatures increase by 2°C, the irrigation water demand would increase by 40% (assuming the same number of hectares under cultivation) and demand for cooling water might increase by 1% to 5%. Nonstructural adaptation options were suggested, including water conservation, integrated ground and surface water management, improved water supply, drought controls, and improved operation of reservoirs.

Water Resources in Slovakia
Slovakia is in the preliminary stages of their analysis. The team provided information on the past and present state of water resources. Under the anticipated climate change conditions, preliminary findings suggest that there will be approximately a 20% decrease in water resources by 2030. The WATBAL model (Benioff et al., in press) and statistical models are being used to generate more specific results. A set of possible adaptive measures is suggested.

Water Resources in Ukraine
Ukraine assessed the vulnerability of water resources in the Dnieper river basin. The following models were used in the assessment: a model of natural flow recovery developed at the Ukrainian Research Hydrometeorological Institute; a model of river runoff hydrograph developed by the Russian National Hydrological Institute; and a water resources management model developed at the Institute of Mathematical Machine and Systems Problems, Cybernetic Center of Ukrainian National Academy of Sciences. An estimate of vulnerability of the Dnieper river basin to climate change conditions was developed using data from three main gauging sites. Estimated changes in runoff varied widely under different climate scenarios. For example, results for the total water supply show that, when compared with the baseline, total water supply would increase under the GFD3 and UKMO scenarios (more than twice under the UKMO scenario). In contrast, runoff is estimated to decrease by as much as 20% under the GF01 scenario. A study of the national impacts derived from the Dnieper analysis is expected to be performed in the near future. A preliminary supply and demand assessment suggests that in 2075, under all climate change scenarios, water supply would exceed demand at all locations, with the exception of the Kahovka hydrojunction under transient climate change

scenarios. An economic analysis of the impact on six cascade hydropower stations indicates that economic losses could range from $US10 million to $US17 million.

Coastal Resources

Several countries in the region are land-locked, and those countries with coastal areas are located along either the Baltic Sea or the Black Sea and the Azov Sea. The size of the populations that might be affected by sea level rise in Eastern Europe is small relative to other portions of the world. It has been estimated that the populations at risk from a 1.0 m sea level rise along the Baltic coast, including Poland and Estonia, account for less than 0.5% of the worldwide population that would be at risk, and the population at risk along the Black Sea and Azov Sea coasts, including Bulgaria, Romania, and Ukraine, accounts for only 3% of the worldwide population at risk (Delft Hydraulics, 1993). Nonetheless, sea level rise may have serious effects for some of these countries.

Estonia, Poland, and Ukraine have reported on their assessments for the coastal resources sectors. Nicholls and Leatherman (1995), a report compiling recent sea level rise research, includes a report on Poland, but does not include information on the potential impacts of sea level rise on Estonia and Ukraine.

Coastal Resources in Estonia

Estonia is currently assessing the impact of a 1.0 m sea level rise by 2075 on the country's coastal resources at two study sites. Because of a long coastline and the predominance of low-lying coastal areas, a sea level rise associated with a change in climate conditions may cause serious damage. On the other hand, an isostatic land uplift would reduce the consequences of the sea level rise. Both qualitative and quantitative estimates of impacts on land, ecosystems, and dwellings are provided. A 1.0 m sea level rise would inundate 38 km^2 of the Hiiumaa study site, and a number of valuable coastal ecosystems would be destroyed. A 1.0 m sea level rise is also estimated to inundate 22 km^2 of the Pärnu-Ilka study site, which is one of the most popular recreation areas in the country. The land uses and structures that would be lost were identified, but the assessment of economic damages has not been completed. Two other study sites are currently under investigation.

Coastal Resources in Poland

Poland's assessment of the coastal sector is based on long-term research and contains condensed information about impacts that can be expected in four well-defined areas. Sea level rise scenarios of 10 cm and 30 cm by 2030, and 30 cm and 100 cm by the year 2100, are assumed for the entire Polish coast. Three strategies (retreat, limited protection, and full protection) are assumed and compared in physical and socioeconomic terms. Possible development patterns for the Polish coast were projected, with emphasis on recreation and tourism, harbors and shipyards, industry, transportation, and other human activities, to allow the development of the economic analyses. Over 2,200 km^2 and 230,000 people are estimated to be vulnerable to a 1.0 m sea level rise by 2075, and the total cost of land at loss is estimated to be approximately US$30 billion. The cost of full protection is assessed to be US$6 billion.

Coastal Resources in Ukraine

Ukraine is developing an assessment of the effects of a 0.5 m and 1.0 m sea level rise on the Black and Azov sea coasts within the Ukrainian territory. The final analysis will include an assessment of economic losses and a more complete evaluation of adaptation options. The Adaptation Strategy Evaluator was used for a preliminary evaluation of various adaption options. For this preliminary assessment, estimated flooding and soil salinization data were used, but the primary losses are expected to be from erosion and submergence. It is estimated that, without protection, wetlands loss for the Black Sea coastline under a 0.5 m and a 2.0 m sea level rise scenario would be 32 km^2 and 370 km^2, respectively.

Summary and Discussion

The ongoing economic transformation in the Central and Eastern European region makes it difficult to estimate how socioeconomic trends may interact with the biophysical impacts of climate change. This may become a particularly important issue as the countries of the region work to assess the economic importance of the changes in agricultural production and water resources, shifts in forest types, and land loss due to sea level rise.

The estimated impacts of climate change on the agriculture sector for this region vary between increases and decreases across the region and across climate scenarios, indicating that this sector may be quite sensitive to climate change.

- The results of impacts to the agriculture sector provided by the Czech Republic appear to vary between an increase or decrease in winter wheat (-3% to +16%), depending on the scenario used. Similarly, a study in Russia and the former Soviet Union (Menzhulin and Koval, 1994, as cited in IPCC, in press) and two studies in Russia (Sirotenko and Abashina, 1994; Sirotenko et al., 1991; as cited in IPCC, in press) estimate that crop yields may increase or decrease.

The assessments of the forestry sector presented in this report rely on a range of methodologies. The Czech Republic used a nondynamic correlative model, Estonia used a forest gap model, and Ukraine used a Holdridge model (Slovakia is also currently using the Holdridge model). Each of these models provide insight into potential changes, yet the limitations of each model suggest that each assessment could be strengthened by further work using an alternative model type.

- The assessments of the forestry sector for the Czech Republic, Estonia, and Ukraine all indicate that significant shifts will occur in type and distribution of forests. In each of these assessments, the Norway spruce forest is estimated to decline and the oak or beech or pine forest is estimated to increase. This is consistent with study findings indicating that the temperate forest may advance poleward, displacing boreal forest (IPCC, in press).

The impact of climate change on water resources is highly dependent on local or regional conditions and, therefore, it is difficult to draw regional conclusions from a limited number of studies. As an example, the results for the Czech Republic indicate that while some sites may be very sensitive to climate change, others may be able to withstand even relatively large changes.

- The results of impacts to the water resources sector provided by the Czech Republic and Ukraine appear to vary between an increase and decrease in runoff, depending on the size of the basin or the scenario used (-45% to +15% for the Czech Republic; more than double increase to -20% for Ukraine). These results provide an initial assessment of potential biophysical impacts; potential changes in water demand or water resource management have yet to be assessed. The nonclimatic factors that influence water resources may be more influential than potential climate changes in determining vulnerability in this sector (IPCC, in press).

The countries that report on their assessments for the coastal resources, Estonia, Poland, and Ukraine, used sea level rise scenarios ranging from 0.3 m to 2.0 m. These scenarios extend well below and above the current estimate of potential sea level rise, and thus the estimated impacts may not be the most likely, but they assess the full range of sensitivity to climate change.

- The coastal resources assessments completed by Estonia and Poland suggest that the effects for these countries may be serious, but are not likely to threaten as large of a population as may be affected in other regions. The Estonia assessment estimated that a 1.0 m sea level rise would inundate 50 km^2 at two study sites. Poland estimated that over 2,200 km^2 and 230,000 people would be vulnerable to a 1.0 m seal level rise by 2075, with damages of approximately US$30 billion. These estimates are similar to previous estimates of a loss of 1,700 km^2 and 235,000 people under a 1.0 m sea level rise with damages of approximately US$24 billion (Nicholls and Leatherman, 1995).

In addition to the results summarized here, Slovakia and Estonia reported on assessments that are currently underway, and a number of the reporting countries have assessments of additional sectors that are not included in this report. Completion and documentation of these results in the future will allow a more comprehensive assessment of the region.

Summary for the Latin American Region

Region-Specific Issues

Geography and Climate

The Latin American region consists of Mexico, Central America, and South America (Figure 4.4). The climate varies greatly throughout the region and within the countries of the region. In South America, elevations range from sea level to the high Andes mountains, second in height only to the Himalayas. Consequently the climate of the region ranges from humid tropical areas with no dry season such as the Amazon river basin, to arid and semi-arid areas along the coast of Chile and in Mexico, to temperate areas throughout the region. Mexico, Central America, and South America have extensive coastlines along both the Pacific and Atlantic oceans that are affected in varying degrees by erosion pressures and

Figure 4.4. Map of the Latin American Region

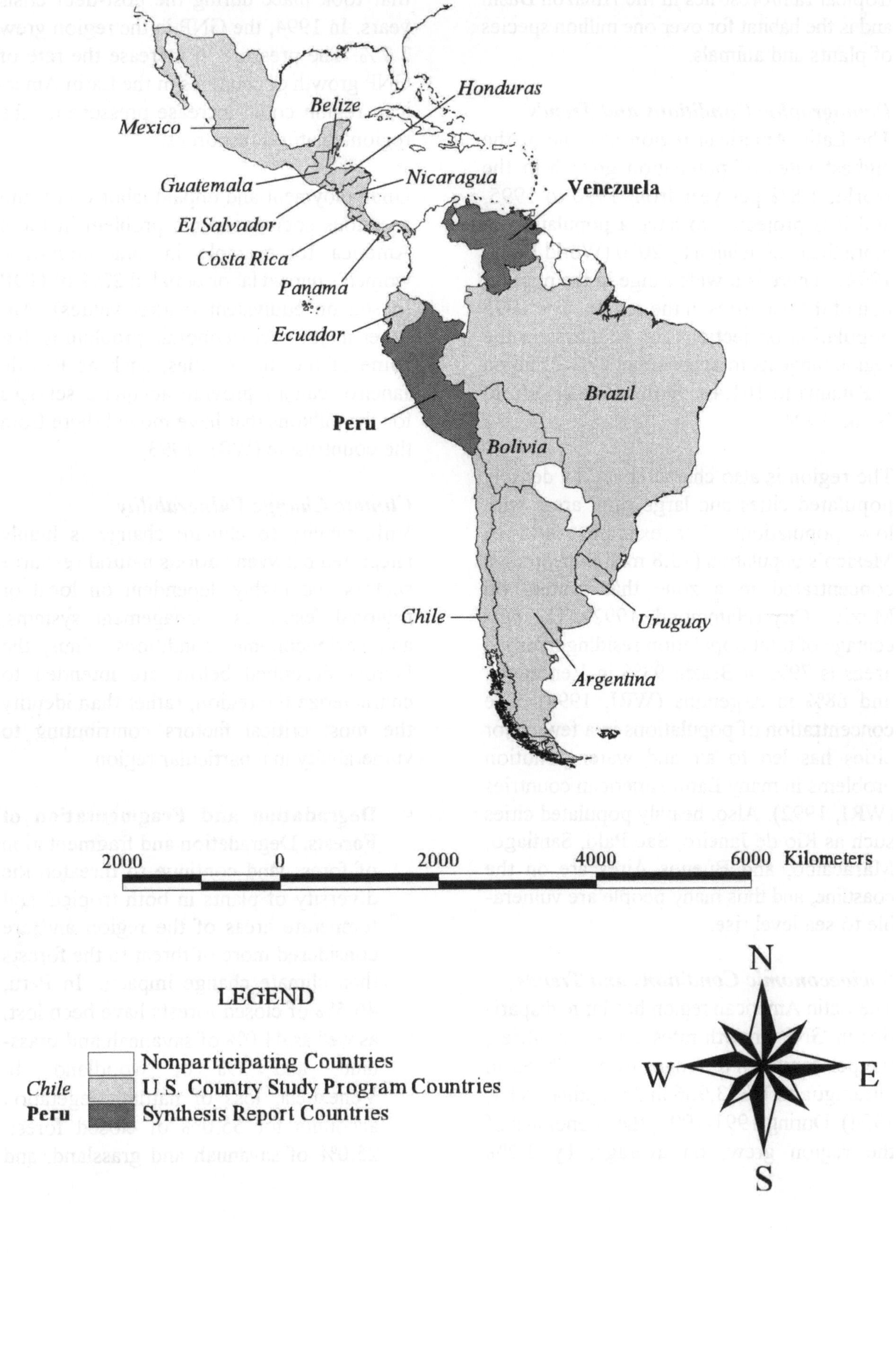

extreme weather events. The world's largest tropical rainforest lies in the Amazon Basin and is the habitat for over one million species of plants and animals.

Demographic Conditions and Trends

The Latin American region has one of the highest rates of population growth in the world, 1.8% per year from 1990 to 1995, and it is projected to have a population of more than 720 million by 2030 (World Bank, 1994). There is a wide range in the population of the countries in the region. The 1995 population projections for countries in the region range from approximately 2.7 million in Panama to 161.4 million in Brazil (World Bank, 1994).

The region is also characterized by densely populated cities and large rural areas with low populations. For example, 50% of Mexico's population (13.8 million people) is concentrated in a zone that centers on Mexico City (Hammond, 1992). The percentage of total population residing in urban areas is 79% in Brazil, 93% in Venezuela, and 88% in Argentina (WRI, 1994). The concentration of populations in a few major cities has led to air and water pollution problems in many Latin American countries (WRI, 1992). Also, heavily populated cities such as Rio de Janeiro, Sao Palo, Santiago, Maracaibo, and Buenos Aires are on the coastline, and thus many people are vulnerable to sea level rise.

Socioeconomic Conditions and Trends

The Latin American region has large disparities in GNP growth rates across countries; the per capita GNP ranges from US$283 in Nicaragua to US$3,966 in Argentina (WRI, 1994). During 1991-1993, the economies of the region grew, on average, by 3.2% annually, almost double the average growth that took place during the post-debt crisis years. In 1994, the GNP in the region grew 3.7%. The pressure to increase the rate of GNP growth of countries in the Latin American region could increase pressure on the region's natural resources.

Unemployment and unpaid labor constitute a serious socioeconomic problem in Latin America; for example, in some countries, women's unpaid labor equaled 22% of GDP (based on equivalent market values). Another major socioeconomic problem is that some of the major cities, such as Rio de Janeiro, cannot provide adequate services for the millions that have moved there from the countryside (WRI, 1993).

Climate Change Vulnerability

Vulnerability to climate change is highly integrated between various natural resource sectors and highly dependent on local or regional resources, management systems, and socioeconomic conditions. Thus, the factors described below are intended to characterize the region, rather than identify the most critical factors contributing to vulnerability in a particular region.

- **Degradation and Fragmentation of Forests.** Degradation and fragmentation of forest land continue to threaten the diversity of plants in both tropical and temperate areas of the region and are considered more of threat to the forests than climate change impacts. In Peru, 46.5% of closed forests have been lost, as well as 41.0% of savannah and grassland, and 3% of woodland. In Venezuela, loss of natural vegetation accounts for 55.0% of closed forest, 25.0% of savannah and grassland, and

2.5% of woodland (WRI, 1994). Between 1980 and 1990, forest areas in the region shrank by approximately 6.29 million ha (WRI, 1994). Disturbance of the tropical forest through land conversion and resource use due to population and economic pressures is expected to continue to be a significant threat to tropical forests (IPCC, in press). This fragmentation of forest land threatens the biodiversity in many locations such as the Amazon river basin, and can reduce global evapotranspiration and increase runoff, thus affecting the amount and distribution of precipitation over wide areas (Salati and Jose, 1984, as cited in IPCC, in press).

- **El Niño-Southern Oscillation.** International and regional studies have shown that natural and anthropogenic climate variations ranging from short-term events such as the seasonal and interannual variation due to the El Niño-Southern Oscillation, to long-term changes such as temperature shifts and sea level rise associated with greenhouse warming, may have significant impacts on the Latin American region. In Bolivia, Peru, and Colombia, for example, temperatures are modified by positive temperature anomalies in the east Pacific Ocean. These countries are also affected by floods and droughts that are influenced by El Niño-Southern Oscillation. It is not yet possible to project the impact of climate change on the El Niño-Southern Oscillation; however, long-term studies are under way (IPCC, in press).

Participating Countries

The countries in the Latin American region that are participating in the U.S. Country Studies Program are shown in Figure 4.4. Of the participating countries, two have submitted chapters to this report: Peru and Venezuela. Both are Round 1 countries. Peru reported results from their assessment of the coastal resources sector, and Venezuela reported results from their assessments of the forestry and coastal resources sectors. Table 4.4 lists the sectors analyzed by the countries, the scenarios used, the assessment methods used, a summary of results, and the status of their work.

Table 4.4. Latin American Countries Contributing to the Interim Synthesis Report

Country	Sectors	Scenarios	Vulnerability Assessment Method	Summary of Results	Progress
Venezuela (Round 1)	Forestry	GFD3, GISS, OSU, UKMO	Holdridge model	About 35 million ha of subtropical forest converted to tropical forest; shift of 40-50 million ha from subtropical moist or wet forest to tropical dry and very dry forest	Biophysical impacts
	Coastal	SLR 0.5 m and 1.0 m	IPCC	Land loss at five study sites: 85.54 km^2 under 0.5 m SLR 124.74 km^2 under 1.0 m SLR	Biophysical impacts and socioeconomic impacts
Peru (Round 1)	Coastal	SLR 1.0 m	IPCC	Land loss at selected areas: 78.32 km^2 under 1.0 m SLR	Biophysical impacts

Forestry

The forests of the Latin American region account for approximately 26% of all forests in the world. The immense tropical forest of the Amazon River basin was estimated to cover 1,756 million ha at the end of 1990. Almost half of the tropical forests in the Latin American region are located in Brazil, Peru, and Venezuela (WRI, 1992). Although the global vegetation models provide conflicting estimates of whether the total area of tropical forest will increase or decrease, large gross changes in tropical dry forest are expected (IPCC, in press).

For this interim report, Venezuela presented information on its forest sector. The Venezuelan forest area is roughly 58 million ha, which represents more than 60% of the country's territory. About 70% of the forest land is found in the south of the country, in the Amazon Basin. Venezuela used four GCMs [GFD3, GISS, Oregon State University (OSU) (Benioff et al., in press), and UKMO] under current and 2×CO_2 conditions in conjunction with the Holdridge model to estimate current life zones and how those life zones may shift under climate change conditions. Under current conditions, tropical dry forest (savannah) and subtropical moist forest are the largest life zones, covering roughly 52 million ha; under 2×CO_2 conditions using the UKMO model, the tropical very dry forest and tropical dry forest would increase by roughly 40 to 50 million ha, while the tropical and subtropical moist forest would no longer be sustainable and would decrease by roughly the same amount. Estimates developed from all of the GCM scenarios indicate a tendency for drier patterns, and several critical zones were identified. The Venezuelan chapter indicates that some work on adaptation assessment is being carried out using forest gap models.

Coastal Resources

The total length of the Latin American region's coastline is approximately 59,700 km (WRI, 1992). It contains a wide diversity of morphologies, types, and land use activities. Central America is a natural divide between two major oceans; it therefore contains one of the richest and most diverse coastal and marine systems in the world. This system includes coral reefs, mangroves, and one of the longest continuous beaches in the world. Relative to other regions of the world the Latin American region is less vulnerable to potential sea level rise. A recent global study of sea level rise impacts found that the South American Pacific coast is not very vulnerable to sea level rise with respect to both socio-economic systems and ecosystems, and the South American Atlantic coast was found to be vulnerable with respect to wetlands only (Delft Hydraulics, 1993). Despite these estimates of the relative vulnerable of the region as a whole, local and regional impacts may be significant.

Two countries in the Latin American region reported findings on the coastal resources sector: Venezuela and Peru. Both of the assessments examined the potential loss through erosion and inundation as a result of varying estimates of sea level rise. Nicholls and Leatherman (1995) presents results from Venezuela, but not for Peru.

Coastal Resources in Venezuela
Venezuela's coastline covers approximately 4,000 km², of which 52% is on the Caribbean Sea and 48% is on the Atlantic Ocean. There are many types of coastal environment: sandy marshland, beaches, cliffs, deltaic plans, coastal lagoons, estuaries, barrier islands, and bays. The assessment presents results based on the IPCC common methodology (Benioff et al., in press), which includes scenarios of 1.0 m and 0.5 m sea level rise for 2100. Using information from a study of the entire Venezuelan coast, five study sites were selected for the assessment. Of a total study area (131.43 km²), it is estimated that a 0.5 m and 1.0 m sea level rise would result in losses of land by erosion of 34.24 km² and 47.0 km², respectively. In addition, land losses due to inundation are estimated to be 52.63 km² under a 0.5 m sea level rise and 77.74 km² under a 1.0 m sea level rise. Preliminary estimates indicate that economic damages could be US$10.2 billion from lost land and infrastructure; one of the most important industries, oil, could be significantly affected and important tourism developments and natural areas could be threatened as well. In addition, a preliminary adaptation assessment was undertaken using the Adaptation Strategy Evaluator. In the future, more work will be completed using this model.

Coastal Resources in Peru
The Peruvian coastline is 3,079.5 km long and consists mainly of alluvial and fluvioalluvial deposits and some marine, aeolian, and gravity deposits. The coastline is mostly cliffs, but there are some cities along the coast in low-lying areas, and thus potentially threatened by flooding. The assessment examined eight study sites and presented results based on two scenarios: 0.30 m and 1.0 m. Potential land loss due to inundation at the eight study sites is estimated to be 78.3 km². Examples of the potential cost of damage from the country's preliminary economic evaluation and a brief evaluation of adaptation options are also included.

Summary and Discussion

The degradation and fragmentation of the tropical forests in the Latin American region are currently threatening the forest resources and have the potential to alter the regional climate. The interaction of these changes with potential climate change due to an increase in greenhouse gases will require additional research beyond the assessment presented in this report.

- The assessment of the forest in Venezuela indicates that under climatic change conditions there would be a change toward drier patterns with a shift of approximately 40 to 50 million ha from tropical and subtropical moist forest to tropical dry forest. This shift from subtropical to tropical forests is consistent with findings from a recent assessment (Smith et al., 1995).

The significance of sea level rise may vary within the region, depending on both local morphologies and land uses. Although it is difficult to compare the magnitude of the impact between various studies, the results reported here illustrate the variation in potential impacts to this sector.

- For Venezuela, of the total study area (131.43 km²), it is estimated that a 1.0 m sea level rise would result in a loss of land by erosion of 47.0 km² and a loss by inundation of 77.74 km^2. For Peru the potential land loss due to inundation at eight study sites is estimated to be 78.3 km^2 under a 1.0 m sea level rise scenario. Even though the potential land loss at these sites is significantly lower than estimates of potential land loss for all of Peru (Nicholls and Leatherman, 1995), the economic damages in this study are significantly higher than those reported by Nicholls and Leatherman.

This discussion is based on preliminary assessments from only two countries and two vulnerability sectors; additional analyses are necessary to identify regional impacts.

CONCLUSION

The countries reporting in this document are at various stages in their assessments. A few of the reporting countries have just initiated the screening for their assessment, many have completed the screening for their assessment, and several have completed the assessment of biophysical impacts for one or more sectors. However, most of the country studies presented in this report have only begun their assessment of socioeconomic impacts and evaluation of adaptation options. Both the socioeconomic analyses and adaptation assessments are important steps because they place the biophysical impact assessments in the context of potential changes in the socioeconomic baseline and the potential for adaptation. Thus, although much work has been completed, much work remains to be done.

The results presented in this report are preliminary and the results from many countries are still to come, but a number of conclusions can be drawn:

Successful Use of Methodologies. The methodologies for conducting vulnerability and adaptation assessments are widely available, and as shown in this chapter, are being used by developing and transition country researchers to address a range of vulnerability sectors. The results presented in this report also demonstrate that, with training and efforts to adapt the methodologies, assessments can be made for a wide variety of circumstances.

Contribution to Scientific Understanding. This report also includes results from many countries for which there was no (or limited) prior research on the potential impacts of climate change or for which the research has not yet been widely available. In many cases the findings are consistent with estimates from recent literature. As the 13 countries that have included reports in this document continue their research and as additional countries complete their studies, the literature on vulnerability to climate change will be greatly enriched.

Need for Capacity Building. Many developing and transition countries have been strengthened in their ability to conduct vulnerability and adaptation assessments through financial and technical support from the U.S. Country Studies Program. However, scarcity of human and financial resources continues to limit the ability of some countries to carry out vulnerability and adaptation assessments. For these countries, the efforts of international institutions to bring resources to these regions, to improve communication, and to build the necessary capacity should improve the countries' ability to conduct vulnerability and adaptation assessments.

REFERENCES

African Development Bank. 1991. Environmental Policy Paper, 56 pp.

Allersman, E. and W.K. Tilmans. 1993. Coastal Conditions in West Africa, A Review. *Ocean and Coastal Management* 19: 199-240.

Awosika, L.F., A.C. Ibe, and P. Schroeder, eds. 1993. Coastlines of West Africa. A compilation of papers presented at the Coastal Zone '93. New York: American Society of Civil Engineers.

Benioff, R., S. Guill, and J. Lee (eds.) (In press). *Guidance for Assessing Vulnerability and Adaptation.* Dordrecht, The Netherlands: Kluwer Academic Publishers.

Biodiversity Support Program. 1993. African Biodiversity: Foundation for the Future, A Framework for Integrating Biodiversity Conservation and Sustainable Development, 149 pp.

Crosson, P. 1995. Future Supplies of Land and Water for World Agriculture. In *Population and Food in the Early 21st Century: Meeting Future Food Demand of an Increasing World Population* (Islam, N., ed.). Occasional Paper, International Food Policy Research Institute (IFPRI), Washington, DC, pp. 143-160.

Delft Hydraulics. 1993. Sea Level Rise, A Global Vulnerability Assessment, Population, Coastal Wetlands and Rice Production on a Global Scale. Tidal Water Division, Ministry of Transport, Public Work and Water Management, the Netherlands, 184 pp.

Downing, T.E. 1992. *Climate Change and Vulnerable Places: Global Food Security and Country Studies in Zimbabwe, Kenya, Senegal, and Chile.* Research Report No. 1, Environmental Change Unit, University of Oxford, Oxford, UK, 54 pp.

FAO. 1992. *AGROSTAT Digital Data.* FAO, Statistics Division, Rome, Italy.

Hammond. 1992. Atlas of the World. Hammond Incorporated. Maplewood: New Jersey.

Heath, L.S., P.E. Kauppi, P. Burschel, H.-D. Gregor, R. Guderian, G.H. Kohlmaier, S. Lorenz, D. Overdieck, F. Scholz, H. Thomasius, and M. Weber. 1993. Contribution of Temperate Forests to the World's Carbon Budget. *Water, Air, & Soil Pollution* 70: 55-69.

Horie, T. 1993. Predicting the Effects of Climatic Variation and Effect of CO_2 in Rice Yield in Japan. *Journal of Agricultural Meteorology* (Tokyo) 48: 567-574.

Huq, S., S. Ali, and A. Rahman. 1995. Sea Level Rise and Bangladesh: A Preliminary Analysis. In Nicholls, R.J. and S.P. Leatherman Potential Impacts of Accelerated Sea-Level Rise on Developing Countries. *Journal of Coastal Research* Special Issue No. 14, Spring.

Ibe, A.C. and R.E. Quelennac. 1989. Methodology for Assessment and Control of Coastal Erosion in West Africa and Central

Africa. UNEP Regional Sea Reports and Studies, No.107.

IPCC (in press). Climate Change 1995: The IPCC Second Assessment Report, Volume 2: Scientific-Technical Analyses of Impacts, Adaptations, and Mitigation of Climate Change [Watson, R.T., M.C. Zinyowera, and R.H. Moss (eds.)]. Cambridge University Press, Cambridge and New York.

Kaczmarek, Z. and J. Kindler. 1995. National Assessment — Poland. In *Water Resources Management in the Face of Climatic and Hydrologic Uncertainties*, Z. Kaczmarek et al. (eds.), Kluwer Publ. House.

Kaczmarek, Z., M. Niestepski, and M. Osuch. 1995. Climate Change Impact on Water Availability and Use. International Institute for Applied Systems Analysis, WP-95-48, 18 pp.

Lin, E. 1994. The Sensitivity and Vulnerability of China's Agriculture to Global Warming. *Rural Eco-Environment* 10(1): 1-5.

Magadza, C.H.D. 1994. Climate Change: Some Likely Multiple Impacts in Southern Africa. *Food Policy* 19(2): 165-191.

Matthews, R.B., M.J. Kropff, D. Bachelet, and H.H. van Laar. 1994. The Impact of Global Climate Change on Rice Production in Asia: A Simulation Study. Report No. ERL-COR-821, U.S. Environmental Protection Agency, Environmental Research Laboratory, Corvallis, OR.

Mearns, L.O., C.R. Rosenzweig, and R. Goldberg. (in press). "The Effect of Changes in Daily and Interannual Climatic Variability on CERES-Wheat: A Sensitivity Study." *Climatic Change*.

Menzhulin, G.V. 1992. The Impact of Expected Climate Changes on Crop Yields: Estimates for Europe, the USSR, and North America Based on Paleoanalogue Scenarios. In *Economic Issues in Global Climate Change: Agriculture, Forestry, and Natural Resources* (Reilly, J.M. and M. Anderson, eds.). Boulder: Westview Press, pp. 353-381.

Menzhulin, G.V. and L.A. Koval. 1994. Potential Effects of Global Warming and Carbon Dioxide on Wheat Production in the Former Soviet Union. In *Implications of Climate Change for International Agriculture: Crop Modeling Study* (Rosenzweig, C. and A. Iglesias, eds.). U.S. Environmental Protection Agency, FSU Chapter, Washington, DC, pp. 1-35.

Muchena, P. 1994. Implications of Climate Change for Maize Yields in Zimbabwe. In *Implications of Climate Change for International Agriculture: Crop Modeling Study* (Rosenzweig, C. and A. Iglesias, eds.). U.S. Environmental Protection Agency, Zimbabwe Chapter, Washington, DC, pp. 1-9.

Neilson, R.P. 1993. "Vegetation Redistribution: A Possible Biosphere Source of CO_2 during Climate Change." *Water, Air, and Soil Pollution* 70: 659-673.

Neilson, R.P., G.A. King, and J. Lenihan. 1994. Modeling Forest Response to Climatic Change: The Potential for Large Emissions of Carbon from Dying Forests. In *Carbon Balance of World's Forested Ecosystems:*

Towards a Global Assessment (Kanninen, M., ed.). Academy of Finland, Helsinki, Finland, pp. 150-162.

Nicholls, R.J. and S.P. Leatherman. 1995. Potential Impacts of Accelerated Sea-Level Rise on Developing Countries. *Journal of Coastal Research* Special Issue No. 14, Spring.

Roostalu, H., A. Piho, and P. Saarnak. 1980. Complex Zoning of Soil-Climatic Resources. *Transactions of Estonian Agricultural Academy* (Tartu, Estonia), 122: 101-116.

Rosenzweig, C. and A. Iglesias (eds.). 1994. *Implications of Climate Change for International Agriculture: Crop Modeling Study*. EPA 230-B-94-003, U.S. Environmental Protection Agency, Washington, DC.

Salati, E. and P.B. Jose. 1984. Amazon Basin: A System in Equilibrium. *Science* 225: 129-138.

Seino, H. 1994. Implications of Climate Change for Japanese Agriculture: Evaluation by Simulation of Rice, Wheat, and Maize Growth. In *Implications of Climate Change for International Agriculture: Crop Modeling Study* (Rosenzweig, C. and A. Iglesias, eds.). U.S. Environmental Protection Agency, Japan Chapter, Washington, DC, pp. 1-18.

Sirotenko, O.D. and E.V. Abashina. 1994. Impact of Global Warming on the Agroclimatic Resources and Agriculture Productivity of Russia (result of the simulation). *Meteorology and Hydrology*, Vol. 4.

Sirotenko, O.D. et al. 1991. Global Warming and the Agroclimatic Resources of the Russian Plain. *Soviet Geography*, 32(5): 337-384.

Smith, T.M. and H.H. Shugart. 1993. The Transient Response of Terrestrial Carbon Storage to a Perturbed Climate. *Nature* (London) 361: 523-526.

Smith, T.M., P.N. Halpin, H.H. Shugart, and C.M. Secrett. 1995. "Global Forests." In *As Climate Changes: International Impacts and Implications*. K.M. Strzepek and J.B. Smith (eds.). Cambridge: Cambridge University Press.

Strzepek, K., J. Niemann, L. Somlyody, and S. Kulshreshtra. 1995. *A Global Assessment of National Water Resources Vulnerabilities: Sensitivities, Assumptions, and Driving Forces*. IIASA Working Paper (in press).

Sugihara, S. 1991. A Simulation of Rice Production in Japan by Rice-Weather Production Model. *Kikogaku-Kishogaku Kenkyu Hokoku*, 16: 32-37.

Tao, Z. 1992. Influences of Global Climate Change on Agriculture of China. In *Climate Biosphere Interactions*. New York: John Wiley and Sons.

VEMAP Participants. 1995. "Vegetation/Ecosystem Modeling and Analysis Project (VEMAP): Comparing Biogeography and Biogeochemistry Models in a Continental-Scale Study of Terrestrial Ecosystem Responses to Climate Change and CO_2 Doubling." *Global Biogeochemical Cycles* 9: 407-437.

World Bank. 1994. World Population Projections 1994-1995. Estimates and Projections with Related Demographic Statistics. Baltimore and London: The Johns Hopkins University Press.

(WRI) World Resources Institute. 1992. *World Resources 1992-1993. A Guide to the Global Environment toward Sustainable Development*. New York: Oxford University Press.

(WRI) World Resources Institute. 1993. *Environmental Almanac*. Boston: Houghton Mifflin Company.

(WRI) World Resources Institute. 1994. *World Resources 1994-1995. A Guide to the Global Environment, People and the Environment*. New York: Oxford University Press.

Zhang, H. 1993. The Impact of Greenhouse Effect on Double Rice in China. In *Climate Change and Its Impact*. Meteorology Press, Beijing, China.

Chapter 5

Vulnerability and Adaptation Assessments for The Gambia

Bubu P. Jallow
Department of Water Resources
Banjul, The Gambia

SUMMARY. The Gambia assessment of vulnerability and adaptation to climate change covers the development of scenarios and the assessment of the agriculture (crop production), water resources, forestry, grasslands and livestock, and coastal resources sectors. This chapter presents the development of scenarios, and the analyses of the agriculture and coastal resources sectors. (The other sector analyses are still incomplete.) For the agriculture sector in The Gambia, under $2\times CO_2$ *scenarios, production of cereal crops is estimated to decrease, and production of groundnuts, an export crop, is estimated to increase. Second-order impacts such as the need to import more food are possible. Because most of the developed area of The Gambia, including the capital city of Banjul, is along the topographically flat coast, a 1.0 m sea level rise would have devastating effects. Also lost would be the mangroves and their fish spawning areas, and approximately US$217 million worth of land. Implementation of adaptation strategies and policies to both protect the coastline and discourage development on the coast could alleviate much of the damage. An economic analysis of crop production has not yet been performed. Also, adaptation options have not yet been completely assessed. These issues will be addressed as the study is completed.*

INTRODUCTION

The Gambia, a British colony until 1965, is located between the 13th and 14th north latitudes along the Atlantic coast of Africa. It is the smallest country (about 11,000 km²) in Africa (see Figure 5.1). The coastline is where most of the development in the country lies, including the capital city of Banjul. The coast also supports mangrove systems on the Kombo Peninsula and associated fish spawning and breeding grounds.

Based on the 1982 forest inventory of The Gambia (Foster, 1983), the total area of closed and open broadleaf forest was 438,400 ha. Based on the moderate deforestation rates in Pepper et al. (1992), 17,157 ha were lost by 1994 and 123,710 ha are expected to be lost by 2100. Thus, by 2100 only 314,690 ha of the 1982 broadleaf

Note: This chapter does not contain a detailed description of the assessment methodologies that were recommended by the U.S. Country Studies Program. Refer to Chapter 2 for an overview of these methodologies.

J. B. Smith et al. (eds.), Vulnerability and Adaptation to Climate Change, 109–128.

Figure 5.1. Map of The Gambia

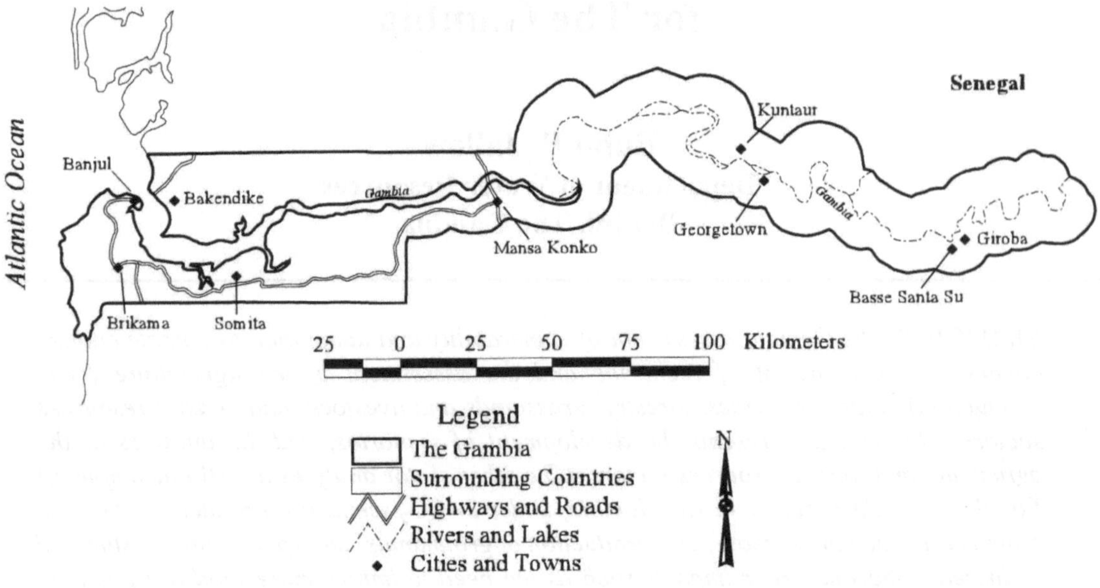

forest cover will be left. Most of this area is assumed to be lost to agriculture and some may be barren flats.

The climate is sudano-sahelian, with a short rainy season from June to October and a long dry season from November to May. Average annual temperature is about 28°C and average annual rainfall is about 880 mm. During the dry season, the relative humidity is about 68% along the coast and 41% inland, but generally above 70% throughout the wet season. An analysis of the historical climate data (1951-1990) of The Gambia indicates a slight warming and about a 25% to 30% decrease in rainfall.

According to the 1993 population census, the population of The Gambia is 1,025,867 (CSD, 1993). The population change during the inter-censal period of 1983-1993 shows a steep and startling increase of about 49% and an annual growth rate of 4.1% per annum. The population density of approximately 96 persons per square kilometer is among the highest in Africa. It is estimated that more than one third of the population is concentrated on the coastal zone.

From 1986 to 1992, the economy of The Gambia was dominated by trade. Trade, mainly re-export of manufactured goods and commodities to neighboring countries, has become a major determinant of economic activity in terms of output accounting for 23% of 1991-1992 GDP (UNDP-The Gambia, 1993).

The agriculture sector contributes about 27% of GDP and employs about 75% of the labor force. Development in the agriculture sector is constrained by a long tradition of dependence on a single cash crop, groundnuts, which is also the main source of foreign exchange earnings. However, the production of food crops, mainly coarse grains and rice, has steadily increased in recent years. The liberal policy of the government has further reduced the distortions in the agriculture market, and farmers are diversifying away from groundnuts. Although agriculture's share of GDP fell consecutively for three years before 1991, output of crops other than groundnuts increased by 20% in 1991-1992. The value of the marketable surplus of foodcrops now exceeds that of groundnut products. This increase has been obtained primarily through better marketing channels and cultivation of additional land, with little productivity gains.

Recently tourism has become an important economic sector. The mild climate, long stretches of sandy beaches, and close proximity to Europe have made The Gambia an attractive destination for tourists, especially from Europe. The number of visitors has increased from about 70,000 in the 1985-1986 season to well over 100,000 in 1990-1991. Available statistics and surveys show that tourism contributes about 10% of GDP and a foreign exchange earnings of about US$50 million in 1991-1992 (UNDP-The Gambia, 1993). The industry also employs about 7,000 people, either directly or indirectly, during the height of the season.

Imports and consumption of fossil fuel products dominate the energy sector, and most of the commodity is consumed by the transport sector. However, until the ban on the production of charcoal in the early 1980s, biomass fuel production and consumption featured high in the energy sector. Despite the ban on charcoal, biomass fuels still meet about 85% (EU/MNRE, 1992) of the energy needs of the country.

The Gambia study on assessment of vulnerability and adaptation to climate change covers the agriculture, water resources, forestry, grasslands and livestock, and coastal resources sectors. This chapter presents the development of scenarios and the analyses of the agriculture and coastal resources sectors. The Gambia's studies for the other sectors have not yet been completed. In addition, further economic analysis and analysis of adaptation options are planned for the agriculture (crop production) sector.

SCENARIOS

Baseline Climate Scenario

Forty years (1951-1990) of historical climate data were used to develop a baseline climate scenario (Table 5.1) for The Gambia. The annual mean temperature is 27.5°C, annual mean minimum temperature is 20.9°C, and annual mean maximum temperature is 34.1°C. Mean annual rainfall is 878.7 mm, mean annual potential evapotranspiration is 144 mm, mean annual solar radiation received is 255 W/m^2, and mean annual relative humidity is 61%. The monthly variations of these parameters are given in Table 5.1. The behavior of the climate of The Gambia, and the Sahel in general, shows almost equal distribution of wet/cool and dry/warm years. The 1951-1980 period was wet and relatively cooler, and the 1961-1990 period was drier and warmer. It is for this reason that the 40-year period has been used instead of the 30-year (1951-1980) period. Selection of this period is expected to minimize bias.

Climate Change Scenarios

In this study only the equilibrium general circulation model (GCM) outputs are used to create climate change scenarios. Five

Table 5.1. Baseline Climate Scenario of The Gambia

	Jan	Feb	Mar	Apr	May	Jun	Jul	Aug	Sep	Oct	Nov	Dec	Annual
Mean Temp. (°C) 1951-1990	24.8	26.6	28.4	29.6	29.6	29.3	27.1	27.1	27.2	28	26.8	24.8	27.5
Min. Temp. (°C) 1951-1990	16.3	17.6	19.6	20.9	22.6	23.9	23.7	23.4	23.1	22.8	19.8	16.7	20.9
Max. Temp. (°C) 1951-1990	33.1	35.4	36.4	36.8	36.4	34.5	32.3	31.4	31.9	33.5	34.5	32.8	34.1
Rainfall (mm/month) 1951-1990	0.2	0.0	0.0	1.2	12.1	87.0	202.2	280.7	215.0	73.8	6.3	0.4	878.7
CCCM Rainfall (mm/day)	0.39	0.00	0.41	0.38	1.44	3.70	6.88	9.93	6.60	3.37	1.34	0.55	2.95
Solar Rad. (W/m^2) 1970-1985[1]	232	272	289	295	286	263	235	233	244	256	241	218	255
Rel. Hum. (%) 1951-1990	43	42	44	48	55	67	78	82	82	76	61	50	61
Winds (knots) 1951-1990	4	4	5	5	5	5	4	4	3	3	3	4	4
Pot. Evap. (PET, mm) 1951-1990	138	146	193	189	175	145	134	126	118	124	118	121	144

[1] Data only available for this period.

GCMs were run to obtain $1\times CO_2$ and $2\times CO_2$ output for temperature, rainfall rates, and solar radiation that were used for creating $2\times CO_2$ equilibrium scenarios. The models run are the Canadian Climate Centre Model (CCCM), Goddard Institute for Space Studies (GISS), United Kingdom Meteorological Office 1987 (UKMO), the Geophysical Fluid Dynamics Laboratory equilibrium run (GFD3), and Geophysical Fluid Dynamics Laboratory 1% transient model (GF01) (Benioff et al., in press).

To choose the models that best estimate the climate of The Gambia, the $1\times CO_2$ temperature output from all five models was compared with the 1951-1990 temperature data. The observed data are compared to the GISS, UKMO, and GFD3 models in Figure 5.2. The GISS, UKMO, GFD3, and GF01 models best approximated the climatology. However, since none of the sectoral study teams used the transient scenarios, these are not reported here. Also, some sectoral study teams (e.g., agriculture and forestry) were supplied model outputs from the Oregon State University (OSU) model (Benioff et al., in press) and the CCCM. These teams used these model outputs for their sectoral assessments, in addition to the models that best approximated climatology.

The equilibrium scenarios for monthly mean temperature, precipitation, and solar radiation were created by combining average monthly model output values with observed climate data from the 1951-1990 baseline

Figure 5.2. 1×CO_2 Scenarios with 1951-1990 Average Monthly Mean Temperature

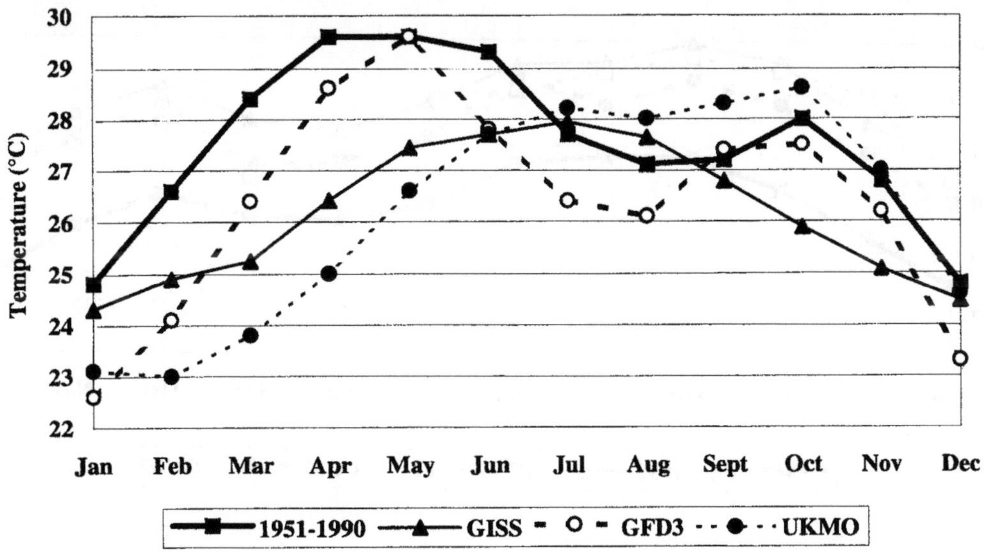

climate period. The changes in temperature (2×CO_2-1×CO_2) were added to the 1951-1990 average for every month and the annual mean was evaluated. On average, by 2075, mean temperatures are expected to increase by 3°C (GFD3), 4.1°C (GISS), and 5.0°C (UKMO). The ratio of 2×CO_2 to 1×CO_2 for the rainfall and solar radiation output were multiplied by the 1951-1990 average to obtain the scenarios. Average rainfall in The Gambia is estimated to decrease from the 1951-1990 average by 15% (GFD3) to 21% (GISS). In contrast, rainfall is estimated to increase by approximately 86% above the 1951-1990 average under the UKMO scenario. Little change is estimated in solar radiation (-0.4% GISS, -5% GFD3, and 6% UKMO) while all the models show an increase in PET (5% GISS, 4% GFD3, and 7% UKMO). Figures 5.3 and 5.4 show the scenarios for mean monthly temperature and rainfall.

Baseline Socioeconomic Scenarios

Baseline socioeconomical scenarios are designed taking into account changes in population levels, national and per capita income, and productivity. The average growth rate of population in The Gambia has been 2.9% since the beginning of this century, with a faster rate in the last 30 years of about 4%. To develop the scenarios for the next century it is assumed that the growth rate will remain at 4% for the first decade of the analysis period and will then decline. This assumption, when applied to the 1993 population, yields a population of about 1.3 million by 2000 and over 9.7 million in 2100.

The average GNP for The Gambia from 1986 to 1992 is 2.871 Dalasis (US$319,000). GNP growth rates from Pepper et al. (1992) were applied to the 1993 value to yield a value of 55.124 million Dalasis (US$6.12 million) by 2100.

Figure 5.3. 2×CO_2 Scenarios of Mean Monthly Temperature of The Gambia in 2075

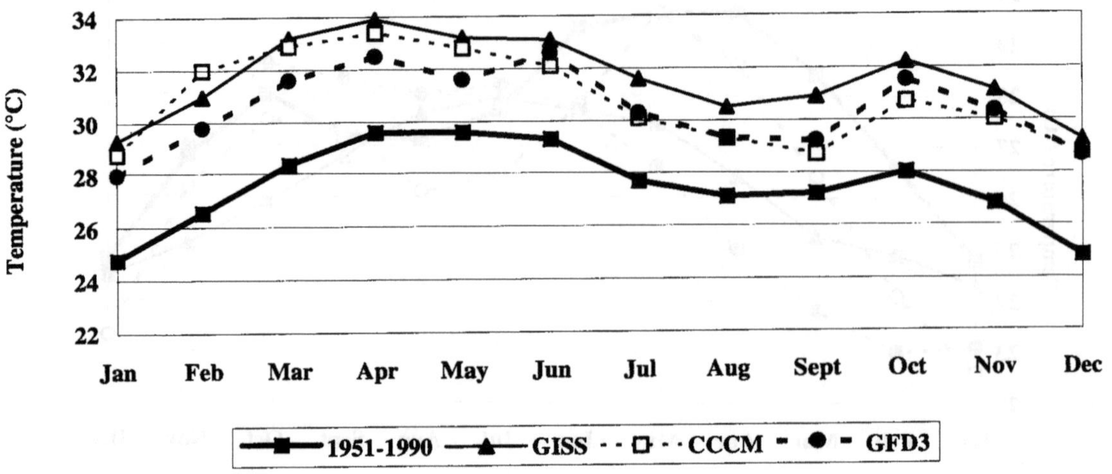

Figure 5.4. Model Simulation of Mean Monthly Rainfall of The Gambia in 2075

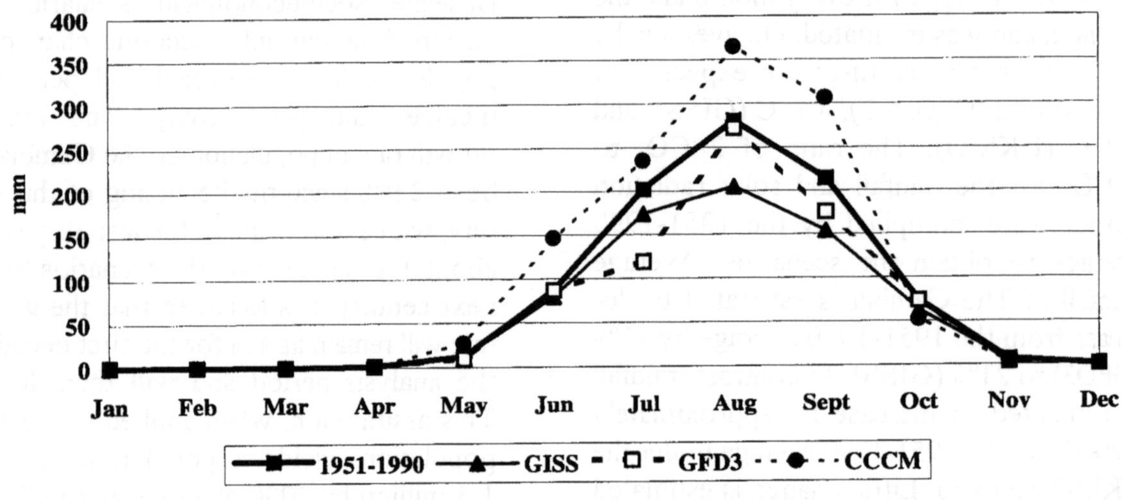

Baseline Sectoral Scenarios

Although the baseline sectoral scenarios have not been directly incorporated into this assessment, the following information provides a context for integrating environmental analysis with socioeconomic pressures on the sectors.

Data on crop cultivation, harvest, yield, and production for the period from 1980 to 1994 were obtained from the 1990 and 1993 Year Books of Gambian Agriculture (DOP, 1990; 1993) and then were combined with annual growth rates from FAO (1993) to develop cropland area and production assumptions to

2100. The 1994 values of 68,826 ha for coarse grains, 20,640 ha for rice, and 87,808 ha for groundnuts are averages for the 1980-1994 period. Based on the analysis, the area of land under cultivation of coarse grains, rice, and groundnuts is expected to increase by 2100 to 112,900 ha, 33,643 ha, and 143,131 ha, respectively. Similarly, production is expected to increase by 2100 to 185,989 tons for coarse grains, 151,619 tons for rice, and 443,259 tons over the 1994 baseline (1980-1994 average) values of 57,863 tons for coarse grains, 31,720 tons for rice, and 93,943 tons for groundnuts.

The total forest area of The Gambia (excluding mangroves) is estimated to be 438,400 ha (Foster, 1983). Applying estimates in Pepper et al. (1992) and based on a moderate deforestation assumption of a 0.2% per year forest clearing rate, this area is combined with population assumptions to estimate the area cleared (deforestation) to 2100. From the analysis, 17,157 ha would be lost by 2000, leaving a balance of 421,243 ha. By 2100, 123,719 ha would be cleared, leaving a balance of 314,681 ha. This analysis considers only the natural broadleaf forest cover.

Sea Level Rise Scenarios

The sea level rise scenarios adopted in this study are based on the IPCC first assessment report (IPCC, 1990a). These are 0.2 m as baseline, and 0.5 m, 1.0 m, and 2.0 m by 2100.

Agriculture Sector

Methods Used

Selection of Crops and Study Sites

The major crops analyzed in this study are maize, millet (early and late variety), and groundnuts (export crop). Although rice is the staple food in The Gambia, the model used for the simulation did not perform well for rice, and this crop had to be dropped from the analysis. Maize is a cereal crop that was initially grown as a cash crop, filling the gap between the previous year's harvest and the current year's harvest of the major food crops (NARI, 1994). Early millet is one of the most widely consumed cereals in The Gambia next to rice and also one of the most drought-tolerant cereals next to sorghum. Groundnuts do well in sandy soils and are the main export crop of The Gambia. The crop needs sufficient water at all its growth and development stages, but more water is needed during the period of useful flowering (Alimi, 1990).

Four sites were selected, each representing an area where one of the selected crops is cultivated. Bakendik was chosen as the site where groundnut production could be simulated; in Somita early millet was simulated; in Kuntaur late millet was simulated; and Girobakunda was selected for maize production simulation. These sites also represent major centers for agricultural extension services and farm trials.

Crop Models Used

The methodology used involves the applications of the Decision Support System for Agrotechnology Transfer version 3 (DSSAT 3) and the CERES-Maize model (Benioff et al., in press).

Weather, soil, and crop parameters are needed for crop growth simulation. The climate data used in this study are daily maximum air temperature, daily precipitation, and daily solar radiation. All these data were available in the meteorological data appropriate format to be directly utilized by the DSSAT 3.

Soil characterization for the study sites was carried out by a soil specialist from the Soil and Water Management Unit of the Department of Agricultural Services under a contract with The Gambia National Climate Committee (NCC). Data on the amount, type, and depth of fertilizer application were collected. Crop management data, including planting dates, plant population, cultivar, and irrigation practices were also collected.

The scenarios used to assess the impacts of the predicted climate change on crop production were based on GCM outputs of simulations at $2\times CO_2$. Using the Weather-Man Utility program, it was possible to create daily weather scenario data files from the monthly climate change scenarios. The daily weather scenarios created using this bank of The Gambia, though not in the approach were based on the current (baseline) and the climate change scenarios from the CCCM, GFD3, and GISS models. It was also assumed that technology and management of crop production such as fertilizer application and irrigation practices will remain the same.

Results

Production of Maize

Tables 5.2, 5.3, and 5.4 give the potential impacts on the production of maize in The Gambia of the predicted climate change

Table 5.2. Growth Parameters for Maize at Giroba

Parameter (kg dry matter/ha)	Baseline	CCCM	GFD3	GISS
Leaf Weight	578	472	376	559
% change		-18%	-35%	-3%
Stem Weight	1,357	1,057	900	1,203
% change		-22%	-34%	-11%
Root Weight	1,048	741	765	901
% change		-29%	-27%	-14%
Grain Weight	2,108	1,520	1,270	1,664
% change		-28%	-40%	-21%
Total Dry Matter Produced	5,091	3,790	3,311	4,327
% change		-26%	-35%	-15%

Table 5.3. Water Balance Parameters for Maize at Giroba

Parameter (mm)	Baseline	CCCM	GFD3	GISS
Evapotranspiration	87	23	21	70
% change		-74%	-76%	-20%
Precipitation	868	812	1,316	1,028
% change		-6%	52%	18%
Runoff	169	153	442	226
% change		-9%	162%	34%
Extractable Water	154	155	159	159
% change		1%	3%	3%
Drainage	350	389	598	468
% change		11%	71%	34%

Table 5.4. Nitrogen Utilization Parameters for Maize at Giroba

Parameter (kg N/ha)	Baseline	CCCM	GFD3	GISS
Nitrogen Uptake	69	51	44	61
% change		-27%	-36%	-12%
Total (stem + leaf) Nitrogen	505	396	300	400
% change		-22%	-41%	-21%
Nitrogen Leached	43	52	68	56
% change		19%	57%	28%
Soil Nitrogen	2,432	2,331	2,188	2,234
% change		-4%	-10%	-8%

Figure 5.5. Dry Matter Production of Maize at Giroba

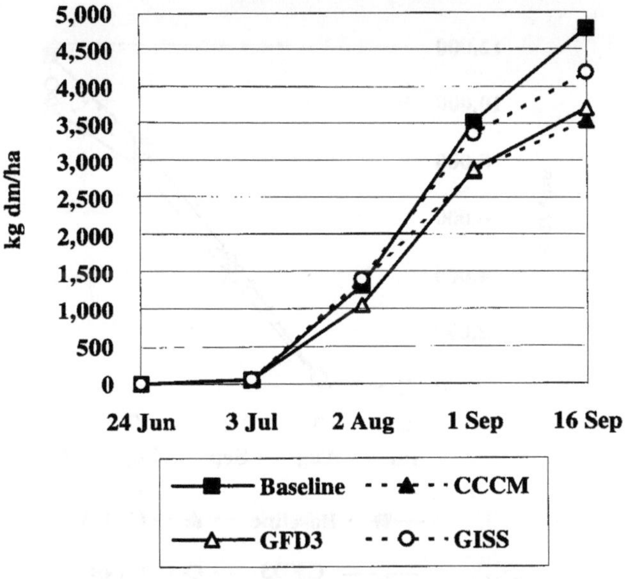

Figure 5.6. Dry Matter Production of Early Millet at Somita

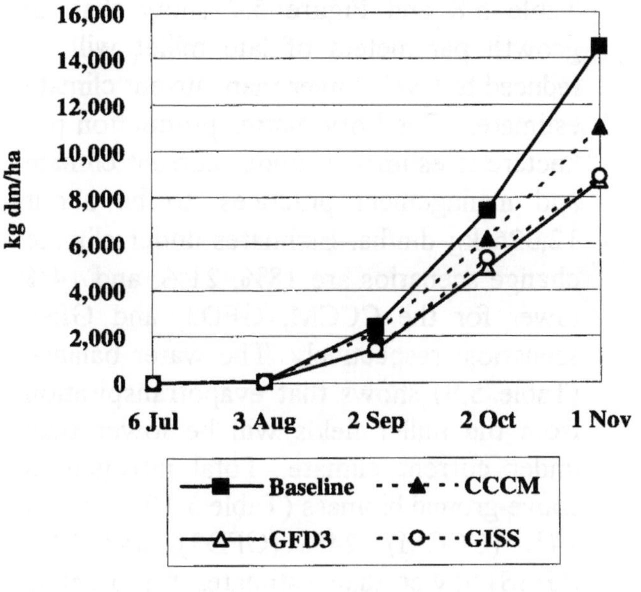

under a 2×CO_2 scenario and current management practices. On average, total bio mass dry matter (dm) production is estimated to be 26% (CCCM), 35% (GFD3), and 15% (GISS) below current climate estimates (Table 5.2 and Figure 5.5). Under current climate, a seasonal average of 154 mm of water is available in the soil for extraction by the maize crop, and that amount does not change greatly under climate change. The CCCM scenario estimate is 1% above current climate value, and the GFD3 and GISS models estimates are about 3% above current climate. Model estimates of nitrogen utilization (Table 5.4) show that under climate change, nitrogen uptake is reduced by about 27% (CCCM), 36% (GFD3), and 12% (GISS). Also nitrogen content of above-ground biomass (leaf, stem, and fruits) is reduced by about 22% (CCCM), 41% (GFD3), and 21% (GISS).

Production of Early Millet

All the model scenarios show a reduction in the growth parameters of early millet under current climate conditions and current management practices (Table 5.5 and Figure 5.6). Total dry matter (dm) production (root, leaf, stem, and grain) is estimated at 15,562 kg dm/ha under current climate conditions; the model estimates show a reduction of 29% (CCCM), 44% (GFD3), and 43% (GISS) below current climate estimates. Table 5.6 shows that precipitation and extractable water are all expected to increase over the estimates under current climate. Evapotranspiration is estimated to decrease below the current climate estimate by 21% (GFD3) and 22% (CCCM and GISS). Also, Table 5.7 shows that nitrogen content of above-ground biomass (leaf, stem, and fruits) is reduced by about 29% (CCCM), 39% (GFD3), and 37% (GISS).

Table 5.5. Growth Parameters of Early Millet at Somita

Parameter (kg dry matter/ha)	Baseline	CCCM	GFD3	GISS
Leaf Weight	4,402	2,823	2,092	2,058
% change		-36%	-52%	-53%
Stem Weight	8,395	6,385	5,114	5,374
% change		-24%	-39%	-36%
Root Weight	744	536	424	406
% change		-28%	-43%	-45%
Grain Weight	2,021	1,318	1,095	1,071
% change		-35%	-46%	-47%
Total Dry Matter Produced	15,562	11,062	8,725	8,909
% change		-29%	-44%	-43%

Table 5.6. Water Balance Parameters of Early Millet at Somita

Parameter (mm)	Baseline	CCCM	GFD3	GISS
Evapotranspiration	152.1	118.5	120.5	118.6
% change		-22%	-21%	-22%
Precipitation	770.0	807.0	981.0	1,015.0
% change		5%	27%	32%
Runoff	35.9	53.5	74.5	102.3
% change		49%	108%	185%
Extractable Water	176.8	183.9	194.7	195.5
% change		4%	10%	11%
Drainage	384.1	475.3	565.6	568.7
% change		24%	47%	48%

Table 5.7. Nitrogen Utilization Parameters of Early Millet at Somita

Parameter (kg N/ha)	Baseline	CCCM	GFD3	GISS
Nitrogen Uptake	111.1	78.0	65.9	67.2
% change		-30%	-41%	-40%
Total (stem + leaf) Nitrogen	92.1	65.6	56.4	58.0
% change		-29%	-39%	-37%
Nitrogen Leached	32.7	54.0	56.2	51.0
% change		65%	72%	56%
Soil Nitrogen	67.1	71.0	72.9	75.8
% change		6%	9%	13%

Figure 5.7. Dry Matter Production of Late Millet at Kuntaur

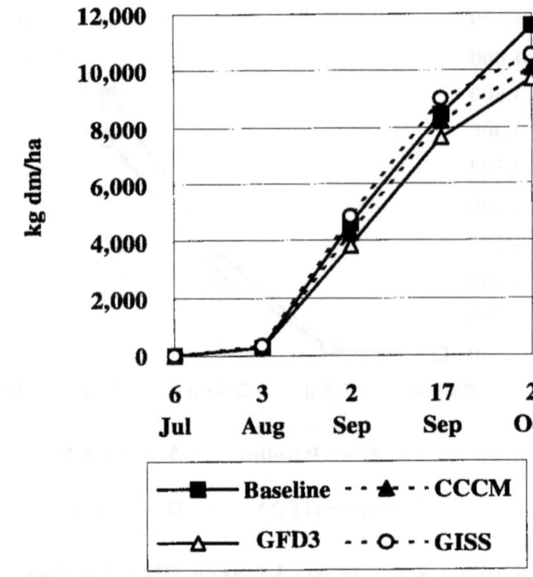

Production of Late Millet

Table 5.8 and Figure 5.7 show that all growth parameters of late millet will be reduced to levels lower than current climate estimates. Total dry matter production per hectare is estimated, under current climate and management practices, to be about 12,326 kg dm/ha. Estimates under climate change scenarios are 18%, 21%, and 14% lower for the CCCM, GFD3, and GISS scenarios, respectively. The water balance (Table 5.9) shows that evapotranspiration from the millet fields will be lower than under current climate. Total nitrogen in above-ground biomass (Table 5.10) is about 14% (CCCM), 24% (GFD3), and 18% (GISS) lower than estimates for baseline climate and management conditions.

Production of Groundnuts

All the climate change scenario estimates of growth parameters of groundnuts are higher than estimates under current climate and management conditions. Table 5.11 shows

Table 5.8. Growth Parameters of Late Millet at Kuntaur

Parameter (kg dry matter/ha)	Baseline	CCCM	GFD3	GISS
Leaf Weight	5,940	5,110	5,001	5,522
% change		-14%	-16%	-7%
Stem Weight	4,141	3,272	3,145	3,359
% change		-21%	-24%	-19%
Root Weight	635	493	481	502
% change		-22%	-24%	-21%
Grain Weight	1,610	1,256	1,123	1,175
% change		-22%	-30%	-27%
Total Dry Matter Produced	12,326	10,131	9,750	10,558
% change		-18%	-21%	-14%

Table 5.9. Water Balance Parameters of Late Millet at Kuntaur

Parameter (mm)	Baseline	CCCM	GFD3	GISS
Evapotranspiration	86.6	64.2	69.9	74.9
% change		-26%	-19%	-13%
Precipitation	417.0	431.0	580.0	443.0
% change		3%	39%	6%
Runoff	9.7	18.7	51.3	18.9
% change		93%	429%	95%
Extractable Water	131.9	138.1	144.9	142.4
% change		5%	10%	8%
Drainage	97.8	167.2	263.2	153.4
% change		71%	169%	57%

Table 5.10. Nitrogen Utilization Parameters of Late Millet at Kuntaur

Parameter (kg N/ha)	Baseline	CCCM	GFD3	GISS
Nitrogen Uptake (cumulative)	110.7	94.7	84.7	90.3
% change		-14%	-23%	-18%
Total Nitrogen (highest value)	96.5	83.3	73.2	79.0
% change		-14%	-24%	-18%
Nitrogen Leached (cumulative)	15.7	29.1	45.6	32.0
% change		85%	190%	104%
Soil Nitrogen (daily average)	58.0	60.4	58.7	63.8
% change		4%	1%	10%

Figure 5.8. Dry Matter Production of Groundnuts at Bakendik

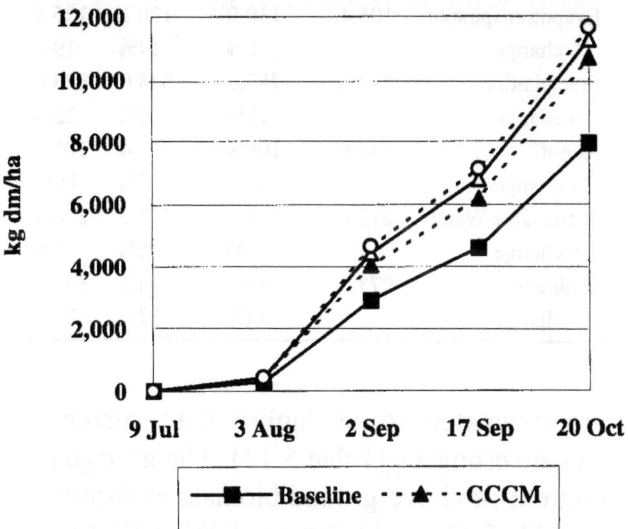

Table 5.11. Growth Parameters of Groundnuts at Bakendik

Parameter (kg dry matter/ha)	Baseline	CCCM	GFD3	GISS
Leaf Weight	1,867	2,735	2,845	2,982
% change		46%	52%	60%
Stem Weight	3,857	5,829	6,000	6,195
% change		51%	56%	61%
Root Weight	474	661	697	730
% change		39%	47%	54%
Grain Weight	1,752	1,911	2,146	2,156
% change		9%	22%	23%
Total Dry Matter Produced	7,950	11,136	11,688	12,063
% change		40%	47%	52%

that total dry matter production is estimated to be 7,950 kg dm/ha under current climate and management conditions. Model scenario estimates under 2×CO_2 climate conditions are 11,136 kg dm/ha (CCCM), 11,688 kg dm/ha (GFD3), and 12,063 kg dm/ha (GISS). These figures represent about 40% (CCCM), 47% (GFD3), and 52% (GISS) increases in total dry matter production (Figure 5.8). All water balance parameters

Table 5.12. Water Balance Parameters of Groundnuts at Bakendik

Parameter (mm)	Baseline	CCCM	GFD3	GISS
Evapotranspiration	105.4	110.9	120.1	125.6
% change		5%	14%	19%
Precipitation	672.0	787.0	848.0	823.0
% change		17%	26%	22%
Runoff	58.8	109.9	115.4	105.7
% change		87%	96%	80%
Extractable Water	212.8	214.7	222.5	214.9
% change		1%	5%	1%
Drainage	273.5	366.8	374.3	346.4
% change		34%	37%	27%

are estimated to be higher than current climate estimates (Table 5.12). The nitrogen content of above-ground biomass as shown in Table 5.13 would increase 50% (CCCM), 55% (GFD3), and 63% (GISS) above current climate conditions.

Higher-Order Impacts

Results of the analysis for The Gambia show that dry matter production in cereals would be less under climate change than estimates under current climate and management practices. The nitrogen content of the above-ground biomass of these cereals is also estimated to be lower. These decreases could lead to unfavorable higher order impacts on the economic and social well-being of The Gambia. The decrease in cereal production is likely to increase production costs because farmers may wish to boost food availability by increasing inputs such as fertilizers. The higher demand for inputs could lead to frequent shortages and thus increased prices.

Any increase in the price of inputs could cause economic concern because the output prices may not be as high as the input prices.

Table 5.13. Nitrogen Utilization Parameters of Groundnuts at Bakendik

Parameter (kg N/ha)	Baseline	CCCM	GFD3	GISS
Nitrogen Uptake (cumulative)	91.5	94.5	98	104.5
% change		3%	7%	14%
Total Nitrogen (highest value)	118.6	177.9	183.4	193.1
% change		50%	55%	63%
Leached Nitrogen (cumulative)	22.2	21.6	24.3	23.0
% change		-3%	9%	4%
Soil Nitrogen (daily average)	56.3	53.4	55.0	57.6
% change		-5%	-2%	2%
Fixed Nitrogen (cumulative)	141.9	226.8	241.5	248.1
% change		60%	70%	75%

It has been estimated that a yield decrease of 25% in Africa could increase agricultural prices by 20% (IPCC, 1990a). Net returns would be lower and subsistence farming would continue to be unprofitable, as it has been since the start of the Sahelian drought in the late 1960s and early 1970s.

The level of food security would also be reduced. The present rural-urban drift would be exacerbated, increasing the levels of social problems in the urban areas.

Reduced nitrogen content in the above-ground biomass of cereals will lead to reduced availability of palatable feed for livestock. Animals obtain protein from nitrogen-rich leaves of cereals, and livestock in The Gambia feed on the leaves of these crops.

All the models estimate that the dry matter production and nitrogen content of groundnuts will increase under a $2\times CO_2$ climate change. This would be beneficial to both the human and animal populations of the country. At the national level, farmers experiencing reduced yields of cereal crops may switch to groundnut cultivation. However, because groundnuts are produced mainly for export, there will be a greater demand for imported food to feed the population. The question of the availability of foreign currency for imports may then arise, and the related consequences may be a concern. The food supply and security would also remain threatened and could become worse.

Discussion and Conclusions

The model simulations have shown that precipitation may increase by approximately 50% at most, but increases in runoff and drainage will be much higher. The amount of water available within the root zone for extraction by the plants is estimated to change comparatively very little. The highest expected increase in extractable water is for millet, with about a 10% increase (GFD3) over current climate estimates. This is consistent with information in the IPCC First Assessment Report (IPCC, 1990b) that in tropical regions, particularly areas of monsoon rainfall, increases in total rainfall may not imply increases in availability of extractable moisture. The increase in rainfall could be due to increased storm intensity, which could result in more runoff but no increase in percolation and so less available water. From the results of the crop simulations it is concluded that climate change could lead to a decrease in dry matter production and nitrogen parameters in all the cereal crops studied in The Gambia. An overall increase in all the growth and nitrogen parameters of groundnuts is also envisaged.

The Gambian population prefers meals prepared from maize and millet. A decrease in the quantity produced and the nitrogen content of the dry matter produced will reduce the food supply from these cereals and the protein derived from them. The leaves of millet are used as feed for livestock and a reduction in the nitrogen content of the dry matter produced will also reduce the protein availability in the animal feed. Also, farmers experiencing reduction in yields from cereal crops may switch to groundnut cultivation, creating greater demand for imported food to feed the population, and a subsequent economic problem of the need for foreign currency to buy it.

Coastal Resources Sector

Methods Used

Selection of Study Sites

The study area considered (Figure 5.9) is the 70 km-long Gambian coastline of open coast and the sheltered coast of the mangrove system of The Gambia River estuary, which extends about 200 m inland. For specific analysis such as socioeconomic impacts, only parts of unit 2 and the whole of unit 3 (identified on Figure 5.9) were selected because they are the most developed areas on the coastal zone. The capital city of Banjul is located in unit 3.

Sea Level Rise Methodologies Used

Various data sets on the coastal zone of The Gambia were assembled to inventory and characterize the study area. Black and white aerial photography for 1972 and 1982, color infrared photography for 1993, topographic maps for 1976, contoured orthophoto maps of the 1982 aerial photography, bathymetric maps of part of the coastline, geology maps and documentation on The Gambia (Whyte

Figure 5.9. Detail of Kombo Peninsula

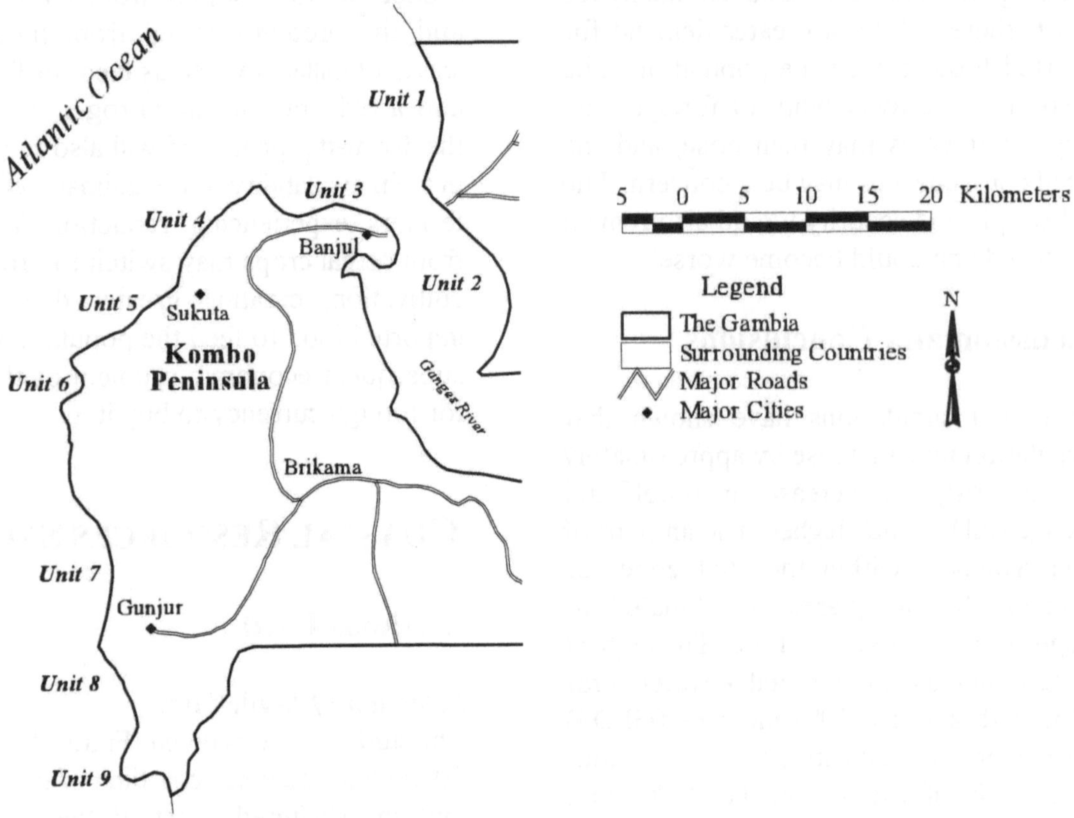

and Russel, 1988), and multiyear and multi-location still photographs of the coastline were acquired from archives or purchased from the Department of Surveys in The Gambia and AERODATA in Luxembourg. Socioeconomic data were obtained on gross domestic product, gross national product, population and population density, annual number of tourists visiting The Gambia, out-of-pocket tourist expenditures, planned tourism development, wetland habitants, and historic and cultural assets.

Land loss due to a 1.0 m sea level rise was estimated using the drowning or inundation concept and the Bruun Rule for estimation of retreat through beach and dune erosion. It was assumed that sea level rise is the only forcing factor, and land loss from a 1.0 m sea level rise was estimated.

For the estimation of land loss due to inundation, a 1.0 m contour line was drawn on the topographic maps. The area bounded by the contour line was defined as the area that will be inundated by a 1.0 m sea level rise. Land loss due to the lower scenarios was estimated by linear interpolation. The area to be lost from a 2.0 m level rise was also estimated for information purposes only.

Dennis et al. (1995) reasonably assumed low and high estimates of closure depth of 3.4 m and 5.9 m respectively, for the low wave energy sector from Dakar to Bissau. Since The Gambian coastline is within this southern sector, these values are used in the applications of the Bruun Rule in this study. The low and active profile widths were determined for different locations along The Gambia coastline.

In areas where the coastline exhibits spits/barriers backed by marshes, mangroves, and strandplain, the drowning (inundation) concept of land loss was applied as the initial potential impacts for the first half of the period (50 years) in question. After inundation, the remaining land would be subject to wave action, and so the Bruun Rule was applied to estimate impacts for the next 50 years.

To estimate the size, pattern, and value of development of the coastal zone, the aerial videotape-assisted vulnerability analysis (AVVA) (Benioff et al., in press) technique (Leatherman et al., 1995) was used. A low-level videotape of both the open and sheltered coasts was acquired using a pilot and a small aircraft. Two technicians provided verbal commentary.

Tide gauge records available in The Gambia cover only a short period and are therefore statistically insignificant. Long-term geological data also suggest insignificant tectonic variations. Hence, the global sea level rise scenarios, widely adopted in the IPCC documentation (IPCC, 1990b; Pepper et al., 1992), were used. The scenarios are the current rate of sea level rise of 0.2 m, 0.5 m, and 1.0 m per century. The 0.2 m rise per century is taken to be the baseline scenario.

Methodology for Economic Analyses

The analysis of the economic consequences of a 1.0 m sea level rise for unit 3 was based on socioeconomic and demographic data collected within The Gambia, particularly from the Departments of State Lands, Central Statistics, and Tourism.

The Department of State Lands provided limited market value assessments of land and infrastructure in units 2, 3, and 5. The cost of land per square kilometer was multiplied by the total land loss to arrive at the monetary value of the lost land. Because the sample of infrastructure given by the Department was unrepresentative, the cost of the infrastructure expected to be lost was not evaluated further.

Results

Table 5.14 shows the potential area to be lost due to inundation under the various sea level rise scenarios due to predicted climate change.

Table 5.15 shows the results of the application of the Bruun Rule to the second half of the 100 year period being studied. This analysis accounts for wave action after initial inundation.

In the sheltered coast of The Gambia (unit 2), only the low-lying mangrove systems and adjacent marshland and agricultural lowlands bordering Banjul, Bakau, Old Jeswang, Eboe Town, Talinding Kunjang, Fagikunda, and Abuko were included in the analysis. These mangrove systems have all their hinterland reclaimed for settlements, and thus the mangrove fringe and the adjacent agricultural lands have become narrower and will eventually disappear as sea level rises.

Table 5.14. Estimates of Land Loss due to Inundation under Various Sea Level Rise Scenarios (km²)

	Sea Level Rise Scenarios			
Coastal Unit	0.2 m	0.5 m	1.0 m	2.0 m
Senegal Border (north) to Barra Point	0.56	14.06	28.13	56.25
River Gambia Estuary	0.94	23.44	46.88	93.75
Banjul Point to Cape St. Mary	1.56	3.90	7.80	17.63
Cape St. Mary to Fajara	0.00	0.00	0.00	0.00
Fajara to Bald Cape	0.01	0.03	0.05	0.20
Bald Cape to Solifor Point	0.41	1.02	2.03	2.89
Solifor Point to Bench Mark (125 km)	0.27	0.67	1.34	1.81
Bench Mark (125 km) to Kartong Point	0.03	0.09	0.17	0.17
Kartong Point to Senegal Border (south)	1.19	2.70	5.93	8.85
Total Land Loss	4.96	45.89	92.32	181.55

Table 5.15. Land Retreat and Erosion Rate in The Gambia under the Bruun Rule

			Active Profile Width				
Coastal Unit	Depth of Closure (m)	Overfill Ratio	Map (cm)	Actual (m)	Dune or Cliff Height (m)	Sea Level Rise Scenario	Retreat[1]
Senegal Border (north) to Barra Point River Gambia Estuary	5.9	1.0	3.5	2,611	2.0	1.0	330
Banjul Point to Cape St. Mary	5.9	1.0	3.0	2,284	2.7	1.0	265
Cape St. Mary to Fajara	5.9	1.0	1.6	1,233	6.1	1.0	102
Fajara to Bald Cape	5.9	1.0	2.8	2,081	2.7	1.0	241
Bald Cape to Solifor Point	5.9	1.0	8.6	6,476	1.8	1.0	840
Solifor Point to Bench Mark (125 km)	5.9	1.0	6.8	5,098	3.0	1.0	570

[1] R = G(L/(B+h*))S (m); where R is the retreat (erosion rate) due to a sea level rise, h is the depth of closure, B is the dune or cliff height, L is the width of the active profile, G is the inverse of the overfill ratio of material being eroded, and S is the scenario.

The mangrove systems provide socioeconomic, ecological, environmental, and cultural value to the people in these settlements. They provide habitat as well as spawning and nursery grounds for fish and other aquatic species such as prawns, which are important for the sustainability of both the artisanal and the commercial fishing industry. The trees serve as substrate for marine organisms such as sponges, molluscs, and algae, which are important sources of protein for the local people. Mangroves also support a variety of birds, and serve as resting and feeding stations for migratory birds. This is particularly true for the mangroves between Banjul and Sarro, and the mangrove system at Tanji River, which has become known as the Tanji Birds Sanctuary. These mangrove systems also provide fuel and building materials for the inhabitants of the adjacent land.

A 1.0 m sea level rise could threaten the loss of the capital city and all its infrastructure. Other capital structures and cultural and heritage sites such as the Banjul Cemeteries, Radio Syd, Utilities Holding Corporation water tanks, Palm Grove and Wadner Beach hotels, the infrastructure at Sarro, the ice and cold storage plants at Mile 5, and the hotel complex around Cape Point are already at risk of being lost due to erosion. The hotel complex (Kairaba, Senegambia, Kombo Beach, etc.) in unit 5 will also be affected by sea level rise through the exacerbation of present erosion rates.

Adaptation Alternatives

The present degradation of the coastal zone in The Gambia is due to both historical sea level rise and resultant forcing agents such as coastal erosion. Considering the current economic situation of The Gambia, in which most development activity is based on technical assistance, the most appropriate response would be to protect only the important areas in units 2, 3, and 5 of the coastal zone. In unit 3, this would mean protecting the coastline of Banjul and the infrastructure described above by stabilizing sand distribution, building, and repairing groin systems.

Innovative sand management and the repair of the groin system are important in the short term. In the long term it may be necessary to employ better and longer lasting shoreline stabilizing techniques. These would include the construction of a 7 km revetment at an estimated cost of 35 million Dalasis (US$3.9 million). This estimate is based on a GAMWORKS/World Bank Project being implemented at the Banjul Cemeteries. Alternatively, a low-cost seawall or a bulkhead could be built. IPCC (1990b) estimates the cost of such a wall as US$440,000 per kilometer for a 1.0 m sea level rise. For 7 km this translates to about US$3.1 million.

For other areas, including the areas bordering the mangrove systems from Bakau to Abuko (unit 2), dikes of about 1.5 m to 2.0 m of sand and vegetation cover would suffice. Assuming that a truck load (5 m^3) of sand costs about 300 Gambian Dalasis, the cost of construction of 16 km of dikes will be about 4,875,000 Gambian Dalasis (US$542,000).

Evaluation of Anticipatory Adaptation Policies

All the alternative adaptation strategies discussed above are enumerated based on the availability of the required materials and their relative cost. These may change with the dynamics of the market and hence will need to be adjusted accordingly.

Policies and regulations concerning the use of the coastal zone for any form of human activity should include consideration of sea level rise. Physical planning and building control measures and regulations should be instituted and implemented. Allocation of land for any economically useful purpose in areas likely to be flooded should be avoided. The public should be informed of the risks of living in coastal and low-lying land areas. Timely public education about erosion, sea level rise, and flooding risks could be a cost-effective means of reducing future expenditures.

Where the building of coastal infrastructure is approved, the authorities and owners should make sure that marginal increases in the height of the structures are included to account for sea level rise and other related phenomena. People located in high risk areas could be offered incentives to relocate. Marginal increases in the height of the infrastructures during construction phase and redirecting growth away from sensitive lands are relatively inexpensive options for reducing the impacts of sea level rise and risks of flooding.

All the adaptation strategies, options, and policies discussed above will reduce the risks from current climate variability and protect against potential climate change and sea level rise. These could be put together in the form of a program that will constitute a Coastal Zone Management Plan (CZMP). A CZMP brings together all actors in the coastal zone to address coastal problems. The program should consist of a set of principles and plans to guide the use of coastal land for conservation, recreation, and development.

Discussion and Conclusions

The assessment of the vulnerability of the coastal resources of The Gambia to sea level rise shows that the whole of the coastal zone is currently affected by erosion, which will be exacerbated by climate change and sea level rise. All the mangrove systems and adjacent marsh and croplands on the Kombo Peninsula could be inundated. The largest loss of land (US$217 million) and infrastructure (not quantified) will most likely be in units 2 and 3. The capital city of Banjul could be lost within the next 50-60 years if there is a 1.0 m sea level rise in the next century. About 15 hotels on the coast could be lost. This would destroy the tourist industry, since tourists come to The Gambia because of the beautiful sandy beaches and luxurious hotels.

Results from the coastal resources assessment indicate that sea level rise will lead to inundation of most of the mangroves, destroying large areas of fish spawning and breeding grounds. Since the population in these areas rely on fish for their daily protein supply, the loss could increase susceptibility to disease, particularly in children, because of a poor diet. Sea level rise will also lead to salinization of some of the water supply systems in the Kombo Peninsula, reducing the availability of fresh water. With sea level rise, the saline front (an imaginary line dividing the upper fresh water and the lower saline water of the River Gambia) could move about 200 km farther upstream. This movement of the saline front and the consequential salinization will render the large area on both banks of the river, now under irrigation, unsuitable for crop cultivation.

Adaptation strategies and policy options evaluated include innovative sand management, repair of the groin systems, construction of a revetment or low-cost sea wall/bulkhead, and the construction of a dike system. Policies should be instituted that will discourage the construction of capital structures within about 150 m of the mean high water. Structures that have to be constructed should have marginal increases in the height of the structures to account for sea level rise and other related phenomena. Legislation should be enacted and implemented, and institutions responsible for the implementation of the regulations should be given authority that they can exercise without fear of being undermined.

Summary Conclusions

Based on these preliminary results and results of other studies conducted by the National Climate Committee. The Gambia Country Study team plans to develop a Climate Change Action Plan and National Communications for submissions to the

United Nations Framework Convention on Climate Change. Work has already started toward the development of a National Action Plan. With assistance from American Forests, an American nongovernmental organization, greenhouse gas mitigation projects have been identified and are being developed. The U.S. Country Study Program also approved and provided funds for a mitigation assessment in the major economic sectors of The Gambia. The study started in October and is expected to end in late 1995. The results of these assessments will be used in conjunction with the vulnerability and adaptation assessments to develop policies to mitigate greenhouse gas concentrations in the atmosphere and adapt to the predicted climate change effects.

REFERENCES

Alimi, N.D. 1990. Determination of Favorable Crop Planting Dates and Rainfall Probability in The Gambia. Department of Water Resources Technical Note, The Gambia.

Benioff, R., S. Guill, and J. Lee (eds.) (In press). *Vulnerability and Adaptation Assessments: An International Guidebook.* Dordrecht, The Netherlands: Kluwer Academic Publishers.

CSD. 1993. Population and Housing Census. 1993: Provisional Report (1). Central Statistics Department, The Gambia.

Dennis, K.C., I. Niang-Diop, and R.J. Nicholls. 1995. Sea Level Rise in Senegal: Potential Impacts and Consequences. *Journal of Coastal Research*, Special Issue, no. 14.

DOP. 1990. *1990 Statistical Year Book of The Gambia Agriculture.* Department of Planning, The Gambia.

DOP. 1993. *1993 Statistical Year Book of The Gambia Agriculture.* Department of Planning, The Gambia.

EU/MNRE. 1992. The Gambia Environment Action Plan. Environment Unit. 1991.

Food and Agriculture Organization (FAO). 1993. *Agriculture: Towards 2000.* The United Nations. Rome, Italy.

Foster, H. 1983. Evaluation of the National Forest Inventory of The Gambia. Technical Report, no. 10, Gambia-German Forestry Project, Banjul.

(IPCC) Intergovernmental Panel on Climate Change. 1990a. *Climate Change: The IPCC Scientific Assessment.* Houghton, J.T., G.J. Jenkens, and J.J. Ephramus, (eds). Report of Working Group I of the Intergovernmental Panel on Climate Change.

(IPCC) Intergovernmental Panel on Climate Change. 1990b. Response Strategies and Adaptation Assessment. Report of Working Group III of the IPCC.

Leatherman, S.P., R.J. Nicholls, and K.C. Dennis. 1995. Aerial Videotape-Assisted Vulnerability Analysis: A Cost-Effective Approach to Assess Sea Level Rise Impacts. *Journal of Coastal Research*, Special Issue, no. 14.

NARI. 1994. Cereal Research Programme Report. National Agricultural Research Institute.

Pepper, W., J. Legget, R. Swart, J. Wasson, J. Edmonds, and I. Mintzer. 1992. Emissions Scenarios for the IPCC Update. Assumptions, Methodology, and Results.

UNDP-The Gambia. 1993. The 1992 Development Co-Operation Report-The Gambia. UNDP, Banjul.

Whyte, W.J. and T.S. Russel. 1988. Geology and Mineral Resources of The Gambia.

ACKNOWLEDGMENT

This study was funded by the United States Country Studies Program. The Author acknowledges the support from the U.S. Country Studies Team, Programme Officer, Sandy Guill, and The Gambia National Climate Committee for their technical and moral support.

Chapter 6

Vulnerability and Adaptation Assessments for Zimbabwe

C.H. Matarira,[1] W. Kamukondiwa,[1] F.C. Mwamuka[1]
J.M. Makadho,[2] L.S. Unganai[3]

[1] Scientific and Industrial Research and Development Center (SIRDC) Environment and Remote Sensing Institute
[2] Department of Agricultural Technical Extension Services
[3] Drought Monitoring Center

SUMMARY. Zimbabwe's vulnerability and adaptation assessment examines the potential effects of climate change on the production of maize in Zimbabwe combining a crop model (CERES-Maize) with climate scenarios derived from two general circulation models. Currently, low rainfall and high temperature threaten agricultural production in Zimbabwe. Simulated maize yields decreased considerably under dryland conditions based on the climate change scenarios. Reductions in simulated maize yields were attributed to shorter growing seasons driven by increased temperatures. In some regions, simulated yield decreases were partly offset by the direct effects of CO_2 on photosynthesis and water use. Switching to drought-tolerant varieties of small grains and applying appropriate management practices would counteract the negative impacts of climate change. Rapid geographical shifts in the agricultural land base could disrupt rural communities and their associated infrastructure. More research is called for to generate technologies that enable farmers to adapt to the effects of climate change.

INTRODUCTION

Zimbabwe is a landlocked country in southern Africa (Figure 6.1). It can be divided into four topographic regions: the eastern highlands (above 1,500 m altitude), the high veld (1,200-1,500 m altitude), the middle veld (600-1,200 m altitude), and the low veld (less than 600 m altitude). Zimbabwe has a tropical climate with two broad seasons, cool dry winters (May-August) and warm wet summers (November-April). September and October are relatively warm and dry. Climate varies widely within the country as a result of the differences in topography.

Rainfall decreases from annual means of 1,500 to 2,000 mm in the highland areas of eastern Zimbabwe to less than 400 mm in the southern low veld areas of Beit Bridge. The eastern highlands receive orographic rainfall

Note: This chapter does not contain a detailed description of the assessment methodologies that were recommended by the U.S. Country Studies Program. Refer to Chapter 2 for an overview of these methodologies.

J. B. Smith et al. (eds.), Vulnerability and Adaptation to Climate Change, 129–140.

Figure 6.1 Map of Zimbabwe

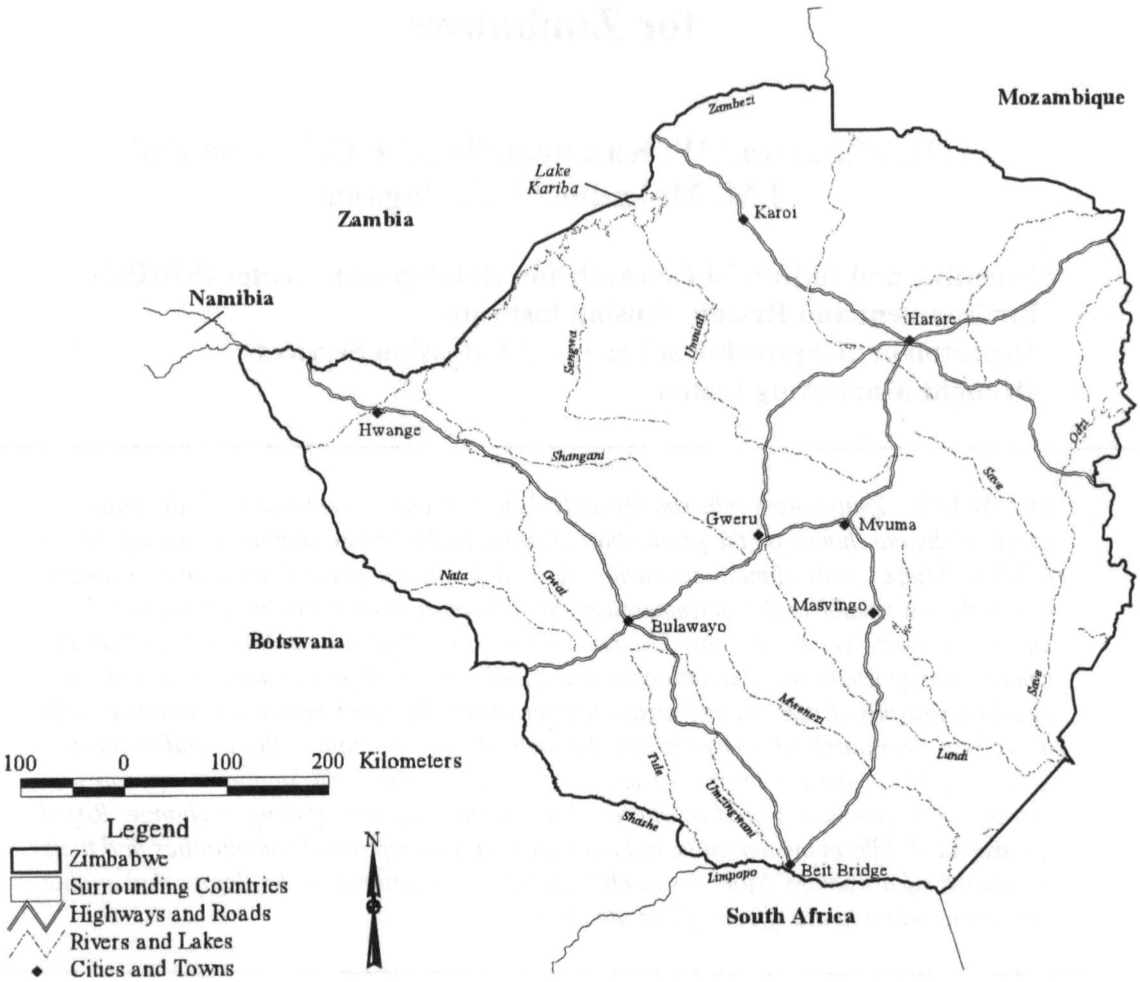

almost throughout the year. The high veld, which runs some 560 km across the country (northeast to southwest), also receives high rainfall and is the source of most of the major river systems in the country.

Land adjacent to and on either side of the 80-km-wide high veld is referred to as the middle veld and receives 400-800 mm rainfall. The hot dry plains below 600 m altitude are basins for the Zambezi, Limpopo, and Save rivers. These often receive less than 400 mm annual rainfall. These topographic features similarly affect temperature, which is lowest in the eastern highlands and highest in the low veld areas.

Agricultural suitability in Zimbabwe can be divided into five agro ecological zones whose demarcation is largely based on the amount and distribution of rainfall. The five zones and the locations of the study sites used in this vulnerability and adaptation assessment are shown in Figure 6.2. The four study sites, Beit Bridge (22°13'S, 30°00'E), Gweru (19°27'S, 29°51'E), Karoi (16°50'S, 29°37'E), and Masvingo (20°04'S, 30°52'E), represent four of the agro-ecological zones.

Figure 6.2. Study Sites and Natural Regions of Zimbabwe

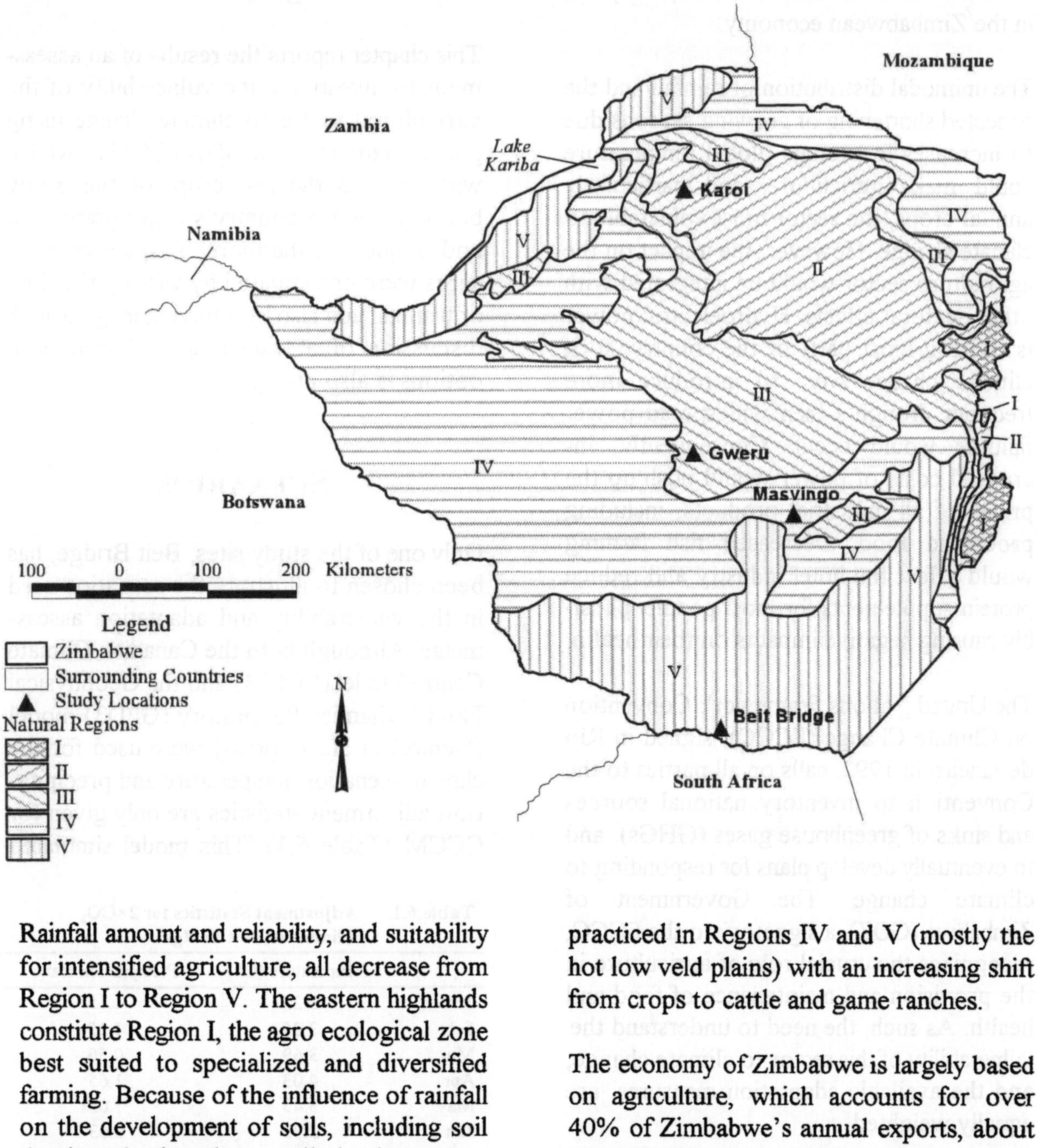

Rainfall amount and reliability, and suitability for intensified agriculture, all decrease from Region I to Region V. The eastern highlands constitute Region I, the agro ecological zone best suited to specialized and diversified farming. Because of the influence of rainfall on the development of soils, including soil depth and mineralogy, soils in the eastern highlands are deep and fairly weathered. Thus, despite the steep slopes of the eastern highlands, the deep soils allow intensive cultivation, including that of tea and coffee. Semi-extensive to extensive farming is practiced in Regions IV and V (mostly the hot low veld plains) with an increasing shift from crops to cattle and game ranches.

The economy of Zimbabwe is largely based on agriculture, which accounts for over 40% of Zimbabwe's annual exports, about 11% to 14% of the gross domestic product, and 50% of the industrial raw materials (GOZ, 1991). About 80% of the country's 10.4 million people are directly or indirectly dependent on agriculture, which also accounts for 30% of formal employment. The agricultural sector is therefore an area of major concern because of the role it plays in the Zimbabwean economy.

The unimodal distribution of rainfall and the expected shortening of growing seasons due to increases in average global temperature could make agriculture, and particularly annual crops, the sector most vulnerable to climate change. However, the impact on the agriculture sector would be integrated with other sectoral effects. Hydroelectric power is used in most parts of the country, so if climate change were to introduce more frequent droughts then widespread power failures would occur. Consequently, increased costs of power would push up the prices of all industrial products, including processed food. Decreased fish farming would affect the hotel industry and reduce protein supplements for most people, possibly causing negative impacts on their health.

The United Nations Framework Convention on Climate Change (FCCC), signed in Rio de Janeiro in 1992, calls on all parties to the Convention to inventory national sources and sinks of greenhouse gases (GHGs), and to eventually develop plans for responding to climate change. The Government of Zimbabwe (GOZ), a signatory to the FCCC, recognizes the crucial role of agriculture in the provision and maintenance of food and health. As such, the need to understand the vulnerability of this sector to climate change, and the available adaptation measures, are equally weighted.

Preliminary work on the vulnerability and adaptation assessments of Zimbabwe has been focused on climate change detection, agriculture, and forestry. The climate change detection and agricultural sector assessments have been completed, and the forestry sector assessment is ongoing.

This chapter reports the results of an assessment to investigate the vulnerability of the agricultural sector to climate change using general circulation models (GCMs). Maize was used as the test crop for this study because it is the country's major grain crop and staple. Furthermore, adaptation measures were investigated by varying the time of planting and also by applying irrigation. A discussion of a wide range of adaptation options is also provided.

SCENARIOS

Only one of the study sites, Beit Bridge, has been chosen to illustrate the scenarios used in the vulnerability and adaptation assessments. Although both the Canadian Climate Centre Model (CCCM) and the Geophysical Fluid Dynamics Laboratory (GFD3) model (Benioff et al., in press) were used for the climate scenarios, temperature and precipitation adjustment statistics are only given for CCCM (Table 6.1). This model simulated

Table 6.1. Adjustment Statistics for 2×CO_2 Climate in Beit Bridge[1]

	Temperature (C°)	Precipitation Ratio
Jan	3.25	1.15
Feb	3.25	1.25
Mar	3.69	0.56
Apr	4.03	1.65
May	3.45	1.03
Jun	2.79	0.63
Jul	3.45	0.85
Aug	4.32	0.66
Sep	3.78	0.46
Oct	3.92	0.76
Nov	4.00	0.77
Dec	3.56	0.56
Annual	3.62	0.83

[1] Generated by CCCM.

Figure 6.3. Comparison of Instrumental Based Observed Climate with the CCCM 1×CO_2 Scenario for Beit Bridge

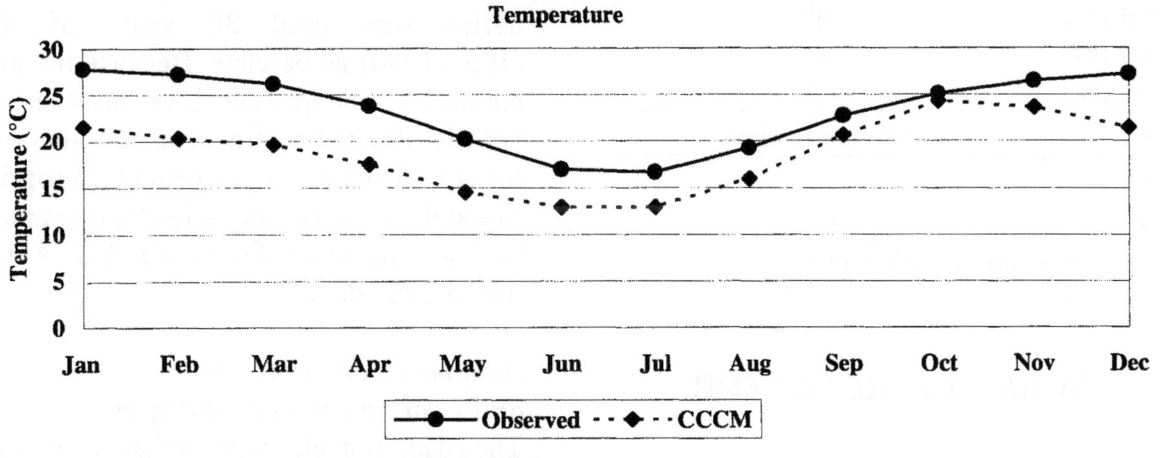

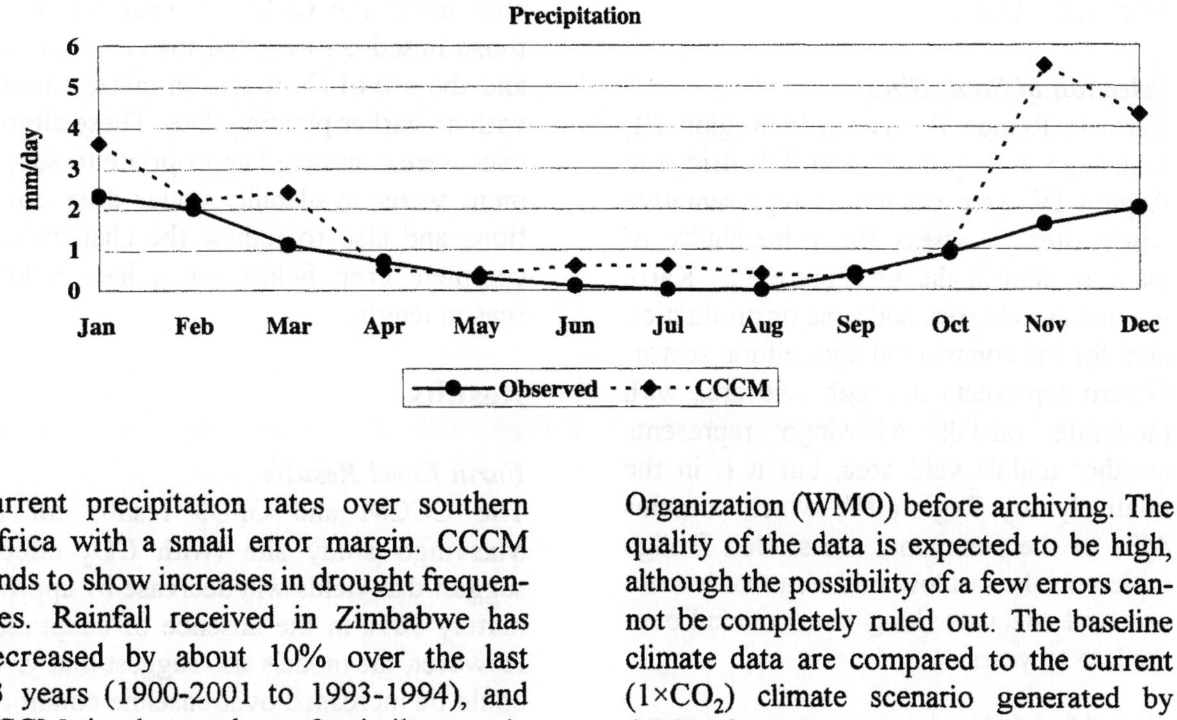

current precipitation rates over southern Africa with a small error margin. CCCM tends to show increases in drought frequencies. Rainfall received in Zimbabwe has decreased by about 10% over the last 93 years (1900-2001 to 1993-1994), and CCCM simulates a drop of a similar margin over the next 80 years. The baseline climate scenarios were developed using 1951-1980 data obtained from the Zimbabwe Meteorological Department's archives. The data had been techniques and other techniques recommended by the World Meteorological Organization (WMO) before archiving. The quality of the data is expected to be high, although the possibility of a few errors cannot be completely ruled out. The baseline climate data are compared to the current (1×CO_2) climate scenario generated by CCCM for Beit Bridge in Figure 6.3.

As the national population continues to swell (Table 6.2), pressure on land and other resources also increases. Thus, the impact of anthropogenic factors on climate change increases with increasing population.

Table 6.2. Baseline Growth in Population

Year	Average Annual Percentage Growth
1901-1921	2.4%
1921-1941	3.4%
1941-1961	3.4%
1961-1982	3.2%
Year	Population (millions)
1901	0.7
1992	10.4

Source: Central Statistical Office, 1993.

AGRICULTURE SECTOR

Methods Used

Selection of Study Sites

Karoi in Region II, Gweru in Region III, Mazvingo in Region IV, and Beit Bridge in Region IV were chosen as representative study sites to assess the vulnerability of maize to climate change (Figure 6.2). Karoi is a middle veld area, and a major production area for the commercial agricultural sector. Gweru represents the high veld area with moderate rainfall. Masvingo represents another middle veld area, but it is in the relatively dry Region IV, which contains 45% of the communal areas. Beit Bridge represents the vast low veld area that has not received adequate (long-term) rainfall over the last 15 years.

Crop Models Used

Maize, the country's staple crop, was analyzed using the CERES-Maize model. Although the GFD3 did not adequately represent the current climate, simulations for the crop model included runs based on the GFD3 model, as well as on CCCM. This was considered necessary because agriculture is expected to be highly sensitive to climate change and even those scenarios that seem less likely had to be considered.

The crop model was based on climate scenarios that used 30 years of data (1951-1980) as baseline. Because the agricultural sector in Zimbabwe has been adversely affected by frequent droughts over the last 15 years, it was deemed appropriate to use those climate scenarios that generated warmer and drier climatic conditions from the baseline data.

Adaptation Alternatives and Evaluation Methodologies

The adaptation alternatives that were evaluated using the CERES-Wheat Model are those based on irrigated maize production and the use of short-season maize varieties with an earlier planting date. These alternatives were considered appropriate to supplement water availability under drier conditions and also to reduce the chances of a complete crop failure when heat reduces season length.

Results

Farm Level Results

The CCCM and GFD3 results for the Masvingo study site (with CO_2 effects) suggest that yields will decrease by approximately 98% in the absence of adaptation. However, the models also suggest that yields could be increased over baseline conditions by 16% and 3% for the models, respectively, when an earlier planting date and short season variety of maize is used (Table 6.3). Season length decreased by about 15% and decreases in growing season precipitation were high (between -8% for the GFD3 model scenario without adaptation and -33% for the CCCM scenario with adaptation).

Table 6.3. Farm Level Results by Climate Change Scenario and Adaptation Alternative for Maize Grown at Masvingo (change from baseline is shown as a percentage)

Adaptation Alternative	Scenario[1]	Yield (t/ha)	Season Length (d)	Growing Season Precipitation (mm)	Growing Season Evapotranspir. (mm)
Without Adaptation Planting date 11/15	Baseline	2.6	121	498	452
	CCCM	-97.8%	-16.5%	-18.7%	-63.3%
	GFD3	-98.2%	-14.9%	-8.5%	-60.6%
With Adaptation Planting date changed to 10/15 and short season variety planted	Baseline	3.0	121	451	448
	CCCM	16.2%	-16.5%	-33.3%	-32.8%
	GFD3	3.0%	-14.0%	-15.9%	26.9%

[1] The CCCM and GFD3 scenarios include the direct effects of CO_2.

Maize production under climate change scenarios was found to be highly vulnerable to increased temperatures, which reduce season length, whereas the physiological effect of increased CO_2 could not fully compensate for decreased yields.

The model was also run for both baseline and climate change scenarios under irrigation. Since rainfall is the most limiting climatic factor for maize production, estimated yields for each scenario did not show significant differences. Under dryland conditions, yields were estimated to decline from a baseline of 2.59 t/ha to 0.06 (CCCM) and 0.05 (GFD3). Under irrigated conditions, yields in both the baseline and under the climate change scenarios increase significantly (Table 6.4).

Table 6.4. Comparison of Maize Yield under Dryland and Irrigated Conditions at Masvingo

Climate Scenario	Yield (t/ha) Dryland	Irrigated
Baseline	2.59	9.14
CCCM 2×CO_2	0.06	9.75
GFD3 2×CO_2	0.05	10.13

The CCCM and GFD3 results for study sites other than Masvingo suggest that yields may increase or only decrease slightly.

Aggregate Crop Model Results

Results from the crop model showed variation in maize yields between sites. No attempt was made to aggregate model results because the sites had been chosen to represent different agroecological zones where the impacts of climate change would be expected to be different. However, for ease of comparison with other country studies, aggregate results for the 2×CO_2 scenarios for CCCM and GFD3 are given in Tables 6.5 and 6.6.

Discussion and Conclusions

Climate change is estimated to decrease maize yields in some regions, whether under irrigation or not. This is due to the shortening of the growing season as average global temperature increases. The positive physiological effect of CO_2 on photosynthesis only partly compensated the decreases in simulated maize yield.

Although irrigation would improve maize yields in all areas, those yields under climate change scenarios would still be less than

Table 6.5. Aggregation of CCCM Results for Maize (with CO_2)

Natural Region	Site	Soil Group	Management Practice	Percent of Total Production	Percent Change in Crop Yield	Weighted Change in Crop Yield (t)
II	Karoi	Sandy Loam	Commercial	50%	-0.5%	-4,507
III	Gweru	Sandy Loam	Small Scale	20%	+37.0%	+128,293
IV	Masvingo	Sandy Clay Loam	Communal	25%	-97.8%	-423,889
V	Beit Bridge	Sandy Clay Loam	Ranching	5%	+112.3%	+97,347
National	N/A	N/A	N/A	100%	N/A	-202,756

Table 6.6. Aggregation of GFD3 Results for Maize (with CO_2)

Natural Region	Site	Soil Group	Management Practice	Percent of Total Production	Percent Change in Crop Yield	Weighted Change in Crop Yield (t)
II	Karoi	Sandy Loam	Commercial	50%	-0.7%	-5,721
III	Gweru	Sandy Loam	Small Scale	20%	+12.3%	+42,649
IV	Masvingo	Sandy Clay Loam	Communal	25%	-98.0%	-424,755
V	Beit Bridge	Sandy Clay Loam	Ranching	5%	+176.5%	+152,988
National	N/A	N/A	N/A	100%	N/A	-234,839

under baseline climate conditions. If climate change was to bring warmer and drier conditions, farmers would not benefit from irrigation because of the reduced availability of water supplies for irrigation. This would negatively affect the country's GNP, because agriculture would then contribute less to the national economy.

Hydrological uncertainty in the GCMs limits their use under Zimbabwean conditions. This is compounded by the absence of local studies relating future competition for water among agricultural, industrial, and domestic consumption. There is a need for the formulation and implementation of strategies and policies to mitigate the effects of climate change on the agriculture sector.

Agricultural activities are extremely vulnerable to climatic change, yet there is still considerable uncertainty in our understanding of climate change and its effects (Smit, 1993).

A reactive approach to adaptation taken after or as a response to climate change may not produce satisfactory results and could prove to be costly. Thus, there is a need to examine anticipatory approaches to adaptation. The goal of anticipatory measures is to minimize the impact of climate change by reducing vulnerability (e.g., sensitivity) to its effects or by enabling reactive adaptation to happen more efficiently, faster, and at lower cost (Smith and Mueller-Vollmer, 1993).

Agriculture can be one of the most flexible societal systems with respect to sensitivity to climate change, with adaptations possible at the farm level or the national level as reflected in government policy.

Farm Level Adaptation

At the farm level, the potential for agricultural adaptation is very high. Farm level adaptations arise from farmers' perceptions of changed or changing conditions. Farmers are already operating in an environment where climatic conditions vary from place to place and from season to season. In the last 15 years, Zimbabwe has experienced three severe droughts, in the 1982-1983, 1987-1988, and 1991-1992 seasons. Thus, history has taught farmers the need to reexamine land use, management practices, and farm infrastructure.

Changes in Land Use. Climate change in Zimbabwe is likely to increase constraints on agricultural production. Marginally productive areas may become nonproductive and revert to nonagricultural use. For areas where cropping becomes infeasible, livestock production for beef and dairy can take over as the major agricultural activity. Game ranching would certainly become an important economic activity. Similarly, forestry would gain importance. However, biodiversity in parks and wildlife may decrease, thereby imposing limitations on adaptation measures.

Farmers wishing to pursue cropping would have to switch to different crop types or change to more drought- and disease-tolerant crop varieties. Farmers may introduce irrigation systems in areas where high temperatures and rates of evapotranspiration rapidly deplete moisture. Switching from monocultures to more diversified agricultural production systems will help farmers cope with changing climatic conditions. Monocultures are more vulnerable to climate change, pests, and diseases. The use of livestock breeds adaptable to drought and the use of supplementary feeds (including tree crop fodder) will give farmers greater flexibility in adapting to climate change.

Changes in Management and Infrastructure. Changes in management practices can offset many of the potentially negative impacts of climate change (Smith and Mueller-Vollmer, 1993). The timing of various farming operations (e.g., planting, fertilizer and pesticide applications) is critical for farmers to reduce their vulnerability to the impacts of climate change. Planting densities and application rates of agrochemicals and fertilizers will also be critical in maximizing water-use efficiency. The use of conservation tillage, intercropping, and crop rotation practices will similarly enhance the long-term sustainability of soils and improve the resilience of crops to changes due to climate change (Strzepek and Smith, 1995). To gain control of the environment under which crops grow, farmers may consider the use of greenhouses for the production of those crops whose economic return warrants the extra cost.

Increased average global temperatures will require that more agricultural land be brought under irrigation. The simultaneous tightening of water management will require the use of more efficient irrigation systems. For orchards and vines, drip-irrigation systems can be used to improve water-use efficiency. In other agricultural systems, water losses through seepage and evaporation in canal- and flood-irrigation systems can be minimized by lining the canals with cement or by switching to pipe-irrigation systems. The high costs of production related to irrigation systems will presumably result in shifts to uses that require less water in areas where there are high rates of moisture loss. Livestock farmers can use supplementary feeds, including fodder trees, as low-cost grazing systems become less sustainable in areas that become marginal. Farmers can also consider improving their pastures by irrigating with municipal wastewater, a move that would possibly reduce costs for recycling sewage water.

Changes in the types of agricultural production and irrigation systems will require significant changes in farm layout and the types of capital equipment required. In areas where there is need to irrigate, there may be need for additional water reservoirs or boreholes. Parry and Duinker (1990) noted that because of the large costs involved in infrastructural changes, only small incremental adjustments can occur without changes in government policy.

National Level Adaptations

Performance of the agricultural sector is affected by government policies that influence input costs, product pricing, and marketing arrangements. Parry and Duinker (1990) observed that relatively minor alterations to these policies can have a marked effect on agriculture. Government policies pertaining to land and water resources should be more explicit, and the implementing agencies should give due consideration to the possible impacts of climate change.

Given the uncertainties about the magnitude and rate of climatic change, the prospects of the government acting directly to promote adaptation to anticipated change are limited. It is imperative that any anticipatory measures considered allow the greatest flexibility in order to allow these measures to be revised as new information about the magnitude and direction of climate change becomes available. Through its policies on infrastructural developments, research and development, education, water resources management, and product pricing, the government can put both reactive and anticipatory adaptive measures into place. Ideally, a policy-relevant research program could help identify appropriate actions as the current state of knowledge evolves (OTA, 1993).

Infrastructural Developments. The GOZ is on a major drive to increase the number of medium to large dams in the country, particularly in the drier regions, III and IV (Figure 6.2). Even though this may be a reaction to droughts of the recent past, this can still be considered anticipatory. Irrigation schemes have also increased with the construction of dams. Rukuni (1994) noted that there is growing evidence of high rates of return on investments in smallholder irrigation schemes, and that large areas of shallow groundwater could be put to use in intensive cultivation. However, research has to focus on some aspects of environmental protection as well as developing low-volume water pumps.

There is need for the government to undertake a major review of land-use planning with due consideration given to an integrated resources management approach. Thus, the current exercise to assess land tenure systems suitable for Zimbabwe should seriously consider conferring ownership to those currently exercising user rights. There should also be formal obligations on the part of the owner to use the land in a sustainable and productive way.

Government agencies in charge of executing Zimbabwe's resettlement program can also take into consideration the anticipated impacts of climate change. Though the resettlement programs have been primarily targeted at relieving population pressure in the communal areas, it is important to note that most of these areas are marginal and the situation there will become more vulnerable to climate change. Thus, if the resettlement programs are executed with due regard to climate change, they can be made more efficient and enhance the sustainability of agricultural production in these marginal areas. As more areas become marginal, there will be a shift to more intensive agricultural production in the more favorable areas. Hence, if such areas can be identified, supporting infrastructure (e.g., transportation and communication networks, and markets) can be improved in these areas.

Research and Development

The availability of new crop varieties, livestock breeds, and agricultural technologies adaptable to climate change can only be guaranteed if government supports research and development. There is a need for research on crops and livestock that are more tolerant to disease and drought conditions. Government expectations to increase wheat production from 300,000 tons in 1990 to about 487,000 tons in 1995, and to a level of national self-sufficiency thereafter (GOZ, 1991), are noble. To realize and sustain these expectations in an environment with a changed climate, the government needs to commit itself to supporting an intensified program of research into higher yielding, drought hardy, and disease and pest tolerant wheat varieties.

The government's aim of promoting increased goat production in Regions III, IV, and V (GOZ, 1991) should consider the marginality of these regions. The government should support research efforts aimed at ascertaining how to effectively increase and sustain goat production in these regions without sacrificing the environment, because goats can be quite destructive to the environment if not managed properly.

The increased frequency of droughts obviously calls for increased research in effective storage systems for agricultural products. The government has made a commitment for 500,000 to 600,000 tons of grain to be kept by the GMB as a strategic reserve for use by the urban population and in times of drought or similar disasters (GOZ, 1991). From events of the recent past, however, it is evident that the rural majority are hardest hit by a drought. The government should therefore consider supporting research into a more decentralized method of keeping these strategic reserves, with increased local participation. An enabling environment and government support would encourage the private sector to invest more resources into these areas of research. Private sector participation would result in more rapid application of research output within the agricultural sector.

There is a need to bridge the institutional separation of research and extension services, which has tended to minimize the responsibility for developing technology that is farmer-based and problem-oriented. It is also important that the government fully utilize information from research and development bodies in its formulation and/or reformulation of policies affecting the agricultural sector. The government should examine inadvertent damage to the capacities of research and development institutions as a result of budgetary and staff cuts under the Economic Structural Adjustment Program. There is also a need for improved incentives to attract and retain outstanding scientists in these research and development institutions.

The simulated yield changes under climate change are indicative of the potential problems ahead. New and fluctuating weather patterns could have severe negative impacts on economic activities, particularly in the natural resources sector. Without appropriate policies or adaptive strategies, smallholder farmers will find it extremely difficult to operate sustainable agricultural production systems under a changed climate. The potential solutions will require increased financial resources and greater commitment in research.

The considerable uncertainties about the magnitude and extent of the impacts of climate change make it relatively difficult to plan appropriate responses (policy formulation and strategy development). This calls for flexibility in all strategies that may be adopted. While more research and extension programs will enhance our capacity to adapt to climate change, there may be room for improvement by merging the research and extension services, which currently are separate in Zimbabwe.

REFERENCES

Benioff, R., S. Guill, and J. Lee (eds.). (In press). *Vulnerability and Adaptation Assessments: An International Guidebook.* Dordrecht, The Netherlands: Kluwer Academic Publishers.

Central Statistical Office. 1993. Census 1992: Zimbabwe Preliminary Report. Harare: Government Printers.

GOZ. 1991. "Sectoral Development — Agriculture." In: *Second Five-Year National Development Plan 1991-1995.* Government of Zimbabwe. Harare. pp. 23-33.

OTA. 1993. Preparing for an Uncertain Climate: Summary. Office of Technology Assessment, U.S. Congress. OTA-O-563.

Parry, M.L. and P.N. Duinker. 1990. "Agriculture and Forestry." In: W.J. McG. Tergart, G.W. Sheldon, and D.C. Griffiths. 1990. *Climate Change — The IPPC Impacts Assessment.* WMO/UNEP Intergovernmental Panel on Climate Change. Canberra: Australian Government Publishing Service.

Rukuni, M. 1994. Getting Agriculture Moving in East and Southern Africa and Framework for Action. Discussion paper for East and Southern Africa Conference of Agricultural Ministers held in Harare, Zimbabwe, 12-15 April 1994. Commissioned by the World Bank for Global Coalition for Africa (GCA).

Smit, B. (ed.). 1993. Adaptation to Climatic Variability and Change. Report of the Task Force on Climatic Adaptation. Climate Adaptation Branch Bulletin 93-04. (Prepublication draft.)

Smith, J.B. and J. Mueller-Vollmer. 1993. Setting Priorities for Adapting to Climate Change. Prepared for Office of Technology Assessment. Oceans and Environment Program. RCG/Hagler Bailly, Inc.

Strzepek, K.M. and J.B. Smith (eds). 1995. *As Climate Changes: International Impacts and Implications.* New York: Cambridge University Press.

Chapter 7

Vulnerability and Adaptation Assessments for Bangladesh

Ahsan Uddin Ahmed,[1] Saleemul Huq[1]
Zahurul Karim,[2] Md. Asaduzzaman,[3] A. Atiq Rahman[1]
Mozaharul Alam,[1] Youssouf Ali,[1] Rawshan Ali Chowdhury[2]

[1] Bangladesh Center for Advanced Studies (BCAS)
[2] Bangladesh Agricultural Research Council (BARC)
[3] Bangladesh Institute for Development Studies (BIDS)

SUMMARY. In Bangladesh, vulnerability and adaptation assessments are being performed for the following sectors: agriculture, water resources, and coastal resources. More limited assessments are also being performed for the fisheries and forestry sectors. These sectoral analyses are in the preliminary stages. This chapter introduces the climate change issues in the country, and the four types of climate change scenarios chosen for the vulnerability analyses: climate change, economic development, sea level rise, and watershed development. Preliminary analysis indicated that for Bangladesh the most affected important sector in terms of climate change effects is the water resources sector because of because possible effects include devastating floods, severe droughts, and changes in salinity levels in the surface and groundwater systems as well as in soil. Further, the vulnerability of the water resources sector would affect the vulnerability of the agriculture production, coastal resources, forestry, and livestock sectors.

INTRODUCTION

Bangladesh, situated at the confluence of the Ganges, the Brahmaputra (Jamuna), and the Meghna rivers, is one of the largest deltas in the world. About 120 million people inhabit a land area of about 144,000 km^2, making it one of the most densely populated areas in the world. Half of the country is 12.5 m below mean sea level. Owing to its low-lying topography, the country is well known for its high sensitivity to natural calamities such as floods, storm surges, and droughts. Geologically, Bangladesh is located at the Himalayan Foredeep or Indo-Gangetic Trough, a tectonically active region. As a result, part of the country is undergoing considerable subsidence. This is, however, counteracted by sedimentation; 1.8 to 2.2 billion tons of sediment flow through the river systems of

Note: This chapter does not contain a detailed description of the assessment methodologies that were recommended by the U.S. Country Studies Program. Refer to Chapter 2 for an overview of methodologies.

J. B. Smith et al. (eds.), Vulnerability and Adaptation to Climate Change, 141–159.

Bangladesh annually (Kuehl et al., 1989; Khan, 1991).

Bangladesh can be divided in three broad physiographic regions: the tertiary hills, the Pleistocene uplands, and the recent alluvial plains (Rashid, 1991). The alluvial plains can be further subdivided into Piedmont plain, floodplain, deltaic plain, tidal plain, and coastal plain. Much of the present plain was originally an estuary that was filled up by the combined drainage system in the later part of the tertiary period. At one time the entire floodplain was forested, but most of the native forest disappeared as a result of population pressure and the subsequent conversion of forest land to other land uses. Approximately 6% to 7% of forest remains (ADB, 1994).

The climate of Bangladesh is controlled primarily by summer and winter winds and partly by pre-monsoon and post-monsoon circulations. The southwest monsoon originates over the Indian Ocean and carries warm, moist, and unstable air. The easterly trade winds are also warm, but relatively drier. The northeast monsoon comes from the Siberian Desert, retaining most of its pristine cold, and blows over, usually in gusts, during dry winter months.

The temperature ranges of the country can be grouped according to three periods: March-May, June-October, and November-February. The maximum temperature, which can exceed 40°C in some places, is generally recorded between late March and mid-April. The mean maximum temperature between June and October is about 30° to 31°C over most of the country, and the highest maximum does not normally exceed 34°C (Rashid, 1991).

Topographic conditions modify the distribution of rainfall in Bangladesh. Moisture-laden wind blows against mountain sides and is forced upward, producing vigorous condensation. This causes rainfall in the regions facing the wind. Annual rainfall of Bangladesh varies widely, ranging from as low as 1,200 mm in the extreme west to 5,800 mm in the east and northeast. Consequently, the mean annual isohyets over Bangladesh run in a semicircular fashion from north to south with a bulge toward the east (Rashid, 1991).

The economy of Bangladesh is agrobased: the agricultural contribution to the GDP is about 35%. The contributions for the other sectors are public administration (about 18%), transport (12%), and industry (11%). The remaining sectors are trade, housing, construction, etc. (BCAS, 1995).

Vulnerability Issues

Since Bangladesh has a relatively flat topography, except for a few hilly areas in the Chittagong region, the entire country will probably be affected by climate change induced impacts (Figure 7.1).

Previous studies on the implications of climate change for Bangladesh have demonstrated that the major issues regarding climate change are inundation, low flow and drought, salt water intrusion, flash flood, storm surge, and river bank and coastal erosion (BCAS/RA/Approtech, 1994; ADB, 1994; BUP/CEARS/UEA, 1994). All these

Figure 7.1. Map of Bangladesh

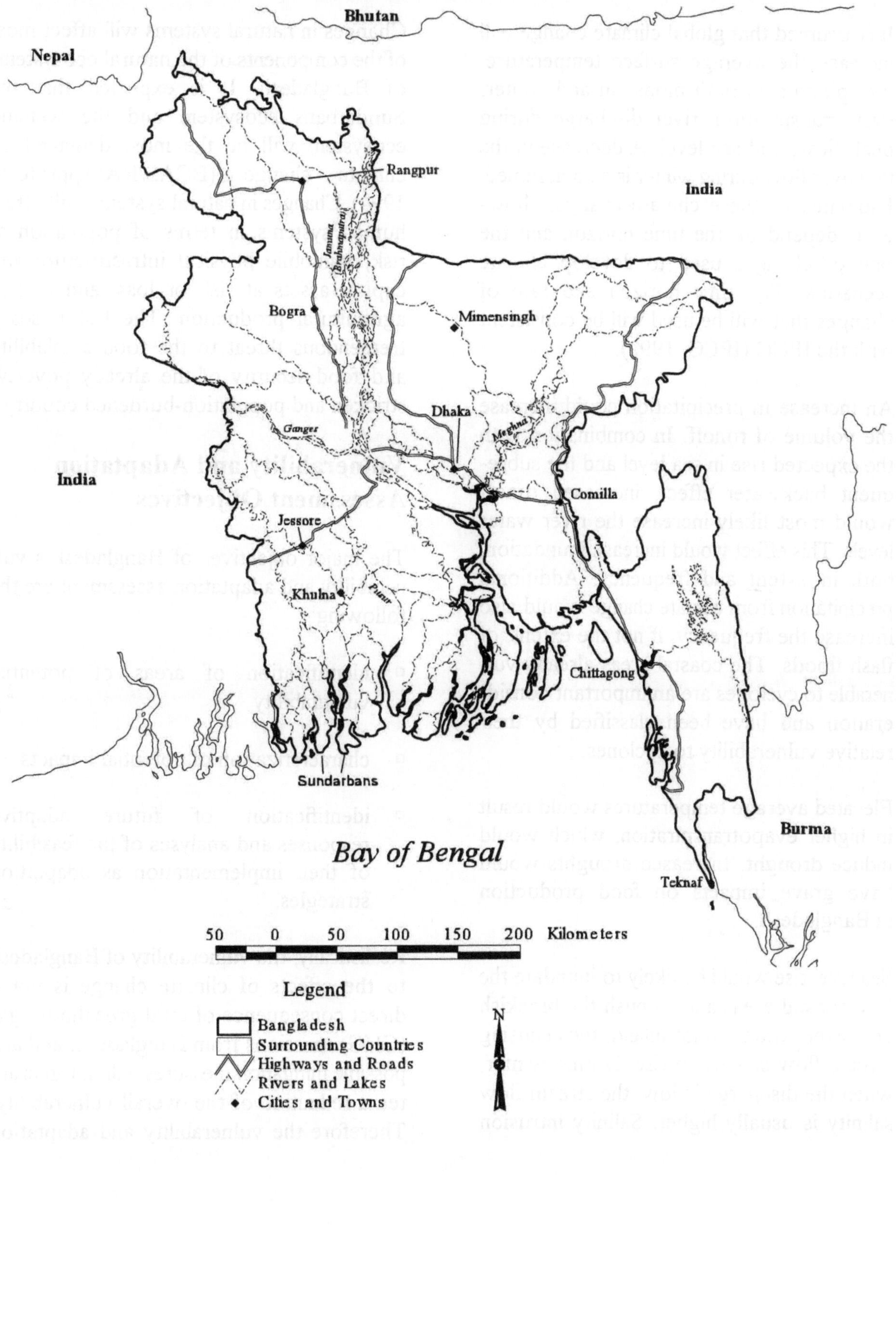

natural phenomena are likely to be enhanced under typical estimates of climate change, either in extent or frequency or both.

It is assumed that global climate change will increase the average surface temperature, precipitation in both monsoon and winter, evapotranspiration, river discharge during peak flow, and sea level. A decrease in the low river flow during winter is also assumed. Estimates of these climate changes, however, depend on the time horizon and the rate of changes used to develop climate scenarios. The time horizon and rate of changes that will be used will be consistent with the IPCC (IPCC, 1992).

An increase in precipitation could increase the volume of runoff. In combination with the expected rise in sea level and the subsequent backwater effect, increased runoff would most likely increase the river water levels. This effect would increase inundation, both in extent and frequency. Additional precipitation from climate change would also increase the frequency, if not the extent, of flash floods. The coastal areas already vulnerable to cyclones are an important consideration and have been classified by their relative vulnerability to cyclones.

Elevated average temperatures would result in higher evapotranspiration, which would induce drought. Increased droughts would have grave impacts on food production in Bangladesh.

Sea level rise would be likely to inundate the low coastal areas and to push the brackish and sweet water zones inland, thus causing stream flow salinity to rise. During winter, when the discharge is low, the stream flow salinity is usually higher. Salinity intrusion currently has considerable impacts on the agricultural and industrial activities of Bangladesh.

Changes in natural systems will affect most of the components of the natural ecosystems of Bangladesh. It is expected that the Sundarbans ecosystem and the wetland ecosystem will be the most damaged by climate change (BCAS/RA/Approtech, 1994). Changes in natural systems will affect human systems in terms of population at risk, immobile physical infrastructure and capital assets at risk or loss, and loss in agricultural production. The last poses a tremendous threat to the food availability and food security of the already poverty-stricken and population-burdened country.

Vulnerability and Adaptation Assessment Objectives

The major objectives of Bangladesh's vulnerability and adaptation assessment are the following:

- identification of areas of potential vulnerability
- characterization of potential impacts
- identification of future adaptive responses and analyses of the feasibility of their implementation as adaptation strategies.

Realistically, the vulnerability of Bangladesh to the effects of climate change is not a direct consequence of total greenhouse gas (GHG) emissions from Bangladesh, and any possible mitigation measures will not guarantee a reduction of the overall vulnerability. Therefore the vulnerability and adaptation assessment component of the Bangladesh Country Study is not directly connected with the emission inventory and mitigation components. However, the impacts of climate change will have grave consequences on the natural system, ecology, and socioeconomic aspects of Bangladesh. The results to be obtained from the different sectoral studies under the vulnerability and adaptation assessment component need to be disseminated amongst the general population and those who will be directly affected by the impacts of climate change. It is also important to inform the national planners about the threats posed by climate change so that they may incorporate the issues into long-term planning exercises.

As evident from the above-mentioned issues, for Bangladesh, the most important sector in terms of climate change effects is the water resources sector, because possible effects include devastating floods, severe droughts, and changes in salinity levels in the surface and groundwater systems as well as in soil. Further, the vulnerability of the water resources sector would affect the vulnerability of the agriculture production, coastal resources, forestry, and livestock sectors. The vulnerability of the fisheries sector is not directly linked with the water resources sector; it is, however, sensitive to water temperature variations and salinity concentrations.

For Bangladesh, grassland vulnerability is irrelevant since there is no grassland or range land in the country. The vulnerability of the forestry sector is confined to the Sundarban mangrove forest, and will be assessed only as a case study. The vulnerability assessment of the fisheries sector will only concentrate on the economically important fish species and only look at the economic vulnerability of the prawn/shrimp fisheries. The vulnerability of the wildlife and human health sectors will not be considered for this study. Maximum emphasis is being given to water resources, agriculture, and coastal resources.

The major studies previously undertaken on climate change issues in Bangladesh served as inventories and provided background information. This analysis provides relevant information on certain sectors in greater detail. However, because of limitations regarding data, time, and resources, some issues are beyond the scope of the present study. This chapter describes the scenarios used. It also presents the methods used and when available preliminary results for the following sectors:

- Agriculture
- Grassland/Livestock
- Forestry
- Water Resources
- Coastal Resources
- Fisheries.

SCENARIOS

Scenarios have been developed to analyze future vulnerability of the country to climate change and sea level rise. For the vulnerability and adaptation assessment of Bangladesh, four types of scenarios are considered: climate change, economic development, sea level rise, and watershed development.

Climate Change Scenarios

Given the uncertainties about regional climate change, regional impacts cannot be predicted. However, sensitivities of the systems to climate change can be evaluated by using scenarios of global and regional climate change (Strzepek and Smith, 1995). The general circulation models (GCMs) used in this study were supplied by the National Center for Atmospheric Research (NCAR). These models include the Canadian Climate Centre Model (CCCM) and the Geophysical Fluid Dynamics Laboratory equilibrium model (GFD3) and 1% transient (GF01) model (GF01)s (Benioff et al., in press). In addition, observed climate data were supplied by the CLIM database. Figure 7.2 shows the comparative status of the model outputs and observed monthly average temperature data (1948-1980) (FAO, 1988).

To develop climate scenarios for particular timeframes, the above-mentioned models may not be used although they may be used to project the change in climate at 2×CO_2. The GF01 model GF01, on the other hand, provides information regarding decadal change in climate variables. The model estimates changes of climate parameters due to a 1% increase of GHG atmospheric concentrations per year. For the Bangladesh study, the GF01 model has been used to gain average monthly temperature data for the reference year (1990, the fourth decade); and for the two projection years (projection year 1, 2030, as the seventh decade, and projection year 2, 2075, as the tenth decade). The results are presented in Table 7.1.

Table 7.1. Estimated Temperature Adjustment Statistics under GF01 1% Transient Scenario

Month	Observed Data[1]	Temp Change 2030 (°C)	Temp Change 2075 (°C)
Jan	18.40	+1.26	+1.44
Feb	21.30	+1.89	+4.47
Mar	26.10	+2.61	+6.09
Apr	28.90	+2.32	+4.79
May	28.90	+2.47	+4.38
Jun	28.60	+2.19	+2.07
Jul	28.50	-0.12	+1.33
Aug	28.60	-0.13	+1.25
Sep	28.60	+0.78	+1.57
Oct	27.20	+1.43	+3.50
Nov	23.50	+1.36	+3.69
Dec	19.40	+3.20	+3.41
Average		+1.61	+3.17

[1] Observed data refers to the observed time series temperature values recorded during 1948-1980 at Dhaka station (90.14°E, 23.47°N) which is located at the center of Bangladesh.

Temperature and Evaporation

As presented in Table 7.1, the average monthly temperature for the base year 1990 and the two projected years 2030 and 2075 were obtained from GF01 model runs. Temperature change was calculated using the approach described in Chapter 2. Table 7.1 shows the increase rate and the calibrated projected temperatures for Bangladesh. The average annual increase of temperature would be 1.6°C and 3.2°C for 2030 and 2075, respectively. The results show the seasonal variation of the temperature. For example, it is estimated that by 2030 the temperature will increase by 2.1°C in the winter and 0.7°C in the monsoon season. For 2070, temperatures in the winter and monsoon are estimated to increase by 3.1°C and 1.6°C, respectively.

Figure 7.2 Comparative Analysis of GCM Model Outputs (1×CO_2) with Observed Data[1]

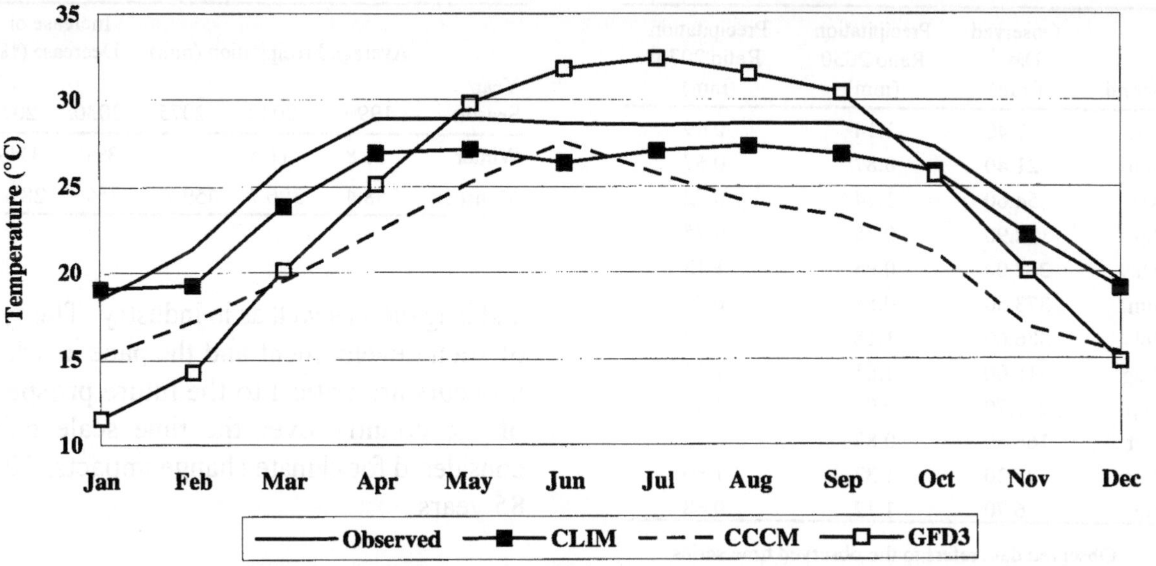

[1] Observed data refers to the observed time series temperature values recorded during 1948-1980 at Dhaka station (90.14°E, 23.47°N), which is located at the center of Bangladesh.

It is assumed that the change of temperature with respect to average winter and monsoon temperatures will be of the same order as that for evaporation. The observed average monthly temperature values for December, January, and February and June, July, and August represent average winter and average monsoon temperatures for the base year. The estimated percent change in evaporation is presented in Table 7.2.

Precipitation

The same method was used to estimate future precipitation change for Bangladesh, and the results are shown in Table 7.3. Here, precipitation ratio (between estimated and baseline) was considered the key value. Table 7.4 shows that winter precipitation will decrease by 3% and 37%, and monsoon precipitation will increase by 11% and 28% for the projected years, respectively.

Table 7.2. Changes in Temperature and Evaporation for 2030 and 2075 Based on GF01 1% Transient Scenario

	Temp (°C)[1]		Temp Increase (°C)[2]		% Change in Evaporation[3]	
Year	Winter	Monsoon	Winter	Monsoon	Winter	Monsoon
Base 1990	19.7	28.6	0.0	0.0	0%	0%
2030	21.7	29.3	2.0	0.7	10%	2%
2075	22.8	30.2	3.1	1.6	16%	5%

[1] Estimated values obtained by correlating model output data with the observed data.
[2] Estimated based on model output data regarding rate of temperature change.
[3] Estimated on the basis of base year average data.

Table 7.3. Estimated Precipitation Adjustment Statistics under GF01 1% per Year Transient Scenario

Month	Observed Data[1] (mm)	Precipitation Ratio 2030 (mm)	Precipitation Ratio 2075 (mm)
Jan	7.40	1.14	0.49
Feb	21.40	0.87	0.67
Mar	54.60	1.14	0.72
Apr	122.90	0.75	0.75
May	262.00	0.95	1.13
Jun	373.50	0.58	0.83
Jul	386.60	1.18	1.53
Aug	315.00	1.65	1.51
Sep	255.70	0.92	1.22
Oct	166.50	0.82	0.73
Nov	25.30	1.33	1.80
Dec	6.70	1.12	0.68

[1] Observed data refers to the observed time series precipitation values recorded during 1948-1980 at Dhaka station (90.14°E, 23.47°N), which is located at the center of Bangladesh.

Table 7.4. Estimated Increase in Precipitation under GF01 1% per Year Transient Scenario

Year/ Season	Average Precipitation (mm) 1990	2030	2075	Increase or Decrease (%) 2030	2075
Winter	11.8	11.5	7.5	-3%	-37%
Monsoon	358.4	396.7	459.7	11%	28%

Economic Development Scenario

The main aspects of development in Bangladesh stem from the high population/land ratio, demographic characteristics, and the major contribution of the agriculture sector to the GDP as well as employment. Thus the efficiency and ability of agricultural output to continue to outpace the increase in population is an important consideration in the future development of the country. Other important considerations include the ability to continue the downward trend in the rate of increase in population, which in turn is tied to improved education, particularly of women, and the employment prospects of the youth, who represent half of the present population. Improving the pace of economic growth is also an important consideration that will require higher investments in infrastructure such as roads, communications, watershed development, and irrigation as well as in industry. The type of such development and the pace at which it occurs are critical to the future prospects of the country over the time scale being considered for climate change impacts, 30 to 85 years.

The baseline development analysis on population, macroeconomics, socioeconomics, and agriculture considered in this study is taken mainly from two projection reports: the Task Force Report (Task Force, 1991) and the World Bank Report (World Bank, 1993).

The population of Bangladesh has grown at a rapid rate, from 50 million in 1961 to 110 million in 1990 (BBS, 1991). Expected future population growth can be captured by the net reproduction rate (NRR), showing the natural increase of the population (ratio of current population and the population that would result if fertility would reach replacement level, morbidity would remain constant, and net migration would be zero). The World Bank's estimate is based on the assumption that NRR will decrease from the 1990 level of 1.9 to 1.0 in 2015. This implies a population of 153 million in 2010 (World Bank, 1993). The projections of the

Table 7.5. Baseline Growth in Population for Bangladesh

Time Period	Average Annual Percentage Growth
1990-1995	1.96%
1995-2000	1.78%
2000-2005	1.59%
2005-2010	1.36%
2010-2015	1.23%
2015-2020	1.16%
2020-2025	1.06%
2025-2075	1.00%

Source: World Bank, 1993.

Table 7.6. Present and Future Population

Year	Population (millions)
1990	109.0
2000	131.2
2010	152.0
2020	171.1
2025	180.3
2030	189.5
2075	296.5

Source: World Bank, 1993.

Planning Commission show a population of 155 million in 2010. The average annual percentage growth in population assumed for the present study is based on the World Bank report, as presented in Table 7.5. Based on this growth rate, the present and future population of Bangladesh is presented in Table 7.6.

The economic development in Bangladesh during the last decade has been characterized by the World Bank as a period of successful stabilization but low growth (World Bank, 1993). As projected by M. Asaduzzaman, senior fellow, Bangladesh Institute of Development Studies, the average growth of GDP was 4% annually, and domestic investment, the motor of economic growth, was 10% of GDP in 1990.

The success of economic policy in Bangladesh depends to a large extent on the development of the world economy. For this reason, the development options have been tuned to world scenario outlooks.

In the *Global Shift* (Chowdhury, 1994) and *Balanced Growth* (Nurkse, 1958) scenarios, the economies of the Asian countries have large opportunities for great economic development. In the first scenario there is a shift in economic activity from Europe to Asia; the second scenario represents a worldwide economic growth with "open" trade borders. The World Bank has recommended a set of structural reform policies that Bangladesh should follow in order to benefit from optimistic scenarios for Asia (World Bank, 1993). These policies make an appeal to free market principles, but at the same time define an important role for the government of Bangladesh.

Since 1990, the new, democratic government has made some policy changes to implement a market economy. These policy changes include providing incentives for industrial development and investment, and liberalizing investment policy. This shift in policy is expected to increase the economic growth in the country up to 2025. The estimated average annual percentage growth rates in GDP put forward by the World Bank are somewhat lower than what would be expected due to these policy changes. Average annual percentage growth rates and GDP are presented in Table 7.7.

Table 7.7. Baseline Growth in GDP for Bangladesh

Time Period	Percentage Growth
1990-2000	5.0%
2000-2025	6.0%
2025-2050	5.0%
2050-2075	4.0%
Year	**GDP (US$)**
1990	210
2000	342
2025	1,468
2030	1,874
2050	4,970
2075	13,250

Sea Level Rise Scenario

The proposed rate of current sea level rise (0.2 m by 2100) given in the methodology presented in Chapter 2 is not being used in Bangladesh because the impacts regarding backwater effect may not be significant enough to be reported. Two accelerated sea level rise scenarios of 0.5 m and 1.0 m sea level rise by 2100 are instead being used. To allow correlation with the projection years, decadal rises in sea levels are used. Table 7.8 provides a summary of the sea level rise scenarios being used by Bangladesh.

Changes in sea level could increase flooding as well as salinity intrusion. The relevant information would be made available by MIKE11 model runs and the most affected regions would be the area covered by the southwest regional model. MIKE11 is composed of one general model which considers the entire country and five regional models considering northwestern, and southwestern regional models considering northwestern, northeastern, northcentral, southeastern, and southwestern parts of Bangladesh (SWMC, 1992). Therefore, the proposed sea level rise scenario I (lower bound rise by 2100) is used along with the sea level rise scenario II (upper bound rise by 2100) for only one regional model, i.e., southwest, for both the projection years. It will allow sensitivity analysis on flooding and salinity intrusion with respect to different rates of sea level rise.

Table 7.8. Sea Level Rise Scenario for the Three Projection Years

Projection Years	Rise in Sea Level Scenario- I[1] (m)	Scenario- II[2] (m)
2030	0.15	0.30
2075	0.35	0.70
2100	0.50	1.00

[1] Scenario- I refers to a 0.5 m in rise by 2100.
[2] Scenario- II refers 1.0 m in rise by 2100.

Watershed Development Scenario

Inside Bangladesh

For simplicity, the base runs will assume that by 2030 about 60% of the proposed embankment projects along the major rivers (e.g., Jamuna and Ganges) will be implemented. Similar assumptions allow 100% implementation of those embankments by 2075.

Outside Bangladesh

There will be further withdrawal of surface water from the major rivers, especially in the dry season. But the questions of volume and the location of withdrawals are not easy to predict. So this issue will not be considered in this study.

AGRICULTURE SECTOR

The key issues addressed in the analysis of the agriculture sector include the impacts of climate change parameters such as drought, inundation, and salinity intrusion on rice, the major crop of Bangladesh.

Methods Used

Selection of Study Sites
The entire country has been considered as a study site, except for the southeast and the northeastern hilly areas. There are eight representative agroecological regions in Bangladesh. Crop phenological data were collected from at least two representative farm sites per agroecological region and subsequently extrapolated for the entire country.

Stepwise Methodology
The following stepwise methodology is being used for the analysis of the agriculture sector in Bangladesh.

Crop model installation and calibration:

- application of crop, soil, land, and agroclimatic data into the agroecological land suitability rating

- installation of the Decision Support System for Agrotechnology Transfer version 3 (DSSAT 3) and preparation of working formats

- calibration of Agro Ecological Zone (AEZ) crop model system (FAO, 1988) with DSSAT 3 and transfer of AEZ database into DSSAT 3 system.

Creation of crop phenological data sets for different crops/crop cultivars:

- Local and modern cultivars of rice are being used in the model.

- Appropriate phenological coefficients for DSSAT 3 models were developed based on on-farm data and nationally available data bases.

- Crop response functions to variation in climatic parameters under different levels of management and inputs availability were developed for DSSAT 3 models to create scenario of crop impacts.

Creation of agroclimatic data sets for GCM scenarios:

- 15-20 years daily climatic data of 14 parameters were processed for GCM scenario development. The CCCM and GFD3 models were being used for this assessment.

Crop impact scenarios:

- AEZ land suitability ratings and DSSAT 3 outputs will be compared and a national production database for different GCMs will be provided. Reduction of yields under different management adaptation levels will also be estimated.

Linkage with water sector:

- An attempt will be made to link water demand for irrigation and water availability in the surface water systems.

- Data on crop water use and irrigation potential under different climatic scenarios will be developed by AEZ region catchment areas.

- Irrigation water demands under alternate crops/cropping pattern will be estimated for projecting impacts of climate change on water availability and crop production status.

Adaptation Alternatives

Adaptation alternatives to be evaluated include:

- rice genotype development to sustain rice yield

- management options and adjustment of agroecological variability of rice yield with the change in climate.

Preliminary Results

The analysis is partially completed. From the phenological response database it was found that the high yielding varieties of rice cultivars are more susceptible to water availability in topsoil, resulting in a considerable decrease in production in drought conditions. In addition, when land is submerged by flooding water, the productivity of the high yielding varieties cultivars also decreases. For lands that are under more than 90 cm of floodwaters, the expected loss would be 100% for high-yielding Boro cultivars. (Boro is a rice crop planted under irrigation during the dry season from December to March and harvested from April to June.) On the other hand, the local varieties are significantly less susceptible to inundation.

The proposed linkage with the water sector and water availability has not yet been completed since it depends on data from surface water MIKE11 modeling outputs.

GRASSLANDS/LIVESTOCK SECTOR

There is no land in Bangladesh that can be termed grassland; however, over 45% of the agricultural land is used for grazing when it is not being cultivated. The livestock of Bangladesh largely depend on agricultural residues (hay, molasses, rice husks, etc.) as feed (Hossain, 1989). Thus a reduction of grain production due to climate change would cause a severe shortage of livestock feed in Bangladesh. This would be estimated in terms of shortfall in diet or nutritional value once the agricultural vulnerability is ascertained. The livestock population at risk due to cyclonic storm surges can be estimated by delineating the coastal areas affected by storm surges and overlaying that with a livestock population database using a Geographical Information System (GIS). From the computer GIS database on livestock population, 2.1 million bovine and 1.4 million goat and sheep are vulnerable to cyclonic storm surges in the coastal areas. Bangladesh will not directly model the grasslands and livestock sector.

FORESTRY SECTOR

Bangladesh's assessment for the forestry sector relates to the coastal zone vulnerability study; it focuses on the vulnerability of the Sundarbans mangrove forests and adaptive measures to reduce the vulnerability.

Methods Used

Selection of Study Sites

The Sundarbans mangrove forest in southwestern Bangladesh was selected as the study site for the following reasons:

- Sundarbans is the single largest continuous area of productive mangrove forest in the world. The forest covers 577,000 ha, of which 401,600 ha are land and the remaining 175,400 ha are under water, in rivers, canals, and creeks of widths varying from a few meters to several miles.
- The forest plays a vital national socio-economic role. Substantial portions of the country's requirements for timber, fuelwood, teaching materials, honey, and wax are met by the forest.
- The forest functions as an essential habitat of wildlife biodiversity, nutrient producer, water purifier, storm barrier, shoreline stabilizer, aesthetic attraction, and energy storage unit. The forest also contains many endangered species of global importance, including the Royal Bengal tiger (*Panthera tigris*).
- Under the International Ramsar Convention, Sundarbans has been selected as a Ramsar site.
- Although mangroves are resilient to fluctuations in mean sea level, the existing physical, ecological, and anthropogenic stresses on the mangroves of Sundarbans are diminishing their natural resilience, leaving them increasingly vulnerable to possible climate and sea level changes. The future health of the mangrove forest is a critical issue: future climate change and sea level rise may modify the composition and location of the predominant mangrove species. The decline of the forest will reduce its economic usefulness, with potentially serious implications for the coastal economy. In addition, depletion of the Sundarbans could expose Khulna City and Mongla to increased risks from storm surges. Economic analysis will be done once the assessment of the Sundarbans is completed.

Forest Ecosystem Models Used

The Holdridge model is predominantly for temperate boreal forests in the Northern Hemisphere and thus is inappropriate for use with the tropical Sundarbans mangrove forest. A methodology has been developed to assess the vulnerability and adaptation of the Sundarbans mangrove forest; this methodology may also be appropriate for other countries with tropical mangrove forests.

This methodology considers that the mangrove forests are already under physical, ecological, and anthropogenic stresses, and may not be able to sustain any additional stresses of higher salinity from sea level rise and increased salinity during low water flow periods in the dry season. A surface water model may be used to project future salinity levels and water discharge at different points in the rivers of Sundarbans for the projection years 2030 and 2075. The response of the dominant mangrove species, sundri (*Heritiera fomes*) and gewa (*Excoeceria agallocha*), to changes in soil and water salinity, fresh water flow, sedimentation, and tidal range may be determined from the trend analysis of the available time series and other data. The response functions for the regenertation and survival, growth, distribution, and composition of sundri and gewa with respect to the said changes may be studied from the time series and other available data. The results could then be presented graphically using GIS techniques.

Adaptation Alternatives and Evaluation Methods

Though adaptation alternatives will be evaluated later, some indications are given at this stage of the study:

- To mitigate loss of forest land due to sea level rise, the process of land accretion may be accelerated by considering appropriate interventions.
- To prevent further intrusion of salinity in the southern region of Bangladesh, it may be necessary to increase the freshwater supply in the Mathabhanga and Gorai-Madhumati river systems. This would require implementation of proposed Ganges and Brahmaputra barrages (dams placed in a water course to divert the water), which will, among other things, allow more freshwater supply in the southwest region, including the Sundarbans.

WATER RESOURCES SECTOR

Methods Used

Assessing the vulnerability of the water resources sector requires examination of the following topics:

- changes in water levels
- salt water intrusion
- drought
- groundwater availability.

Information on the supply side of these topics will be obtained by using a surface water hydrodynamic model, MIKE11, except for information on groundwater availability, which will be obtained using a groundwater model. Both models were developed by the Danish Hydraulic Institute (DHI) and have been calibrated for the river systems of Bangladesh (MIKE11). The input and output data of MIKE11 have repeatedly been validated by many national and international hydrologists and the outputs are already being used widely within the country. The hydrology working group thus chose MIKE11 over the other existing hydrodynamic models.

Information on the demand side will be obtained from published information sources (MPO, 1987). The sources provide a comprehensive picture of water requirement for 58 planning areas, each of which comprise two to three catchment areas distributed over the entire country (except the Chittagong Hill Tracts area). This water requirement information will be compared with the water availability data to reveal water stress and/or water surplus. The future supply will also be compared with future water requirements based on increased irrigation, industrialization, domestic use, and other uses, and the future water availability for the projected years will be estimated to assess water resources vulnerability.

Selection of Study Sites

For inundation analysis, the MIKE11 models will be used to provide information on water levels and discharge. The major river systems flowing through the country are incorporated into the general model, a component of MIKE11, and water level and discharge data are made available for more than 700 grid points distributed along all the rivers. The floodplains are divided into many catchment areas and are distributed into five regional model networks of MIKE11; the northwest (NW) model, the north-central (NC) model, the southeast (SE) model, the northeast (NE) model, and the southwest (SW) model (SWMC, 1992). These regional model networks provide hydrologic information at over 4,000 locations (grid points) distributed over the major and minor river networks.

All five regional models will be run for inundation analysis. Data from regional model runs will be interpreted by overlaying water level data on the DEM model using GIS. Salinity intrusion is important for only the SW region and thus only the SW regional model will be considered for salinity information. The drought model will be considered only for the northwest region where the problem is acute. Groundwater availability will be estimated for all the regions and also be interpreted by the use of GIS. The model runs and subsequent interpretations are expected to be completed in late 1995.

Hydrological and Water Resources Methodologies

From the water levels of the output stations, water surface maps will be produced for the base year 1990 and the projection years 2030 and 2075. The base year water surface map will then be superimposed on the water surface map of each projection year to determine the change in inundation levels. Superimposition will be done using GIS techniques.

A water level change map will then be superimposed on a land type map to ascertain the changes of land type from one class to the other. The calculation of the changes in land types will be obtained primarily from the GIS analyses. Loss of agricultural production, loss of physical infrastructure, and number of threatened population due to flooding will be estimated.

It is expected that the salinity runs of the MIKE11 salinity model will estimate the present and the future salinity impacts, to be represented in GIS maps. Salinity runs will estimate the concentration of salinity in absolute numbers, i.e., micromhos/cm or ppt. A water surface map would be prepared to ascertain the changes of land type and to be compared with the general model output. This information would be analyzed in terms of crop response functions with respect to salinity to reveal possible crop production loss.

Integration of Results with Analyses of Other Sectors

The analysis of the vulnerability of the water resource sector will provide vital information regarding the vulnerability of other sectors. Coastal resources vulnerability due to flood, drought, salinity, etc., will be obtained from model outputs. Moreover, isosaline areas given by the GIS analyses coupled with MIKE11 model runs will provide additional information regarding the impacts on the mangrove forest in the southwest region of Bangladesh.

COASTAL RESOURCES SECTOR

The coastal resources vulnerability assessment will consider impacts on rice production, on primary physical infrastructure and capital assets, and on population. Floods, droughts, cyclonic storm surges, salinity intrusion, and coastal and river erosion will be examined. Most of the groundwork was completed in several previous studies (BCAS/RA/Approtech, 1994; BUP, 1994; ADB, 1994) and will be used. Coastal erosion will be examined using the Bruun Rule. A recently completed coastal survey will provide a contour line 1.0 m above the average sea level along the shoreline in the Chittagong area, from Teknaf to Kumira. Because of a lack of time and resources, shoreline erosion for other areas will not be estimated.

Estimates of quantitative losses due to loss of crop production and infrastructure will be attempted using 1990 values; however, the impact on GDP will not be calculated. It is anticipated that the coastal vulnerability analyses will be completed in late 1995.

FISHERIES SECTOR

Bangladesh has extensive inland water fisheries from which its people have been harvesting living resources (fin fish, prawns, crabs, turtles, etc.) from time immemorial. The inland water habitats are largely freshwater, but the habitat conditions in the rivers and associated water bodies in the south of the country change to brackish conditions with wide range in the levels of salinity both over time and space. The brackish and estuarine water conditions extend into the north Bay of Bengal up to considerable distances from the shoreline.

This gradual change from freshwater to brackishwater and estuarine conditions to pure marine conditions has allowed many of the aquatic living organisms to migrate back and forth between freshwater habitat and estuarine as well as marine habitats in order to complete their life cycles. Of such aquatic animals, the most important one is *Hilsa ilisha*, a fin fish that lives in the sea but annually undertakes anadromous migration into the freshwater rivers to breed. This pelagic fin fish is commercially very important for Bangladesh and is harvested in large numbers from the sea as well as from freshwater rivers. Another group of commercially important aquatic fauna is the Penaeid shrimp. The most commercially important shrimp is tiger shrimp (*Penaeus monodon*), which is followed by the brown shrimp (*Metopenaeus monocoros, M. brevicornis*) and white shrimp (*P. indicus, P. merguiensis, P. semisulcatus*). These marine shrimps live in deeper offshore areas of the sea, but their larvae and postlarva come to the estuarine and brackish shallow waters to feed and grow. Because of the popularity and demand in the export market, brackish water aquaculture of the shrimp is extensively practiced in the tide inundated lands along the coast of Bangladesh, particularly in its southeast and southwest regions. Currently, over 100,000 ha of land along the coast are under brackishwater shrimp culture. These shrimp farming activities would be wiped out by a rise in sea level. Another important group is the freshwater prawn (*Macrobrachium rosenbergii, M. malcolnsonii*) belonging to the family Palaemonidae. These prawns live their adult life in the freshwater rivers and floodplain, but visit the brackish and estuarine water to breed and grow in the semi-saline water for the first few weeks of their life before returning to the freshwaters.

Time series data on the biological parameters on any of the above species, time series data on water temperature, and water salinity data for Bangladesh marine, estuarine, or freshwater habitats do not exist.

The species selected for the vulnerability assessment are *Hilsa ilisha*, the anadromous fish migrating between the Bay of Bengal and the Ganges-Brahmaputra River system, *Penaeus monodon*, *Macrobrachium rosehbergii*, and *Metapenaeus monoceros*.

Probable future sites would be:

- the Meghna estuary near Hatia, Khepupara, and Chandpur for *Hilsa*
- the Satkhira, Bagerhat, and Coxs Bazar regions for the two penaeid shrimp
- the Bagerhat, Patuakhali, and Chandpur regions for *Macrobrachium*.

Very little biological and economic data on any of the freshwater and marine species of fish or prawn are available. Data on estimated total annual yield for all species combined are, however, available for 1983-1984 to the present.

To assess the vulnerability and adaptability of the selected species to climate change, basic information on the biology, migration, behavior in response to temperature and salinity, salinity, population dynamics, annual yield, economic value, and nutritional value is needed.

SUMMARY CONCLUSIONS

The Bangladesh Country Study on vulnerability and adaptation has recently been started with three major subcomponents (agriculture, coastal, and water) and two minor subcomponents (fisheries and forestry). This report identifies the climate change scenarios being examined and methodologies being followed in each of the subcomponents.

The potential vulnerability of Bangladesh in the water resources sector has profound impacts for the coastal, agriculture, and forestry sectors. Flooding will affect development in coastal areas as well as low-lying agriculture. However, low water availability during other seasons will put stress on agriculture and salinity intrusion will severely affect the Sundarbans mangrove forest. Shortfalls in agriculture will in turn affect human settlements and the availability of fodder for livestock.

REFERENCES

ADB. 1994. *Climate Change in Asia: Bangladesh Country Report.* Asian Development Bank.

BBS. 1991. *1991 Statistical Yearbook of Bangladesh.* Bangladesh Bureau of Statistics.

BCAS. 1995. *Financial Indicators Bangladesh*, vol. 5, issue 1. Bangladesh Center for Advanced Studies.

BCAS/RA/Approtech. 1994. *Vulnerability of Bangladesh to Climate Change and Sea Level Rise: Concepts and Tools for Calculating Risk in Integrated Coastal Zone Management.* Technical Reports, vols. I and II, Summary Report and Institutional Report. Bangladesh Center for Advanced Studies (BCAS), Resource Analysis (RA), and Approtech Consultants Limited.

BUP/CEARS/UEA. 1994. *Bangladesh: Greenhouse Effect and Climate Change, Briefing Documents, no. 1-7,* Bangladesh Unnayan Parishad, Dhaka, Bangladesh. Center for Environmental and Resource Studies, University of Waikato, Hamilton, New Zealand and Climate Research Institute, University of East Anglia, Norwitch, United Kingdom.

Benioff, R., S. Guill, and J. Lee (eds.) (In press). *Vulnerability and Adaptation Assessments: An International Guidebook.* Dordrecht, The Netherlands: Kluwer Academic Publishers.

Chowdhury, A. 1994. *Global Shift.* In East Asian Industrialization. Routledge.

FAO. 1988. *Land Resources Appraisal of Bangladesh for Agricultural Development, Report 3: Land Resources Data Base, Volume I: Climatic Data Base,* United Nations Development Programme/Food and Agricultural Organization of the United Nations.

Hossain, G. 1989. *A Review of Livestock Subsector in Bangladesh.* Compendium vol. II, Agriculture Sector Review, UNDP, Dhaka.

IPCC. 1992. *Climate Change 1992: The Supplementary Report to the IPCC Scientific Assessment;* J.T. Houghton, B.A. Callander, and S.K. Verney, (eds.). Cambridge: Cambridge University Press, 1990.

Khan, F.H. 1991. *Geology of Bangladesh.* Dhaka, Bangladesh: The University Press Limited.

Kuehl, S.A., T.M. Harlu, and W.S. Moore. 1989. Shelf Sedimentation of the Ganges-Brahmaputra River System; Evidence for Sediment Bypassing to the Bengal Fan. *Geology*, 17:1132-1135.

MIKE11. Danish Hydraulic Institute, Horsholm, Denmark.

MPO. 1987. *Surface Water Availability, Technical Report No. 10.* Master Plan Organization, Government of Bangladesh, March.

Nurkse, R. 1958. *Balanced Growth.* The Conflict between Balanced Growth and International Specialization. In Lectures on Economic Development. Faculty of Economics. Istanbul University and Faculty of Political Sciences, Ankara University, Istanbul.

Rashid, H.E. 1991. *Geography of Bangladesh.* The University Press Ltd. Second Revised Edition.

Strzepek, K.M. and J.B. Smith, (eds). 1995. *As Climate Changes: International Impacts and Implication, Executive Summary.* Cambridge: University Press.

SWMC. 1992. *General Model Verification Report. Surface Water Simulation Modelling Programme Phase II.* Master Plan Organization, Surface Water Modelling Centre; Ministry of Irrigation, Water Development and Flood Control, Government of Bangladesh, July.

Task Force. 1991. *Report of the Task Forces on Bangladesh Development Strategies for the 1990s, vols. I-IV.* Planning Commission, Government of Bangladesh. Dhaka: University Press Limited.

World Bank. 1993. *Bangladesh; Implementing Structural Reform, South Asia Country Department 1,* World Bank, March.

Chapter 8

Vulnerability and Adaptation Assessments for Kazakhstan

Olga Pilifosova,[1] Irina Eserkepova[1]
Svetlana Mizina,[1] Paishan Kozhahmetov[1]
Grigory Chichasov,[1] Lubov Lebed,[1] Svetlana Dolgih[1]
Vsevolod Golubtsov,[1] Ivan Skotselyas,[1] Valery Lee[1]
Kanat Berentaev,[2] Nadezgda Fedorova,[2] Ervin Gossen[3]

[1] Kazakh Scientific Research Hydrometeorological Institute (KazNIGMI)
[2] Scientific Institute of Kazakhstan Ministry of Economy and Market Relations
[3] Academy of Agricultural Sciences

SUMMARY. The vulnerability and adaptation assessments for Kazakhstan address the agriculture, grassland/livestock, and water resources sectors. As our vulnerability and adaptation assessments show, estimated climate changes will generally have negative impacts on the relevant sectors. Reductions of spring wheat yield are anticipated to be 56% under CCCM, 51% under GFD3, and 12% under GF01 scenarios. Winter wheat yield could increase about 21% under GF01 and 17% under GFD3 and CCCM scenarios. Under the CCCM scenario, wool productivity could decrease 11% to 19% and the number of lambs would be 2% to 26% less than under baseline conditions. The most significant losses of grassland productivity, up to 80%, are projected under the CCCM scenario. The assessment of the water resources sector, conducted for both a mountainous and a plains basin, indicates that water resources throughout the territory of Kazakhstan could be vulnerable. According to the preliminary assessment, climate change under 2×CO_2 levels could cause 20% to 30% reductions in water resources of Kazakhstan.

INTRODUCTION

Kazakhstan is located in the center of the Eurasian continent, nearly equidistant from the Atlantic and the Pacific oceans. The landlocked country has the ninth largest land mass of any country in the world, 2.7 million km^2 (Figure 8.1). Most of its territory is lowlands and plains bordered by high mountain ranges in the east and south-east. The highest point in Kazakhstan is the Khan-Tengry peak (6,995 m above sea level), and the lowest one is the Karagie hollow, which is contiguous with the east coast of the Caspian Sea (132 m below sea

Note: This chapter does not contain a detailed description of the assessment methodologies that were recommended by the U.S. Country Studies Program. Refer to Chapter 2 for an overview of these methodologies.

J. B. Smith et al. (eds.), Vulnerability and Adaptation to Climate Change, 161–181.

Figure 8.1 Map of Kazakhstan

level). Kazakhstan has five natural zones: forest-steppe, steppe, semi-desert, desert, and mountains and foothills (Figure 8.2). The climate of Kazakhstan is sharply continental. Throughout the country, the temperature sums for the growing season are sufficient for the cultivation of all species of cereals grown in a moderate temperature belt. Conditions in the south are favorable for heat-loving crops. However, there is a shortage of precipitation for dryland agriculture in most of Kazakhstan. It is only possible in the north, the northeast, and in a narrow foothills area in the south.

Forests cover only 3.6% of the territory, or 9.65 million ha. Coniferous forests cover 1.8 million ha, and the rest of the forested area is made up of deciduous forests and bushes. The largest area, 4.7 million ha, is made up of saksaul (*Haloxylon*), a deep-rooted, practically leafless tree common in Asian deserts.

The river network has had only limited development. However, rivers are extremely important for irrigation and power generation in the foothills and in the south.

The population of Kazakhstan is above 16 million; in 1989, 40% of the population

Figure 8.2. Map of Kazakhstan's Natural Zones

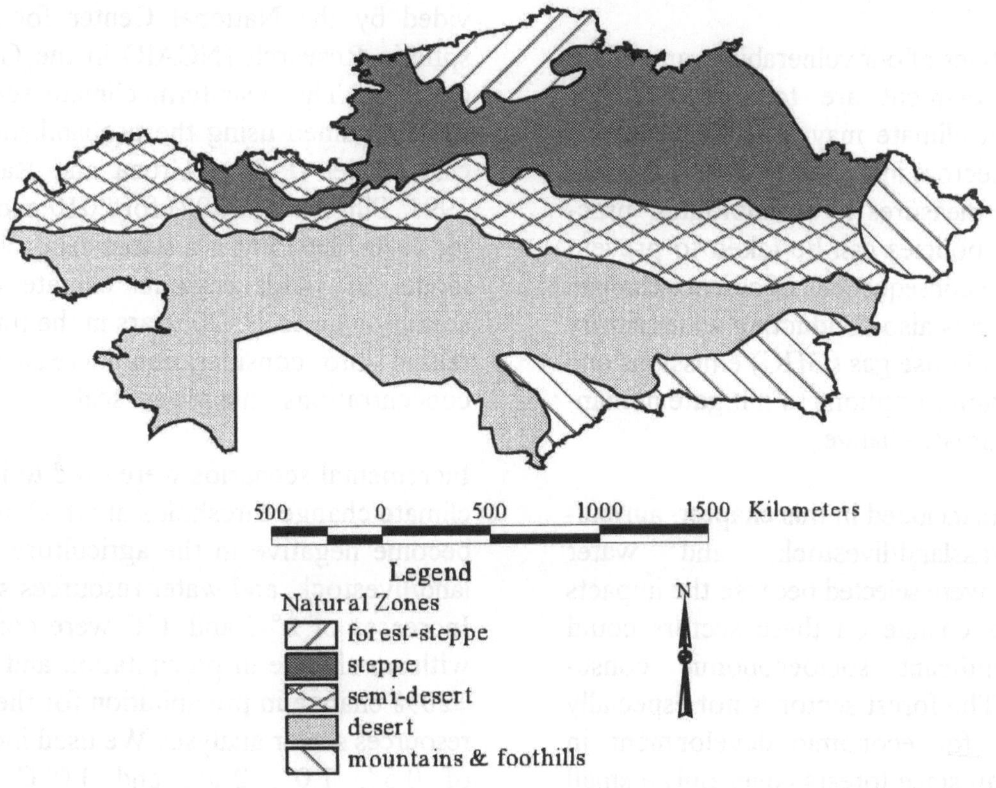

were Kazakh and 38% were Russians. There are 19 administrative regions (oblasts) and more than 80 cities and towns.

Agricultural and industrial production makes up 22% and 69% of the GNP, respectively. Kazakhstan has a relatively diversified economic structure based on its rich endowment of natural resources. In 1992, arable land exceeded 24.1 million ha, including 2.3 million irrigated ha (Conceptual Program, 1994). Cereal harvest is up to 24 million tons per year, some of which is exported. The leading sector of stock-breeding is sheep-breeding on the vast desert and semi-desert grasslands. There were 34.5 million sheep and goats and 9.5 million cattle in 1992. Camels, cows, pigs, and horses are also bred. Now Kazakhstan is facing the typical challenges of a transition process coupled with the negative consequences of the disintegration of the economic structure of the Soviet Union.

Agriculture, natural grasslands, livestock, and water resources are most likely to be vulnerable to climate change in Kazakhstan. The impact on these sectors may be considerable because of their location in the marginal areas of the moderate humid forest-steppe zone and the semi-arid steppe zone. Increased frequency of droughts and dust storms, reduced organic matter in soils, and other negative consequences of global warming can be expected in these marginal areas.

The objectives of our vulnerability and adaptation assessment are to evaluate how changes in climate may affect key natural resource sectors in Kazakhstan and to identify what measures in technologies, practices, and policies can be taken to prevent undesirable consequences of climate change. Kazakhstan is also conducting an inventory of net greenhouse gas (GHG) emissions and an evaluation of options to mitigate human-induced climate change.

The sectors included in this chapter, agriculture, grassland/livestock, and water resources, were selected because the impacts of climate change on these sectors could have significant socioeconomic consequences. The forest sector is not especially important for economic development in Kazakhstan since forests cover only a small part of the country, and therefore this sector was not considered. This chapter presents our preliminary results; economic analyses will be conducted in the future.

SCENARIOS

Climate change scenarios and baseline socioeconomic scenarios were developed for the Kazakhstan vulnerability and adaptation assessment. Climate change scenarios to assess both long-term ($2\times CO_2$ in 2050-2075) and short-term (2010 and 2030) impacts were prepared. The climate conditions under $2\times CO_2$ concentrations were estimated using three global circulation models (GCMs): the Canadian Climate Centre Model (CCCM), the Geophysical Fluid Dynamics Laboratory (GFD3) model, and the 1% transient version of GFDL (GF01) (Benioff et al., in press). The data for the GCM scenarios were provided by the National Center for Atmospheric Research (NCAR) in the GRADS data set. The near-term climate scenarios were obtained using the probabilistic forecast model (PFM) (Gruza and Rankova, 1991; Pilifosova, 1992) for 2010 and GF01 for 2030. The PFM is a statistical probability model. It yields regional climate change scenarios up to 10-20 years in the future by taking into consideration increasing CO_2 concentrations on a global scale.

Incremental scenarios were used to identify climate change thresholds at which impacts become negative in the agriculture, grassland/livestock, and water resources sectors. Increases of 2°C and 4°C were combined with no change in precipitation and with a ±20% change in precipitation for the water resources sector analysis. We used increases of 0.5°, 1.0°, 2.0°, and 3.0°C and a ±10% change in precipitation for the analyses of the agriculture and grassland/livestock sectors.

Climate Change Scenarios

To select the GCMs for creating climate change scenarios, we compared the regional $1\times CO_2$ output with observed climate data. A scenario representing current climate conditions based on observations from 1951 to 1980 was developed. This baseline scenario was used to evaluate which of the GCMs provided the most accurate estimate of Kazakhstan's current climate. This comparison was conducted as follows:

- Within each model, monthly correlation coefficients for the temperature estimates of the GCMs and observed temperature from 25 different sites in Kazakhstan were calculated.

- An analysis was conducted of the differences between the observed and simulated monthly air temperature and daily mean precipitation, averaged across three latitudinal zones within Kazakhstan and over the country as a whole. The analysis was performed using the Student t test.

- The average temperature increase rates derived from the simulation models ($1\times CO_2$ to $2\times CO_2$ levels) were compared to observed data for 1891-1992. This method is based on the assumption that the regional temperature increase rate (in °C per 1% of CO_2 increase) is approximately constant. The regional temperature rise rate over the past 100 years was about 0.03 °C per 1% rise of CO_2 concentration. According to the GFD3 model and the CCCM the rate equals 0.04°C and 0.06°C per 1% increase in CO_2 concentrations, respectively.

This comparison of the current climate scenario (based on observed climate) with the $1\times CO_2$ output from the GCMs showed that the GFD3 model best matched the observed climate.

The adjustment statistics averaged across each of three latitudinal zones within the territory of Kazakhstan and over the country as a whole under each of the GCM scenarios are given in Tables 8.1 and 8.2.

Table 8.1. Adjustment Statistics Generated by the GFD3 Scenarios

Zone		Jan	Feb	Mar	Apr	May	Jun	Jul	Aug	Sep	Oct	Nov	Dec	Annual
First	Temp. (°C)	3.50	3.80	4.50	4.30	4.30	4.50	4.40	4.70	4.80	4.00	3.40	2.30	4.00
(39-45°N)	Precip. Ratio	1.39	1.04	0.86	1.12	0.98	1.68	1.93	2.23	1.10	1.53	1.07	0.94	1.17
Second	Temp. (°C)	3.10	3.50	6.40	6.00	4.90	5.10	5.80	5.80	5.60	5.10	4.30	2.70	4.80
(45-51°N)	Precip. Ratio	1.10	1.14	0.94	0.77	0.89	1.44	1.69	2.01	1.40	1.49	1.03	0.94	1.12
Third	Temp. (°C)	3.50	3.80	7.70	8.20	5.40	5.40	6.40	6.60	5.50	5.70	6.50	4.60	5.80
(51-57°N)	Precip. Ratio	1.18	1.21	1.35	0.93	0.89	1.08	1.12	0.99	1.17	1.48	1.25	1.24	1.17
Averaged for the	Temp. (°C)	3.40	3.70	6.20	6.20	4.90	5.00	5.50	5.70	5.30	4.90	4.70	3.20	4.90
Country	Precip. Ratio	1.22	1.13	1.05	0.94	0.92	1.40	1.57	1.74	1.22	1.50	1.11	1.04	1.16

Table 8.2. Adjustment Statistics Generated by the CCCM Scenarios

Zone		Jan	Feb	Mar	Apr	May	Jun	Jul	Aug	Sep	Oct	Nov	Dec	Annual
First	Temp. (°C)	5.80	7.50	6.60	7.80	7.30	5.70	6.10	5.90	5.30	4.60	5.30	6.80	6.50
(39-45°N)	Precip. Ratio	0.97	1.04	1.03	1.09	0.92	0.67	0.72	1.36	1.13	1.10	0.99	1.04	0.86
Second	Temp. (°C)	8.50	9.80	9.30	11.20	9.30	6.45	6.30	5.90	4.80	3.40	3.90	7.90	7.20
(45-51°N)	Precip. Ratio	1.11	1.20	1.05	1.21	0.94	0.64	0.65	0.71	0.93	1.06	0.94	1.05	0.90
Third	Temp. (°C)	7.80	7.70	6.80	8.40	10.90	5.10	3.70	3.80	3.30	3.10	3.10	8.80	6.00
(51-57°N)	Precip. Ratio	1.17	1.25	1.28	1.51	1.34	1.04	1.08	1.01	1.03	1.08	1.00	1.20	1.12
Averaged for the	Temp. (°C)	7.40	8.30	7.60	9.20	9.10	5.70	5.30	5.20	4.50	3.70	4.10	7.80	6.50
Country	Precip. Ratio	1.08	1.16	1.12	1.27	1.06	0.78	0.82	1.02	1.03	1.08	0.97	1.10	1.00

Probabilistic Forecast Model Near-Term Scenarios

The Probabilistic Forecast Model (PFM) is a modification of the probabilistic weather forecast method worked out by Gruza and Rankova (1991). This method allows the prediction of short-term climate changes as CO_2 concentrations increase. Climate changes caused by factors other than CO_2, of both natural and anthropogenic origins, are also taken into consideration. Their variability within the analogue group (i.e., the time intervals with similar CO_2 concentration variations) makes it possible to give forecasts in a probabilistic form. Temperature changes caused by anthropogenic and natural CO_2 concentration can be expressed by the following relationship (Budyko et al., 1985; Budyko et al., 1991):

$$\Delta_a = \Delta T_c (In \frac{c_1}{c_2} / In\ 2)\mu,$$

where:

ΔT_a = anthropogenic temperature change
ΔT_c = coefficient of temperature sensitivity to $2 \times CO_2$
C_1 and C_2 = CO_2 concentrations at beginning (t_2) and end (t_1) of the period
μ = coefficient of climate system thermal inertia for time period ($t_2 - t_1$).

In the near term, the temperature changes are smaller because the inertia of the climate system dominates the CO_2 effect. Neglecting the dependence of climate system inertia on total temperature changes over the period $t = t_2 - t_1$, can be presented as a sum of two terms:

$$\Delta T(t) = \Delta T_a(t) + \Delta T_N(t) = F(\Delta C_t / C) + N_t,$$

where the first term ($\Delta T_a(t)$ or $F(\Delta C_t / C)$) is the variation of the temperature due to the increase of CO_2 concentration and the second one ($\Delta T_N(t)$ or N_t) is the variation due to other factors not considered here.

In this method we assume that temperature changes are similar to relative variations in CO_2 concentration over the period of instrumental observations of CO_2 concentration in the past and in the future. The statistical characteristics in such a sample of temperature change analogues, i.e., the conditional distribution of climate variables, would represent the probabilistic climate forecast, taking into account anthropogenic changes.

The median or the mean values of the conditional distribution (within the sample) would characterize climate change due to CO_2 increase. Climate oscillations caused by other factors are considered statistically. Their variability within the analogues' group allow the formulation of the forecast in the probabilistic form.

This method is more favorable than extrapolation using regression techniques, because the PFM does not use any hypothesis about statistic relations and distribution functions. We only rely on the assumption that the same "causes" result in the same "consequences." It is not possible to develop climate scenarios beyond 20 years into the future because the period of instrumental observations of CO_2 concentration is too short to derive a sufficient number of analogue periods.

To create the regional climate change scenarios, the change of regional climate variables over the period is presented as the sum of two terms:

$$\Delta X_t = (\Delta X_\tau)_a + (\Delta X_\tau)_n$$

where the first term is the variation of the regional climate variables as a result of CO_2 concentration increases and the second term is their variation due to other factors not under consideration here. The monthly temperature and precipitation amount averaged across the territory were used as dependent variables. The relative variations of CO_2 concentration over the period $t_f - t_0$ was used as a single independent variable. Here t_f is the predicted year and t_0 is the initial year (1992). The observational data obtained at long-term (about 100 years) weather stations in Kazakhstan as well as the CO_2 concentration under a scenario of "business as usual" (IPCC Impacts Assessment Working Group, 1992) were used as the input in the PFM.

The $2\times CO_2$ GFD3 scenario suggests that the minimum increase in temperature is expected in winter, when most of the territory is expected to have temperatures 2.3° to 4.5°C higher. The maximum temperatures (4.3° to 8.2°C) are expected to be in spring. The CCCM scenario estimates an extreme warming above 11°C in spring months. The GF01 outputs provide an "intermediate" scenario (Table 8.3).

In most cases, for all GCMs, the relative changes in precipitation were primarily in the range of 0.8 to 1.2. This range of precipitation change is well within the normal fluctuations found for current climate. According to the GFD3 scenario the maximum precipitation ratios are expected to occur in summer in the first latitude zone (1.68 to 2.23), and in July and August in the second zone (1.69 to 2.01). Precipitation ratios notably above norms are also expected in March (1.35) and October (1.48) in the third zone, and in August (1.74) and October (1.50) for the whole territory.

Under the CCCM scenario the pattern of precipitation ratio distribution is converse, with the minimum in summer and the maximum in spring. Minimum ratios in the CCCM scenario are also lower than under GFD3. For example, under the CCCM scenario the precipitation ratios in June are estimated to be 0.67 and 0.64 in the first and in the second latitude zones, respectively. Under the GFD3 scenario the minimum ratios are estimated to be 0.86 in March for the first zone and 0.89 in May for both the second and third zones. Maximum values under the CCCM are not as large as under

Table 8.3. Adjustment Statistics Generated by the GF01 Scenarios

Decade		Jan	Feb	Mar	Apr	May	Jun	Jul	Aug	Sep	Oct	Nov	Dec	Annual
Fourth (2000)	Temp. (°C)	-0.80	2.60	2.10	1.80	0.20	1.20	0.90	1.40	1.10	2.40	0.90	0.40	1.20
	Precip. Ratio	1.02	1.24	1.00	1.05	1.00	1.20	1.26	1.42	1.28	1.11	1.04	1.08	1.03
Seventh (2030)	Temp. (°C)	2.10	3.80	4.30	2.70	2.40	2.50	2.10	2.50	3.20	4.40	3.10	3.60	3.00
	Precip. Ratio	1.09	1.24	1.17	1.17	1.25	1.19	1.59	1.11	0.97	1.21	1.25	1.14	1.07
Tenth (2070)	Temp. (°C)	4.90	5.10	5.00	4.70	4.50	4.30	4.00	4.70	5.20	6.30	4.20	5.20	4.80
	Precip. Ratio	1.21	1.29	1.19	1.16	1.07	1.09	0.91	0.93	1.10	1.09	0.94	1.32	1.09

Table 8.4. Adjustment Statistics Generated from PFM and Daily Weather Data from 1951-1980 for 2010[1]

	Winter	Spring	Summer	Fall	Annual
Temp. (°C)	3.00	0.60	2.00	1.60	1.60
Precip. Ratio	1.15	0.92	0.67	1.12	0.94

[1] Values averaged for the country and based on increased CO_2 concentrations.

the GFD3; however, they exceed norms. For example, the maximum precipitation ratios for the whole territory are estimated to be 1.27 (April) and 1.74 (August) for the CCCM and GFD3 scenarios, respectively.

The PFM allows the prediction of short-term climate changes based on estimated CO_2 concentrations. The analysis of PFM outputs (Table 8.4) shows that CO_2 concentrations are likely to contribute to increased temperatures and decreased precipitation over all the region.

Baseline Socioeconomic Scenarios

The baseline socioeconomic scenarios presented in Tables 8.5 and 8.6 were developed by experts from Scientific Institute of Kazakhstan Ministry of Economy and Market Relations (State Committee, 1994; National Bank, 1994; National Bank, 1995) in order to provide a context for the estimates prepared for climate change conditions.

The projected socioeconomic trends (taking into account inflation and devaluation of national currency) in the absence of climate change are provided. It is difficult to estimate future GNP because of the many uncertainties in the country's economic development during the present transition. Kazakhstan now is facing a profound crisis in economic, monetary, social, and political terms. A substantial shift in its economic system is expected by 2000.

The population in 2075 is expected to double from 2000. At the same time, GNP is projected to be 10 times higher. However, if climate changes according to the above-mentioned climate change scenarios, the GNP/population ratio might be significantly lower than the baseline scenario reflects.

Table 8.5. Predicted GNP Growth

Period	Growth (avg. annual %)
1990-2000	N/A
2000-2025	3.7
2025-2050	3.3
2050-2075	2.3
Year	**GNP (1994 US$ billions)**
1994	$5.47
2000	$6.17
2025	$20.00
2050	$45.00
2075	$79.65

Source: State Committee, 1994; National Bank, 1994; National Bank, 1995.

Table 8.6. Predicted Population Growth

Period	Growth (avg. annual %)
1990-2000	N/A
2000-2025	1.2
2025-2050	0.9
2050-2075	0.6
Year	**Population (millions)**
2000	16. 9
2025	22.8
2050	28.5
2075	33.1

Source: State Committee, 1994; National Bank, 1994; National Bank, 1995.

AGRICULTURE SECTOR

Methods Used

Selection of Study Sites

Six sites (Kotyrkol, Schuchinsk, Ruzaevka, Kostanai, Pavlodar, and Petropavlovsk) were chosen for simulation with the CERES-Wheat model. In addition, four oblasts (Kostanai, Western Kazakhstan, Semipalatinsk, and Almaty) were chosen for simulation with the KazNIGMI model (Figure 8.3). These sites and oblasts were selected as comprehensive representatives of Kazakhstan.

Crop Models Used in the Assessment

The CERES-Wheat model and the KazNIGMI dynamic and statistical grain yield model (Lebed and Belenkova, 1991) were used for the vulnerability assessment of spring and winter wheat yields in northern Kazakhstan.

The CERES-Wheat model takes into account both the direct effect of $2{\times}CO_2$ and the effects of climate change. The CERES-Wheat model simulations were conducted under three climate change scenarios: CCCM, GFD3, and the tenth decade GF01.

Figure 8.3. Kazakhstan's Study Sites

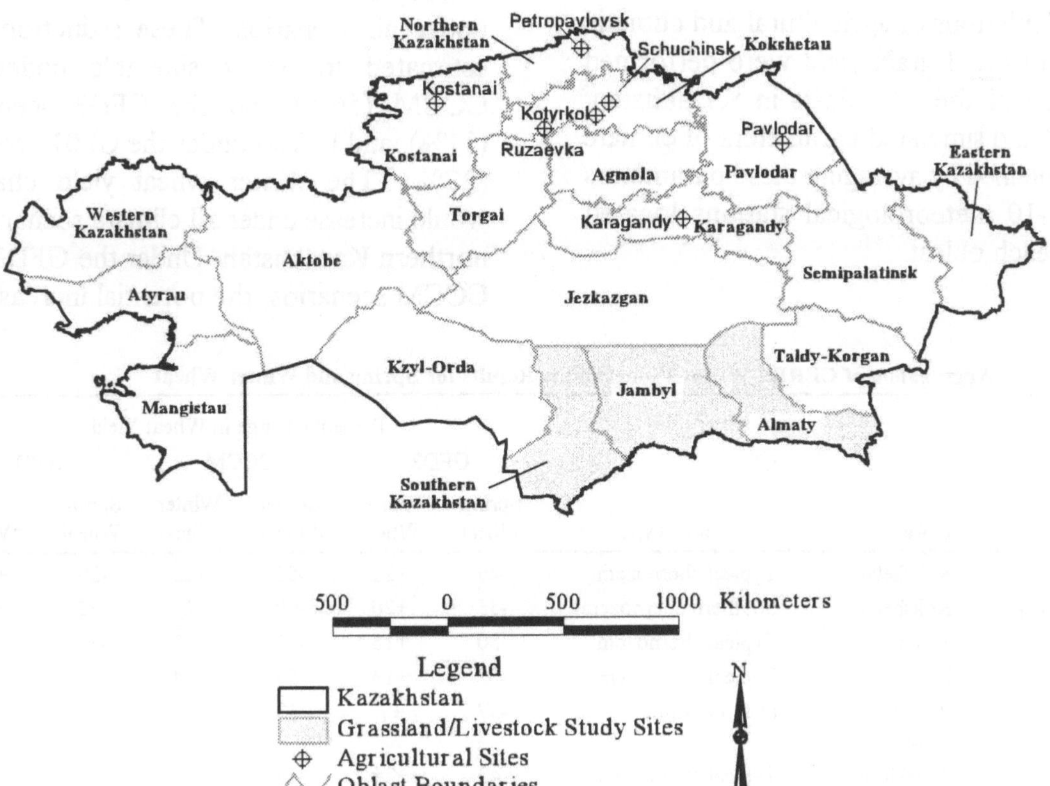

The analyses of temperature and precipitation distribution changes and their impacts on wheat yields were carried out at six study sites. The methodology used is similar to the work carried earlier by Menzhulin et al. (1994). The final results represent mean arithmetic values of estimates obtained for each site.

The dynamic and statistical grain yield KazNIGMI model developed by Polevoi and modified by Lebed and Belenkova (1995) was also applied. The simulations were conducted under the GFD3 climate change scenario and under nine incremental scenarios. This model does not take into account the direct effects of increasing atmospheric CO_2 on plant growth, transpiration, and yield, but only simulates climate change effects.

The calculations of agricultural and climatic conditions and grain yield were performed for each of the 19 oblasts in Kazakhstan. Current and simulated parameters of climate were obtained by averaging these parameters over 5-10 meteorological stations located within each oblast.

The baseline climate scenario represents the current climate for the base period 1951-1980. Other environmental and socio-economic scenarios have not yet been taken into account in the agricultural sector. The observed daily maximum and minimum temperatures and precipitation rate were obtained from the data bank of the Kazakh Scientific Research Hydrometeorological Institute (Antonov and Smirnova, 1995). The solar radiation data were obtained from Hydrometeorological Archives of Kazakhstan. The yield data were provided by the *Kazakhstan State Statistical Account for 1990* (1991).

Results

CERES-Wheat Model

Under the CERES-Wheat model, spring wheat yield would be reduced at all sites and under all scenarios. These reductions are estimated to be considerable under the CCCM (56%) and the GFD3 scenarios (51%), and less so under the GF01 scenario (27%). The winter wheat yield changes would increase under all climate scenarios in northern Kazakhstan. Under the GFD3 and CCCM scenarios, the potential increase

Table 8.7. Aggregation of CERES-Wheat Vulnerability Results for Spring and Winter Wheat

			Percent Change in Wheat Yield					
			GFD3		CCCM		GF01	
Site	Oblast	Soil Type	Spring Wheat	Winter Wheat	Spring Wheat	Winter Wheat	Spring Wheat	Winter Wheat
Kotyrkol	Kokshetau	Typical chernozem	-49	+21	-56	+25	-29	+32
Schuchinsk	Kokshetau	Southern chernozem	-53	+20	-60	+20	-32	+27
Ruzaevka	Kokshetau	Typical chernozem	-50	+16	-62	+21	-32	+25
Kostanai	Kostanai	Typical chernozem	-52	+13	-58	+24	-27	+20
Pavlodar	Pavlodar	Dark chestnut	-57	+7	-63	+10	-37	+15
Petropavlovsk	Northern Kazakhstan	Typical chernozem	-43	+25	-39	+1	-7	+9
Average			-51	+17	-56	+17	-27	+21

would be the same (17%); under the GF01 scenario, the yields would increase by slightly more (21%). The results of the assessment show that the potential vulnerability of spring wheat, which is the main crop of Kazakhstan, could be critical. Table 8.7 shows the CERES-Wheat vulnerability results for spring and winter wheat.

Table 8.8 shows the estimated change in spring wheat generated using incremental scenarios and the CERES-Wheat model.

KazNIGMI Model

The wheat vulnerability assessment using the KazNIGMI model looked at heat, moisture, light, an integrated complex function of these agroclimatic conditions, and wheat yield. The changes under the GFD3 climate scenario are shown in Table 8.9.

Under the GFD3 climate scenario, the KazNIGMI model estimates that spring and winter wheat yields would decrease by 44% to 51% and 12% to 35%, respectively, depending on the region. Because it does not consider the beneficial effect of elevated CO_2 concentrations, the KazNIGMI model overestimates the possible decrease of wheat production. This analysis, as well as analyses using the incremental scenarios, showed that a temperature rise of 3° to 4°C, even when precipitation increases by 25% to 30% above norm, could have a negative effect on spring wheat yield.

Table 8.8. Estimated Change in Spring Wheat Yields under Incremental Scenarios in West-Kazakhstan Oblast

Temp. Change (°C)	Precipitation Change for April to October (mm/day)	Precipitation Change for November to March (mm)	% Change in Yield
+0.5	+0.1	0	-35%
+0.5	+0.1	+10	-29%
+1.0	+0.2	+20	-32%
+2.0	+0.3	+30	-45%
+3.0	+0.4	+40	-28%
+0.5	-0.1	-10	-25%
+1.0	-0.2	-20	-36%
+2.0	-0.3	-30	-65%
+3.0	-0.4	-40	-70%

Table 8.9. Aggregation of KazNIGMI Model Wheat Vulnerability Results under GFD3 Scenario

Wheat	Oblast	Percent of Baseline Agroclimatic Conditions				Estimated Yield (t/ha)	Percent Change in Yield
		Light	Heat	Moisture	ICF[1]		
Spring Wheat	Western Kazakhstan	103	66	81	56	0.46	-44%
	Kostanay	101	72	87	53	0.50	-47%
	Semipalatinsk	99	62	85	52	0.42	-51%
Winter Wheat	Western Kazakhstan	106	77	85	65	0.78	-35%
	Semipalatinsk	99	80	94	76	0.65	-24%
	Almaty	99	86	103	88	1.70	-12%

[1] Integrated complex function of the agroclimatic conditions.

Adaptation Alternatives

The following adaptation alternatives are recommended:

- *Change planting dates.* North Kazakhstan typically has moderate to dry climate conditions in July and therefore precipitation is critical to spring wheat. Under current conditions, planting dates are selected so that the earning phase coincides with the time of maximum precipitation. Under $2\times CO_2$ climate change scenarios, crops should be planted 15 to 20 days earlier.

- *Switch from spring wheat to winter wheat.* Under all climate scenarios, the CERES-Wheat model estimates an increase in winter wheat yield and a decrease in spring wheat yield, so some portion of the cropped area should be switched from spring wheat to winter wheat. Spring wheat is planted in most of the cultivated area in North Kazakhstan (99.83% in 1994). We calculated that 70%, 68%, and 35% of the area should be switched to winter wheat under the CCCM, GFD3, and GF01 scenarios, respectively.

- *Irrigate.* This measure would be very significant for increasing wheat yield. But in north and central Kazakhstan it may be too expensive. KazNIGMI hydrologists estimate that under the GF01 scenario, water resources will be the same as under current conditions. Under the GFD3 and CCCM scenarios, they could be reduced by 20% to 30%. Therefore, irrigation costs would be high under the potential climate change.

The following alternatives could improve agriculture yields and could thereby decrease vulnerability to potential climate change:

- *Increase fallow area.* In North Kazakhstan, increasing the amount of fallowed land of various soil types could increase wheat yield by 0.5 to 0.6 tons per hectare (Abugaliev, 1985).

- *Implement snow reserving.* This is a method of accumulating snow on arable lands to increase soil moisture and keep the soil warm for winter plants. This measure could increase wheat yield by about 0.3 tons per hectare (Abugaliev, 1985). Snow reserving is not a new tillage practice in Kazakhstan, but it has not been used since 1993 because of a lack of funding.

- *Switch to more suitable wheat varieties.* Use of more productive, drought-resistant, earlier maturing, and disease- and pest-tolerant varieties of wheat, could increase wheat yields.

- *Apply fertilizers, pesticides, and weed control.* These measures could increase the wheat yield when combined with proper scheduling of these measures and crop rotation.

- *Apply zonal growing technology.* The technologies used in agriculture should be selected to correspond with the climate conditions of different natural zones.A preliminary economic analysis using recent data on the economic development of agriculture and industry in Kazakhstan (Conceptual Program, 1994) is presented in Table 8.10.

Table 8.10. Characteristics of Selected Intensive Adaptive Measures

Measure	Cost (US$/ha)	Yield Increase (t/ha)	Net Benefits[1] (US$/ha)	Possibility of Use
Change planting dates	—	0.10[2] 0.02[3]		Possible
Switch from spring wheat to winter wheat	—	—	—	Possible
Irrigate	—	0.43	—	None
Increase fallow area	15.62	0.50-0.60[4]	-15.2	Possible
Implement snow reserving	—	0.30[4]	—	Possible, with improved economic conditions
Switch to more suitable wheat varieties	1.19	0.12	+16.63	Possible, with improved economic conditions
Apply fertilizers, pesticides, and weed control	25.00	0.21-0.25	(+8) to (+13)	Possible, with improved economic conditions
Incorporate zonal growing technology	70.10	0.098	+8.18	Possible, with improved economic conditions

1 Calculated as the difference between benefits per hectare and costs per hectare.
2 CCCM, GFD3.
3 GF01.
4 Abugaliev, 1985; Conceptual Program, 1994.

Discussion and Conclusions

Both the CERES-Wheat and the KazNIGMI models show that a potential warming could have a significant detrimental effect on spring wheat yield in Kazakhstan. The estimated decreases of spring wheat yields vary widely, from 62% to 7% under all scenarios. At the same time, the results for winter wheat indicate that crop yields would increase from 1% to 32%. Given that arable land for winter wheat is less than 1% of the total arable land, this increase is not significant, however, switching from spring wheat to winter wheat could be an adaptation alternative.

The adaptive strategies recommended are likely to be more preferable for Kazakhstan than purchasing wheat abroad. This assessment may allow one to draw inferences about the possibilities of adaptation to climate change, but these estimates are preliminary and the continuation of this work is needed.

Grassland/ Livestock Sector

Methods Used

Selection of Study Sites

The Priaralie region was used for the assessment of grassland vulnerability. The region may be divided into North Priaralie (Aktobe and Kzyl-Orda oblasts) and South Priaralie (Almaty and Taldy-Korgan oblasts). Three large sheep-breeding oblasts in the south and southeast of Kazakhstan (South Kazakhstan, Jambyl, and Almaty oblasts) were selected for this assessment (Figure 8.3). Almost one third of the sheep in Kazakhstan are contained in these oblasts. These regions were also chosen because sheep are kept on pastures year round; the viability of sheep and their productivity largely depend on the climate.

Models Used in the Assessment

To estimate possible changes in grasslands under 2×CO_2, the climate change parameters of the GFD3, CCCM, and GF01 scenarios were used. Incremental scenarios were used as well. These climate change scenarios were used with a nonstationary dynamic model of seasonal grassland productivity (Lebed and Belenkova, 1995) to calculate the total grassland productivity for each 10 to 30 days of the growing season and the grassland biomass eaten by livestock in the season. This model is applied to 105 ha areas. The yield calculations take into account the species and structure of grassland vegetation in each area. Temperature and precipitation values were interpolated to these areas by the optimal techniques (Gandin and Kagan, 1976).

We used observed data of meteorological conditions as well as expedition and field station data obtained during airphotometric measurements of grasslands in Kazakhstan in 1989-1991.

Empirical methodologies developed by KazNIGMI were used to estimate changes in summer grazing conditions and in sheep productivity (wool and lambing). In the hot climate of south and southeast Kazakhstan, summer conditions have a critical effect on sheep productivity: sheep lose weight during hot weather (Chekeres, 1973). Using criteria developed by KazNIGMI, we determined the effects of a potential change in the duration of hot weather on sheep grazing. To estimate the vulnerability of sheep productivity, a statistical model was used to look at productivity of grassland vegetation, winter period duration of grassland fodder shortage, and meteorological conditions during shearing, sheep driving, and lambing.

Data from agrometeorological and meteorological stations and the State Committee on Statistical Analysis of the Republic of Kazakhstan were used in the calculations. The statistical data contained information about sheep productivity by region and oblast. The data on grassland vulnerability to potential climate change were obtained from Lebed and Belenkova (1995). Meteorological conditions during winter and summer grazing, shearing, lambing, and driving were estimated using biometeorological criteria developed by KazNIGMI.

Results

Grasslands

Table 8.11 shows the estimates of grassland vulnerability under the GCM climate scenarios for the Priaralie region.

Under the GF01 and CCCM scenarios, grassland productivity in the Priaralie region is estimated to increase 18% to 125% in the spring, mostly due to increased precipitation. Under the GFD3 scenario, only the productivity of such species as *Anabasis salsae* is estimated to increase significantly (25%) in the spring, and biomass of other species would be reduced by 6% to 18% in the spring. Under CCCM scenarios, summer and fall grassland productivity of all considered species could decrease markedly, up to 88%. According to GFD3 and GF01 scenarios, grassland biomass suffers a little in the fall and increases in the summer.

Under the incremental climate scenarios a temperature increase of 2°C and a precipitation increase of 0.2 to 0.3 mm/day is

Table 8.11. Change in Grassland Productivity under GCM Climate Change Scenarios in the Priaralie Region

	Percent Change in Productivity by Species and Season								
	Artemisia terrae albae, etc.			*Anabasis salsae*, etc.			*Artemisia arenaria, Agropirum sibiricum, Calligonum*, etc.		
Scenario	Spring	Summer	Fall	Spring	Summer	Fall	Spring	Summer	Fall
GFD3	-18	+21	-11	+25	+3	-5	-6	-11	+11
GF01	+18	-3	-7	+37	+12	+8	+18	-6	-14
CCCM	+72	-28	-25	+62	-82	-88	+125	0	-14

estimated to increase gross grassland productivity by 20% to 55%. With a 3°C increase, the rise in grassland productivity would be less. The productivity of such species as *Anabasis salsae* and *Holoxylon aphyllum* could be reduced. Under a temperature increase of 0.5°C and reduced precipitation, grassland productivity could be reduced by 6% to 32%. Under increased temperatures of 2° to 3°C and reduced precipitation, grassland productivity could be reduced by 40% to 90%.

Livestock

Table 8.12 shows the results calculated for the duration of hot weather and sheep productivity under the different climate scenarios in the three oblasts studied. Increased temperatures would bring about periods of hot weather 20 days and 30 days earlier than baseline under the GFD3 and the CCCM scenarios, respectively. The duration of these periods would be increased from between 27 and 37 days depending on the region under the GFD3 scenario and from between 37 and 57 days depending on the region under the CCCM scenario. The most unfavorable conditions for sheep-breeding would be under the CCCM scenario: wool productivity would decrease 11% to 19% and the number of lambs would be 2% to 26% fewer than under baseline conditions. Less unfavorable conditions can be observed under the GFD3 scenario.

Adaptation Alternatives

As a result of the grassland/livestock vulnerability assessment, the following adaptation measures will be considered:

- gradually change the periods of spring lambing, sheep driving, and shearing to earlier dates (by 15 to 20 days)
- improve the natural vegetation cover (grasslands) through altering grazing management, reducing undesirable plants, fertilizing, and seeding
- breed only coarse-wool sheep, which are best adapted to hot weather conditions on semi-desert and desert summer grasslands
- intensify veterinary measures to protect sheep, especially during drives
- determine maximal output from mountain pastures to maintain plant community structure
- develop higher summer mountain grasslands to take advantage of an estimated

Table 8.12. Livestock Results by Scenario

Oblast	Scenario	Duration of Hot Weather (days)	Wool Productivity (kg/head)	Lambs Dropped (per 100)
Almaty	Baseline	85	2.7	92
	GFD3	120	2.5	82
	CCCM	132	2.4	87
	GF01	125	—	—
Jambyl	Baseline	90	3.0	92.1
	GFD3	117	2.7	87
	CCCM	127	2.5	90
	GF01	121	—	—
Southern Kazakhstan	Baseline	105	2.6	102
	GFD3	142	2.2	81
	CCCM	162	2.1	75
	GF01	150	—	—

shift in vertical mountain zones due to climate change.

Discussion and Conclusions

The results of the vulnerability and adaptation assessment on grasslands are preliminary and there has not yet been an analysis of adaptation alternatives. However, preliminary results indicate that grassland productivity in the Priaralie region would suffer considerably in the summer under a CCCM climate scenario.

In the south and southeast, estimated climate change under the CCCM scenario could decrease wool productivity by 11% to 19% and decrease the number of lambs by 2% to 26% from baseline conditions. Less unfavorable conditions can be observed under the GFD3 scenario. The region of heat-resistant Astrakhan sheep breeding might be extended to the north, because climate conditions of South Pribalhashie (Almaty and Taldy-Korgan oblasts) may be similar to current conditions in the south of Kazakhstan under $2\times CO_2$ conditions. The adaptation measures offered for sheep breeding are to shift the timing of lambing and other breeding activities by 15 to 20 days earlier, to change grazing systems, and to improve grasslands.

WATER RESOURCES SECTOR

Methods Used

Selection of Study Sites

Assessments of water resources vulnerability to climate change were completed for two watersheds: the Tobol River basin on the plains and the Uba and Ulba watershed in the mountainous area. It should be noted, however, that within Kazakhstan there are a great variety of runoff formation conditions. In particular, this study does not cover watersheds with glacial feed.

The Tobol River basin is located in northwestern Kazakhstan (Figure 8.1). Its watershed area is approximately 13,400 km^2. There are many ponds, small water storages, and artificial reservoirs in the basin, and the largest artificial reservoirs are Bredinskoye and Zhelkuarskoe. The current climate in the Tobol River basin is sharply continental: a hot, dry summer, and a cold but not snowy winter. Temperatures have much variation;

minimum and maximum temperatures can be -40°C and +40°C, respectively, and the average yearly temperature usually falls between 2.2°C and 2.8°C. Average annual precipitation varies from 250 to 400 mm. Between 70% and 85% of the precipitation falls during the warm period of the year, from April to October. The wave of the spring flood, which lasts about a month on average, usually stands out sharply in the water regime of the Tobol River; however, during some years the floods are low.

The Uba and Ulba watershed, located in the Eastern Kazakhstan oblast (Figure 8.3), has an area of 1,484 km^2. The present climate in the Uba and Ulba basin is also sharply continental. In winter the temperature can drop as low as -48°C. In July the monthly average temperature in the foothills is about 20°C. The amount of precipitation varies depending on altitude; at elevations of 200 m and 1,600 m it amounts to 410 mm and 1,220 mm, respectively. Most of the precipitation (up to 70% of the annual value) falls during the warm periods of the year. About 60% of rain adds to the melt water to form the spring floods, which last three to four months and include 60% to 70% of the yearly runoff.

Hydrological Methodology

The KazNIGMI model, developed by KazNIGMI under the direction of Golubtsov (Golubtsov, 1975; Golubtsov et al., 1989), was used in the assessment. The model represents a river basin composed of three successively connected regulating reservoirs, located one above another and identified with the surface, soil-ground, and ground runoff formations within the hydrographic network. Through the use of several special submodels, the model describes particular elementary water balance processes: snow storage formation and water supply to the basin surface; soil moisture content fluctuations; frost-bounding and melting of soil-ground; total evapotranspiration; surface, soil-ground, and ground runoff formations; influx formation to the riverbed network; and runoff at the closure site. A peculiarity of the KazNIGMI model is that, because it is based on limited initial data, it allows simulation of runoff from any height, including parts of the basin with different landscapes and slopes of different exposures. The model parameters are assumed to vary depending on the site elevation, inclination and exposure of the mountain slopes, and characteristics of the landscape (open, forest cover, glacial, etc.).

Improvements to the model to enhance its capability for river basins in the plains have been carried out within the framework of the present vulnerability and adaptation study. To accomplish this, supplements were introduced into the snow storage formation submodel and other submodels. These modifications allow simulation of the runoff formation process, taking into account not only the altitude zones of physical and geographical conditions, but their changes in latitudinal and longitudinal directions as well.

In addition, the model has been adapted to evaluate water resources vulnerability under potential climate change scenarios. All baseline data came from KazNIGMI. The impact of climate change on water resources was assessed using GFD3 and CCCM under 2×CO_2 conditions. Two more intermediate CO_2 concentration levels were assessed using GF01 scenarios.

Results

The water resources vulnerability assessment shows reductions in runoff in the rivers of Kazakhstan under various climate scenarios. According to the climate change scenarios of CCCM and GFD3, reductions in runoff can amount to 30% to 35% and 20% to 25%, in years of water shortages and in years of excess water, respectively. The GF01 model confirms the tendency of water resources reduction during years of plentiful water supply on average for mountainous watersheds.

The estimated changes in precipitation, temperature, and runoff for the Tobol River Basin under different climate change scenarios are shown in Table 8.13.

The estimated changes in precipitation, temperature, and runoff for the Uba and Ulba basin under different climate change scenarios are shown in Table 8.14.

The KazNIGMI model was also used to estimate total water supply for the selected study sites in 2075 (Table 8.15). Under all climate change scenarios, supply in the Tobol River basin decreases from baseline in 2075. Water supply in the Uba and Ulba watershed increases under GF01 S1 and S2, and decreases under GF01 S3, GFD3, and CCCM.

Adaptation Alternatives

The probability of a disturbance in the biological balance of ecosystems and an initiation of critical ecological situations increases with a decrease in water resources. To solve ecological problems it is necessary to address the whole pattern of plant and animal life, partial migration of the community, chemical and biological water treatment, creation of favorable water and temperature conditions for habitation, reproduction of the most valuable living organisms (for example, fish), and water deficit compensation. Under these circumstances ecological expertise should be applied to new construction projects and enhanced control of the water resources condition should be compulsory.

Development of additional water resources is also of great importance in water management: land irrigation, pasture watering, water delivery to residential areas, etc. There are two sources of such resources that could be the most essential: (1) diversion of the portion of runoff of the Volga and Siberian rivers, and (2) groundwater, which has not yet been widely utilized in Kazakhstan.

Table 8.13. Hydrologic Vulnerability Assessment for Tobol River Basin

Scenario	Change in Precipitation (%)	Change in Temperature (°C)	Change in Runoff (%)
GF01 S1	+4%	+1.2	-16%
GF01 S2	+9%	+2.9	-22%
GF01 S3	+13%	+4.5	-29%
GFD3	+12%	+5.7	-26%
CCCM	+21%	+6.7	-24%

Table 8.14. Hydrologic Vulnerability Assessment for Uba/Ulba Watershed

Scenario	Change in Precipitation (%)	Change in Temperature (°C)	Change in Runoff (%)
GF01 S1	+21%	+1.3	+12%
GF01 S2	+22%	+3.2	+6%
GF01 S3	+18%	+5.2	-1%
GFD3	-8%	+4.7	-29%
CCCM	0%	+6.6	-23%

Table 8.15. Water Resource Supply in 2075 (km^3/yr) Estimated Using the KazNIGMI Model

Total Water Supply	Baseline	GF01 S1	GF01 S2	GF01 S3	GFD3	CCCM
Tobol River Basin	0.257	0.216	0.200	0.183	0.190	0.197
Uba/Ulba Watershed	8.889	9.974	9.447	8.782	6.290	6.876

Priority should be given to groundwater since the diversion of runoff involves not only high costs and technical difficulties, but is also complicated by intergovernmental problems between Russia and Kazakhstan and ecological problems inherent to Russia. The resolution of issues concerning water management relations with Kyrgyzstan, Uzbekistan, and China on a mutually beneficial basis is also important. Taken together, these activities could partially reduce the anthropogenic load on surface resources and promote further water management development.

The success of adaptation measures will largely depend on operative actions of managers of different ranks. The conditions necessary to adopt proper and timely decisions regarding adaptation in the water resources sector include legislative acts and interstate agreements regulating water management relations, the reliability and early provision of hydrological forecasts, the availability of well-grounded scientific recommendations and models. These conditions have not yet been completely met in Kazakhstan.

Discussion and Conclusions

Many of the above-mentioned problems are integrated with the impacts of economic activity on water resources. Unfortunately, investigations of some problems are being carried out slowly, while the work on other problems has been suspended as a result of the difficult economic situation in Kazakhstan. The complexity and the number of problems are so great that much time, expenditure, and effort will be needed to adapt water resources to changes in the watershed hydrological regime.

To evaluate adaptation measures in detail it will be necessary to create a system to simulate different situations and to identify the most relevant alternatives for water resources management. However, even without such detailed analyses it is obvious that priority should be given to measures related to water conservation and environmental protection. Such measures include implementation of technologies that conserve water, the reconstruction of irrigation systems, and observance of sanitary norms. Conserved water can be used to maintain normal ecological conditions, to develop economic activities and, hence, to improve the standard of living for the population. In addition, water conservation techniques can solve a number of collateral problems by diminishing secondary salinization of soil and reducing the possible pollution of surface and ground water resources.

- According to assessments of KazNIGMI climatologists, climate change resulting from emissions of GHGs in the atmosphere could reduce water resources in Kazakhstan by 20% to 30%.
- The modeled runoff distribution shows a significant reduction in runoff during the flooding period and an increase in runoff during the autumn/winter period.

Therefore the probability of high flooding should decline.

- The most radical adaptation measure considered is the diversion of the portion of runoff to the Republic of Kazakhstan from the territory of the Russian Federation.

- Priority should be given to water conservation and environmental protection measures.

- Creation of simulation systems should be considered to provide comprehensive assessments and to manage water resources.

This study does not cover the whole territory of Kazakhstan, which has a great variety of runoff formation conditions. Thus, it is assumed that water resources throughout the territory of Kazakhstan will be vulnerable. However, a well-defined solution will require water resources vulnerability and adaptation research using many more study sites.

SUMMARY CONCLUSIONS

Results of the Kazakhstan vulnerability and adaptation assessment include regional baseline and climate change scenarios, an analysis of the vulnerability of winter and spring wheat, grasslands, livestock (sheep-breading) and water resources, and recommended adaptation measures in these sectors. To develop climate scenarios we used two equilibrium versions of GCMs (GFD3 and CCCM) and a transient version (GF01).

We intend to expand our study to include regional interpretations of scenarios from the GISS and UKMO models. We also plan to develop economic analyses of adaptation measures.

REFERENCES

Abugaliev, I.A. (ed.). 1985. *Reference Book of Agronomist*. Alma-Ata, Republic of Kazakhstan: Kainar, 318 pp. (in Russian).

Antonov, A.G. and E.Yu. Smirnova. 1995. General Principles of Organization of the Data Bank "Climate." *Hydrometeorology and Ecology*. No. 2. Almaty, p. 43-53 (in Russian).

Benioff, R., S. Guill, and J. Lee (eds.) (In press). *Vulnerability and Adaptation Assessments: An International Guidebook.* Dordrecht, The Netherlands: Kluwer Academic Publishers.

Budyko, M.I., A.B. Ronov, and A.L. Inushin. 1985. History of the Atmosphere. Leningrad, Hydrometeoizdat, 209 pp. (in Russian).

Budyko, M.I., M.C. MacCracken, A.D. Hecht, Yu.A. Israel (eds.). 1991. Forthcoming Climate Change. Joint Soviet-American Report on Climate Change and its Consequences: Leningrad USSR, 273 pp. (in Russian).

Chekeres, A.I. 1973. *Weather Climate and Pasture Livestock Farming.* Hydrometeoizdat., 175 pp. (in Russian).

IPCC Impacts Assessment Working Group. 1992. *Climate Change 1992.* The Supplementary Report to the IPCC Impact Assessment. The 1992 Report of the IPCC Impacts Assessment Working Group. Australian Government Publishing Service. Canberra.

Conceptual Program. 1994. Conceptual Program of Agriculture and Industry Complex Development of the Republic of Kazakhstan for the 1993-1995 period and till 2000. Kazakhstan, Almaty: Bastau, 314 pp. (in Russian).

Gandin, L.S. and R.L. Kagan. 1976. "Statistical Methods of Interpretation of Meteorological Data." Leningrad, Gidrometeoizdat.

Golubtsov, V.V. 1975. On the Construction of the Mathematical Model for the Runoff Formation in the Mountain Watershed. *KazNIGMI Proceedings*. No. 48 (in Russian).

Golubtsov, V.V., V.I. Lee, and T.P. Stroeva. 1989. The Mathematical Simulation of Formation Processes of Mountain River Runoff in Condition of Limited Information. *Proceedings of the 5th All-Union Hydrological Congress.*

Gruza, G.V. and E.Ya. Rankova. 1991. Probabilistic Forecast of Global Surface Air Temperature up 2005. *Meteorol. Hydrolog.*, 4, 95-103 (in Russian).

Kazakhstan State Statistical Account for 1990. 1991: Kazinform, Alma-Ata (in Russian).

Lebed, L.V. and Z.S. Belenkova. 1991. Agrometeorological Forecast and Computations of Grain Yield in Kazakhstan at Present Economy Conditions. KazNIGMI Proceedings, No. 110, pp. 115-121 (in Russian).

Lebed, L.V. and Z.S. Belenkova. 1995. The Basics of the Nonstationary Model of Seasonal Productivity of Natural Grasslands. Refereed Journal *Deposit Research Works.* Issue 1, 5892-K95-Almaty. KazGosINTI (in Russian).

Menzhulin, G.V., L.A. Koval, and A.L. Badenko. 1994. Potential Effects of Global Warming and Carbon Dioxide on Wheat Production in the Former Soviet Union. In *Implication of Climate Change for International Agriculture: Crop Modeling Study.* U.S. EPA, Policy Planning and Evaluation (2122). EPA 230-B-94-003. FSU pp. 1-36 (in Russian).

National Bank. 1994. Economical Review, 1994. National Bank of the Republic of Kazakhstan. Research Perspective Center, Issue 3, December.

National Bank. 1995. Economical Review, 1995. National Bank of the Republic of Kazakhstan. Research Perspective Center, Issue 1, March.

Pilifosova, O.V. 1992. Probabilistic Climate Scenarios up to 2000-2005 in the Kazakhstan — Middle Asia Region. *Proceedings of KazNIGMI,* 111, pp. 64-72 (in Russian).

State Committee. 1994. Kazakhstan's Statistical Annual Reference Book for the Year 1993. Almaty, Kazinformcenter. State Committee of the Republic of Kazakhstan on Statistics and Analysis.

Chapter 9

Vulnerability and Adaptation Assessments for Mongolia

P. Batima,[1] Sh. Bayasgalan,[2] B. Bolortsetseg[2]
D. Dagvadorj,[2] R. Mijiddorj,[3] L. Natsagdorj[2]

[1] Water Policy Institute
[2] Hydrometeorological Research Institute
[3] Climate and Environmental Changes Research Center

SUMMARY. The results of the vulnerability assessments for Mongolia show that climate change could have a significant impact, in particular on the agricultural, forestry, and water resources sectors. These findings suggest that rising temperatures and reduced precipitation in some regions could increase the amount of arid and semi-arid areas, resulting in increased yields of forage crops and reduced or unchanged yields of grain and seed crops. The severity and frequency of drought may increase in the Gobi Desert area with climate change. This could lead to a greater use of drought-tolerant crops, increased irrigation, or changes in agricultural activity. A warmer climate may also lead to some expansion of agricultural lands, but the nation's soil types would limit expansion. Mongolia's natural zones (life zones) were established using a Holdridge Life Zone Classification Model simulation and the 2×CO_2 climate change scenarios. According to these scenarios, the steppe zone's characteristics will undergo significant change and the zone's boundaries will be shifted. The water resources assessment showed that basin runoff is much more sensitive to temperature and precipitation for the rivers in the northwestern (desert) part of the country than other regions and relatively more sensitive in the steppe region than the mountainous region.

INTRODUCTION

Mongolia is a landlocked country located in Central Asia, bounded by the Russian Federation to the north and China to the east, west, and south (Figure 9.1). Its vast but very sparsely populated territory totals 1,566,500 km^2; its land mass is larger than that of Germany, France, Spain, Belgium, and the Netherlands together. Mongolia's terrain varies considerably in structure of relief and height above sea level; on average, it is 1,580 m above sea level, with its highest elevation at 4,362 m and its lowest at 552 m. The country has three main topographical zones: mountains, mountain basins, and vast

Note: This chapter does not contain a detailed description of the assessment methodologies that were recommended by the U.S. Country Studies Program. Refer to Chapter 2 for an overview of these methodologies.

J. B. Smith et al. (eds.), Vulnerability and Adaptation to Climate Change, 183–206.

Figure 9.1. Map of Mongolia

steppes and semi-desert in the eastern and southern parts of the country.

Such physiographic features give Mongolia a dry continental climate with sharply defined seasons, high annual and diurnal temperature fluctuations, low precipitation, long winters (October to April), and a short growing season (mid-May to mid-September). The frequently unfavorable weather conditions in Mongolia have adverse effects on the national economy.

Mongolia has abundant agricultural land, but intensive crop cultivation is limited by its short growing season and sharply fluctuating temperatures. The country is also rich in mineral resources, including coal, iron, tin, copper, gold, silver, tungsten, zinc, feldspar, and molybdenum, as well as semi-precious stones.

Thermal power, produced from indigenous coal, imported diesel oil, and traditional biomass fuels, is supplemented by electricity from the Russian grid. There is potential for hydro- and wind-generated electricity, which is being investigated, and petroleum exploration is beginning in cooperation with foreign oil companies.

Mongolia's sustainable development depends on close links between the environment and the economy; a significant portion of its economic activity has always centered on natural resources such as agriculture, animal husbandry, mining, and fossil fuel production. One of the reasons Mongolia strongly supports the UN Framework Convention on Climate Change (FCCC), which was ratified by the Great Hural (Parliament) of Mongolia on September 30, 1993, is the possibility that climate change will have a significant negative impact on the country's environment and agriculture.

To implement its FCCC commitments, Mongolia has undertaken a wide range of activities at the national and international levels. These activities focus on developing a national inventory of greenhouse gas sources and sinks for Mongolia, conducting vulnerability assessments, and developing greenhouse gas abatement strategies, concentrating on the power sector. Some scientists suggest that the agriculture, water resources, and forestry sectors may be particularly vulnerable to potential climate change due to Mongolia's physiographic and economic structure, and its climate conditions. In addition, because these sectors play a key role in the country's economic development, they were the focus of this assessment. The location of meteorological stations, agriculture study sites, and the river basins studied are shown in Figure 9.2. A number of major limitations in our knowledge make the findings from this study preliminary and tentative. To obtain more complete and detailed results, particularly for the economic analyses, Mongolia will continue its vulnerability and adaptation assessments for the above-mentioned sectors. Vulnerability and adaptation assessments for the grasslands/livestock sector and soil erosion are still being developed. This chapter describes the scenarios used and presents the methods and results for the following sectors:

- agriculture
- forestry
- water resources.

SCENARIOS

This study examined the general circulation model (GCM) climate change scenarios, a scenario of historical climate change trends in Mongolia, and incremental scenarios. This study also recognizes the importance of integrating environmental considerations into the country's overall economic and social development. However, baseline socioeconomic scenarios have not yet been incorporated into the assessments.

GCM Scenarios

The United Kingdom Meteorological Office (UK89), Oregon State University (OSU), Goddard Institute for Space Studies (GISS), Geophysical Fluid Dynamics Laboratory (GFD3), and the Canadian Climate Centre Model (CCCM) GCMs were used for the vulnerability and adaptation assessments for Mongolia (Benioff et al., in press). It should be noted that the GCM outputs often fail to reproduce the seasonal pattern of climate on a regional scale. This naturally casts some doubt on the ability of GCMs to provide accurate estimates of future regional climate, and consequently on the accuracy of assessments of climate change impacts. Figures 9.3 and 9.4 show how the $1\times CO_2$ scenario from the UK89 model compares to observed current climate (National Standard of Mongolia, 1993).

Estimated temperature increases under $2\times CO_2$ conditions obtained from the GISS model generally fall in the range 2.9°C to 13.4°C for all months. Thus, GCM outputs should be treat The UK89 model suggests that summer month temperatures in

Figure 9.2. Study Sites for Mongolia's Assessment

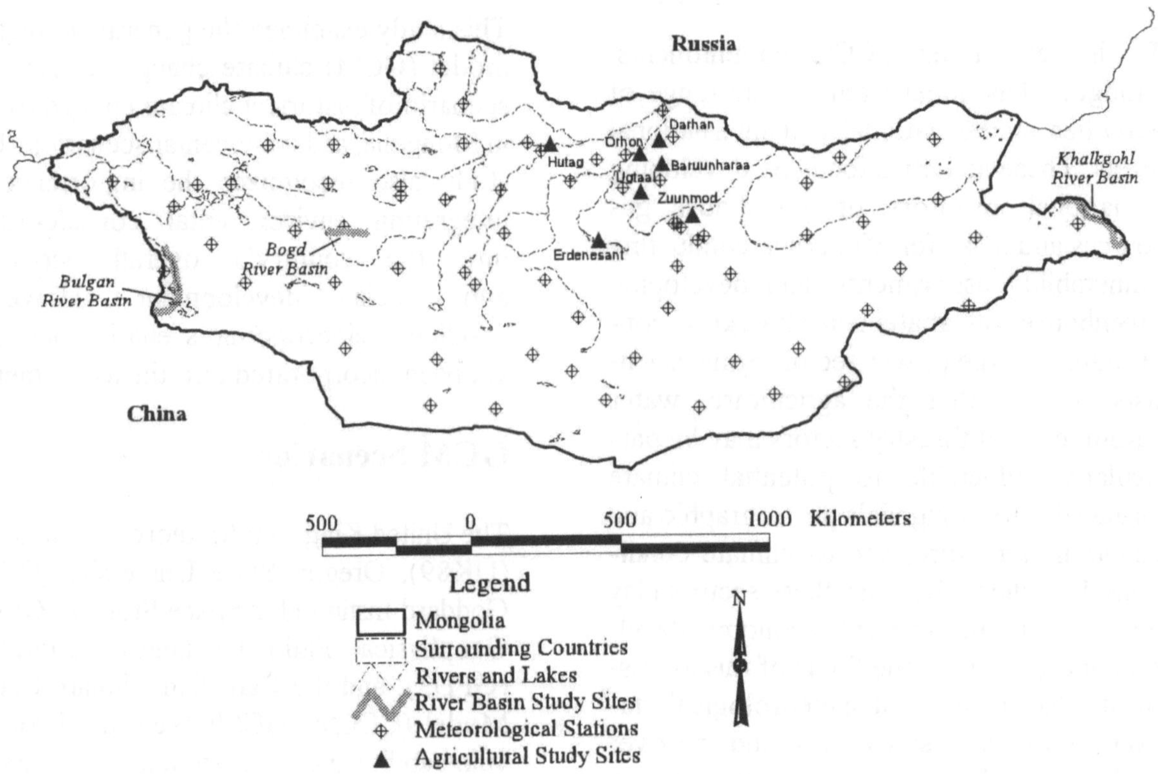

Mongolia overall might increase, which differs from the actual current trend. ed, at best, as broad-scale sets of future climatic conditions and their effects, and should not be regarded as accurate predictions.

Figures 9.5 and 9.6 reproduce UK89 model results under a 2×CO_2 scenario and compare them with the current mean monthly temperature and mean daily precipitation for all territories of Mongolia. The figures suggest that the monthly air temperatures in both winter and summer may warm by 4°C to 5°C. It is also estimated that precipitation may decrease from June to September and increase in the cold months.

Historical Trend Scenario

The historical trend scenario was used in the vulnerability assessment for the agriculture sector. Non-linear models and methods for time series analysis were used to estimate climate change trends in Mongolia (Natsagdorj and Namkhai, 1993).The historical climate data for this analysis were obtained from the National Board of Standards and Meteorology (National Standard of Mongolia, 1993).

A recent study shows that Mongolia is undergoing a change in climate (Dagvadorj et al., 1994). In general, the country's annual mean air temperature increased by 0.7°C for

Figure 9.3. Comparison of Monthly Temperatures under a UK89 1×CO_2 Scenario with Current Observations

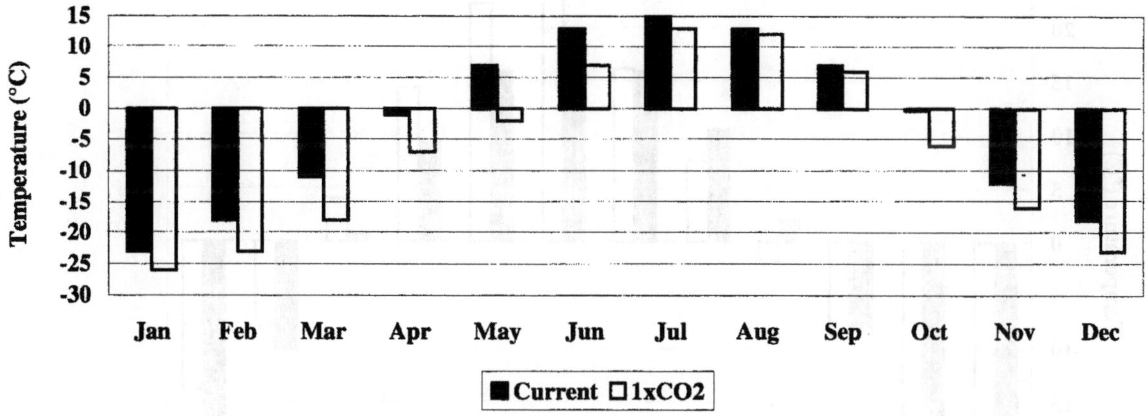

Figure 9.4. Comparison of Daily Precipitation under a UK89 1×CO_2 Scenario with Current Observations

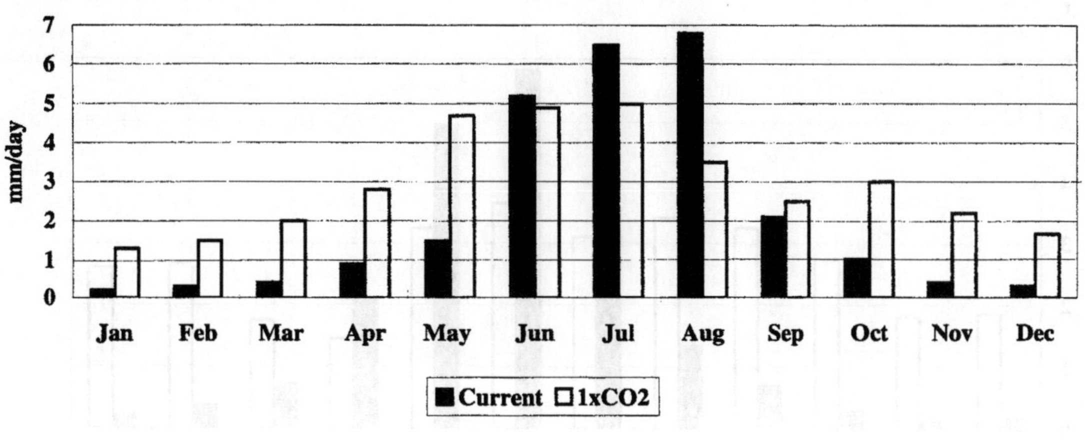

Figure 9.5. Comparison of Monthly Temperatures under a UK89 2×CO_2 Scenario with Current Observations

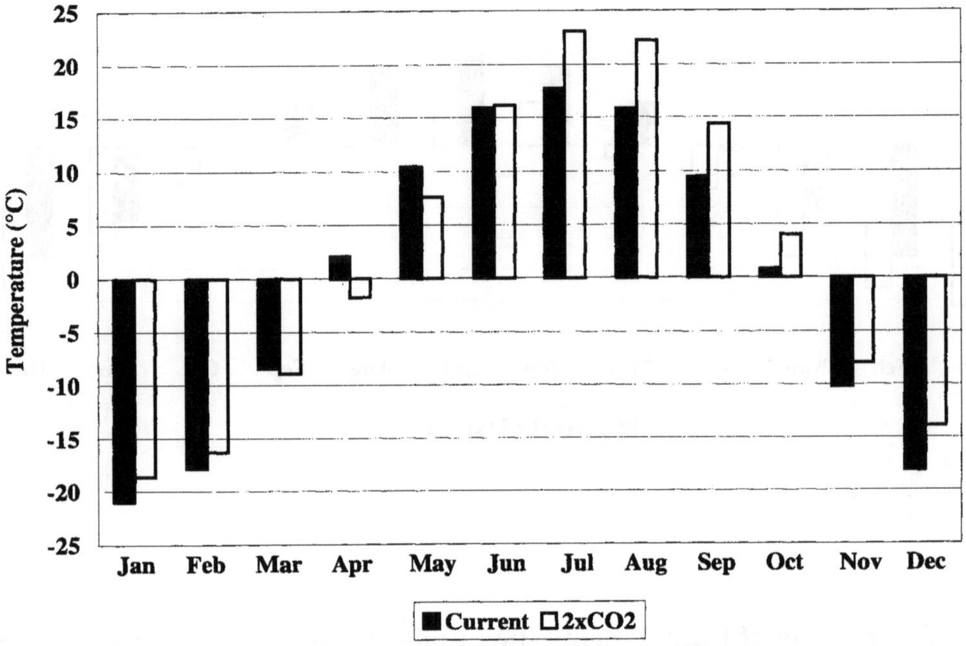

Figure 9.6. Comparison of Daily Precipitation under a UK89 2×CO_2 Scenario with Current Observations

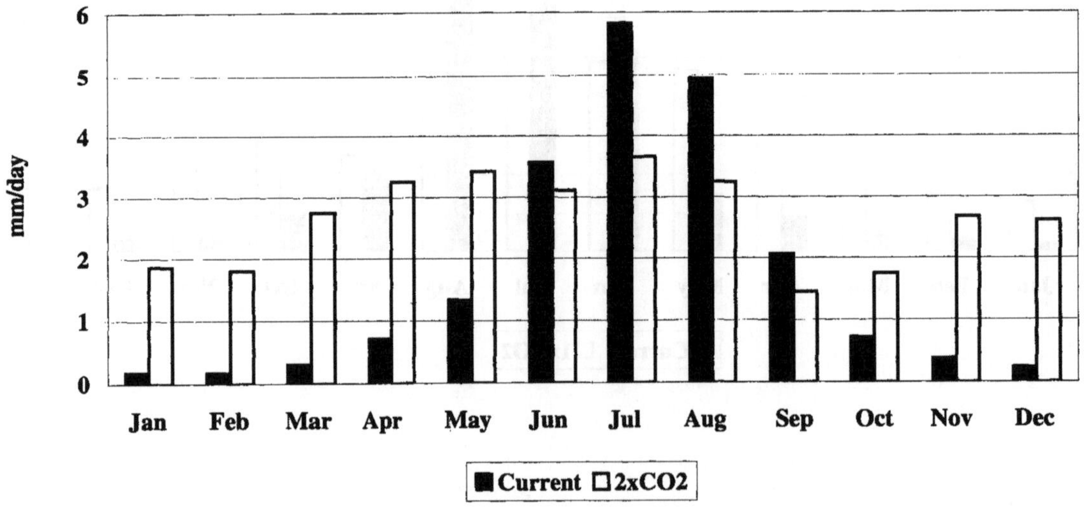

1940 to 1993. The intensity of this warming trend differs among Mongolia's natural and climatic areas. Temperature changes also varies by season, with increases during the winter and decreases during the summer.

Monthly climate change trends were estimated using quadratic polynomial equations for 1940-1994 and trend values were calculated for each month. The main peculiarity of these trends is a temperature decrease of -0.28°C to -0.67°C for June, July, and August, and precipitation increases of 28% and 46% for June and July, excluding the Gobi and desert area. This seems to follow the conclusions of some scientists regarding precipitation decreases in the midlatitudes of the northern hemisphere (IPCC, 1990; Parry, 1990; Parry and Jiachen, 1991).

Incremental Scenarios

Incremental scenarios combining temperature (0°C ± 2°C and ±4°C) and precipitation (±10% and ±20%) were used in the water resources assessment together with GCM climate change scenarios.

Baseline Socioeconomic Scenario

It is important to provide a framework for integrating environmental considerations into a country's overall economic and social development. No detailed socioeconomic scenarios were obtained for this study and we did not investigate how changes in baseline socioeconomic conditions affect the sensitivity of sectors to climate change. However, population and GDP projections through 2015 were obtained from forecasts of GDP and population growth prepared by the National Development Board. The main socioeconomic variables are in Table 9.1.

Table 9.1. Main Socioeconomic Variables

GDP Growth Rate[1]		Population	
Year	%	Year	Millions
1990	—	1990	2.149
1994	2.5	1994	2.200
1995	3.5	2015	3.280
2000	4.8		
2005	5.3		
2010	5.8		
2015	6.4		

[1] Growth rate: +2%/yr (after 2015).

AGRICULTURE SECTOR

Methods

Study Sites Selected

Seven locations were selected to represent the main agroclimatic areas of the country. These locations lie in the central region of Mongolia (Figure 9.2). This region is a major crop-producing center (about 70% of the country's grain is produced there); its arable land accounts for 65% of the country's total arable land of 1,203,000 ha. The staple crop in this region is wheat, whose planted area accounts for 86% of the total planted acreage in the region.

Crop Model

The CERES-Wheat model was used to assess the vulnerability of spring wheat yields and adaptations in the central region of Mongolia (Benioff et al., in press).

Climate Data

The observed data on daily maximum and minimum temperatures and precipitation were obtained from the Climatic Data Bank, which is organized by CLIMCOM system and operated at the computer network of the Hydrometeorological Research Institute. Daily solar radiation data were created from observed daily sunshine durations. The climatological baseline was 1940-1990. Future changes in other environmental and socioeconomic factors were not considered.

Climate Scenarios

Three types of climatic scenarios were used in the study: the GISS and GFD3 2×CO_2 scenarios, and one scenario of historical climate change trends in Mongolia (Mijiddorj, 1994).

Soil Data

Soil chemical and physical characteristics obtained from the Institute of Land Use Policies were input to the model.

Cultivar

We used the "Orhon" spring wheat variety, which is one of the main cultivated varieties in Mongolia.

Adaptation Alternatives and Evaluation Methods

Three adaptation options — changes in planting dates, changes in nitrogen fertilization, and changes in wheat varieties — were assessed using the CERES-Wheat model. Estimated changes in yields under each adaptation option were obtained.

Results

GCM Scenarios

Table 9.2 shows that spring wheat yields for the GISS scenario at five of the seven study areas decreased by 19% to 67% and increased at two locations. For the GFD3 scenario, yields decreased by 29% to 74% at six of the locations. With an increase under both scenarios, Baruunharaa was an exception to the other study locations. The estimated increase in spring wheat yields at Baruunharaa was 32% under the GISS scenario and 13% under the GFD3 scenario. In general, spring wheat yields might be reduced appreciably by the end of the next century as a result of climate change in the region.

"Historical Climate Trend" Scenario

There are many differences between the historical climate change trend in Mongolia and the GCM scenarios. The main feature of this trend is a temperature decrease of 0.28°C to 0.67°C for June, July, and August, and precipitation increases of 28% to 46% for June and July. Simulation results

Table 9.2. Aggregation of Model Results for Wheat

Location	Soil	Current Yield (t/ha)	Percent Change in Crop Yields GISS	GFD3
Baruunharaa	Black brown soil	2.17	+32.0%	+13.0%
Darhan	Black brown soil	2.33	+3.0%	-46.4%
Erdenesant	Black brown soil	2.97	-19.0%	-57.2%
Hutag	Black brown soil	2.59	-29.0%	-39.8%
Orhon	Black brown soil	2.38	-66.8%	-30.3%
Ugtaal	Black brown soil	3.15	-35.9%	-74.3%
Zuunmod	Black brown soil	3.80	-37.1%	-29.2%

from DSSAT 3 under this climate scenario show that the spring wheat yield might be increased if the current climate change trend continues in the future. This result differs from the simulation findings under GCM scenarios. However, the data necessary to make a detailed investigation of Mongolian climate change in the near future and its effect on crop production are currently lacking.

Adaptation Alternatives

According to the projected reductions in spring wheat yields (the country's main grain crop) under a 2×CO_2 scenario, the country will face serious difficulties in providing much of its grain needs.

The following adaptation options were considered as potential responses that might modify any effects of climate change on wheat production.

Changes in Planting Dates

To estimate the optimum planting date for spring wheat under changed climate conditions, we analyzed wheat yield values for different planting dates at the seven study sites. We advanced planting dates (the current average planting date is May 11) since climate scenarios suggest that climate would be warmer in the future. Wheat yields under climate scenarios derived from the GISS and GFD3 models at planting dates of May 11, May 1, and April 25 are given in Table 9.3. Under climate change, the optimum planting date for spring wheat in Mongolia would be earlier than under current climate conditions. If the wheat crop were planted on April 25, then wheat yields would be 9.2% to 79.8% higher than estimated yields without adaptation under the GISS climate scenario (measuring yields as shares of baseline yield), and 8.5% to 55.6% higher than estimated yields without adaptation under the GFD3 scenario.

For example, the May 11 yield for Orhon under the GISS scenario is 33.2% of baseline yield. The April 25 yield is 59.7% of baseline yield, which represents a 79.8% increase over the May 11 yield.

Table 9.3. Spring Wheat Yield Changes by Planting Dates (percent change from current yields)[1]

	Current Planting Dates		Alternative Planting Dates			
	May 11		April 25		May 1	
Location	GISS	GFD3	GISS	GFD3	GISS	GFD3
Baruunharaa	+32.0	+13.0	—	—	—	—
Darhan	+3.0	-46.4	—	-36.9	—	-37.8
Erdenesant	-19.0	-57.2	-5.1	-43.4	-10.8	-51.9
Hutag	-29.0	-39.8	-22.4	-24.7	-20.5	-25.1
Orhon	-66.8	-30.3	-40.3	-23.9	-52.1	-21.0
Ugtaal	-35.9	-74.3	-22.9	-60.0	-27.0	-67.0
Zuunmod	-37.1	-29.2	-28.9	-23.2	-36.6	-24.0

[1] Adaptation options were not analyzed for the study locations with a projected increase in yield under climate change.

Changes in Nitrogen Fertilization

Changes in climate may have large effects on soils (Parry, 1990). Land use policy may need to be reviewed in light of this, with regard to the changing potential of soils to support crops.

Increases in temperature and precipitation will affect the concentration of mineral substances in soils. The nitrogen requirements of spring wheat during the growing season under different climatic conditions are given in Table 9.4, which shows that the requirement for nitrogen would be reduced by 20% to 26% under the GISS and GFD3 scenarios, respectively. We also examined the optimum amount of nitrogen fertilizer and proper date of nitrogen application.

The estimated results show that to obtain maximum yields, the optimum amount of nitrogen fertilizer is 60 kg per ha. We examined different application dates using DSSAT 3. Table 9.5 shows that maximum yields would be obtained if nitrogen were applied on the 70th day following the planting date. If nitrogen were applied at the same time grain is planted, wheat yields would be increased, but not more than with the later application of nitrogen.

Table 9.4. Spring Wheat Requirements for Nitrogen (kg/ha)

Location	Current Climate	GISS	GFD3
Baruunharaa	68.2	70.3	74.7
Darhan	78.2	78.0	60.3
Erdenesant	80.7	70.0	60.0
Ugtaal	94.7	59.6	50.0
Zuunmod	71.7	60.2	70.5
Mean	78.7	67.6	63.1

Table 9.5. Spring Wheat Yields under Different Dates of Nitrogen Application (t/ha)[1]

	Nitrogen Application at Current Planting Date (May 10)			When Grain Is Planted		70th Day after Planting Date	
Location	Current Climate	GFD3	GISS	GFD3	GISS	GFD3	GISS
Baruunharaa	2.17	2.46	2.87	—	—	—	—
Darhan	2.33	1.25	2.41	1.42	—	1.59	—
Erdenesant	2.97	1.27	2.42	1.39	2.52	1.62	2.77
Hutag	2.59	1.56	1.84	1.69	2.04	1.91	2.19
Orhon	2.38	1.66	.79	1.86	1.07	2.01	1.14
Ugtaal	3.15	.81	2.01	1.06	2.11	1.16	2.34
Zuunmod	3.80	2.39	2.69	2.59	2.84	2.74	2.99
Mean	2.77	1.63	2.15	1.67	2.12	1.84	2.24

[1] Adaptation options were not analyzed for the study locations with a projected increase in yield under climate change.

If these nitrogen options were followed, estimated spring wheat yields under climate change could increase by 0.21 t/ha under the GFD3 scenario and by 0.09 t/ha under the GISS scenario.

Changes in Wheat Varieties

The simulation study suggests that yields of the currently planted variety of spring wheat (Orhon) may be reduced under the 2×CO_2 scenarios. Therefore, based on the model simulation, we estimated genetic coefficients for a new, hypothetical variety of spring wheat that might be more adaptable to future climate conditions. The optimum values for genetic coefficients of this variety of spring wheat were estimated by DSSAT 3 simulations (see Table 9.6). This variety gives yields of 200 to 250 kg per ha. The grain starch of this wheat variety is 1.3 to 1.9 times higher than the Orhon variety.

Table 9.6. Genetic Coefficients of a New Variety of Spring Wheat

P1V[1]	P1D[2]	P5[3]	G1[4]	G2[5]	G3[6]
0.5	3.4	4.0	4.6	3.0	4.2

[1] PIV — Relative amount that development is slowed for each day of unfulfilled vernalization, assuming that 50 days of vernalization is sufficient for all cultivars.

[2] PID — Relative amount that development is slowed when plants are grown in a photoperiod one hour shorter than the optimum (which is considered to be 20 hours).

[3] P5 — Relative grain filling duration based on thermal time (degree days above a base temperature of 1 °C), where each unit increase above zero adds 40 degree days to an initial value of 300 degree days.

[4] G1 — Kernel number per unit weight of stem (less leaf blades and sheaths) plus spike at anthesis (1/g).

[5] G2 — Kernel filling rate under optimum conditions (mg/day).

[6] G3 — Nonstressed dry weight of a single stem (excluding leaf blades and sheaths) and spike when elongation ceases (g).

Discussion and Conclusions

The adaptation options discussed above are just a few examples of many possible adaptation options and measures. It is reasonable to expect that a large number of changes in land use (changes in farmed area, crop type and crop location) and changes in management (changes in the use of irrigation and fertilizers, and in crop types) would be adopted over time as the effects of climate change are perceived. It appears that these adaptation measures would be most favorable if climate shifted toward warmer and wetter conditions in the future.

The results obtained to date have been encouraging in beginning to determine the importance of changes in climate variability to crop yields. Future research to determine changes in the photosynthesis rate under 2×CO_2 conditions should help to refine our knowledge of climate change effects.

Clearly, increased atmospheric CO_2 concentrations have a direct effect on the photosynthesis and water use efficiency of plants. As a result of temperature increases, the vegetation would ripen in a shorter time, and consequently, crop ripening time would be decreased. On the other hand, maximum temperature increases have negative effects on the normal phenological phases of wheat maturity, and on the accumulation of organic matter in grain.

Under GCM climate change scenarios for the central agricultural region, the number and area of a leaf in one spike of wheat, the number of grains per unit of area, and grain weights might be reduced due to a decreased photosynthesis process in the crop. However, to determine changes in the photosynthesis rate under $2\times CO_2$ conditions, more detailed information and an advanced methodology are needed (Dagvadorj, 1993).

It would be useful if we could assume technological advances in irrigation efficiency and crop drought resistance as well as improvements in a number of crop-specific characteristics, including harvest index, photosynthesis efficiency, and pest management.

FORESTRY SECTOR

Methods Used

This assessment examined the changes in Mongolia's natural zones under climate change conditions. The baseline natural zones were established using a Holdridge Life Zone Classification Model simulation and climatic data for 1961-1990 from 69 meteorological stations in Mongolia (Figure 9.2) and historical data (Benioff et al., in press). The $2\times CO_2$ climate change scenarios from the UK89 model were used to estimate life zones under climate change conditions. According to these scenarios, the steppe zone's characteristics will undergo significant change and the zone's boundaries will be shifted.

Mongolia contains various overlapping natural zones, including the Siberian *taiga* and the Gobi Desert. On its northernmost edge lies the desert and the southernmost edge is permafrost. Often, Mongolia's natural zones traverse a small spatial area, stretching a longitude of about 1,000 km or 100 km in the mountainous region. Thus, the country's natural zones require further study in terms of more precise spatial factors and more comprehensive surveys.

Baseline Natural Zones

The results of the baseline scenario of natural zones simulated by the Holdridge model using historical climate data are given in Figure 9.7. It shows that a cool steppe is located in the Khangai, Khentii, and Khubsugul mountainous regions; a cool desert scrub lies southward; a warm desert is found in the southern area of Mongolia; and cool desert lies in the hollow of the Great Lakes.

Natural Zones under $1\times CO_2$ Scenarios

To reproduce Mongolia's natural zones on a map, monthly temperatures and precipitation were calculated using the $1\times CO_2$ GCM scenarios on the grid $2°\times2°$ of geographical latitude and longitude. Then, the temperature and precipitation were estimated at selected meteorological stations. Maps of Mongolia's natural zones under the $1\times CO_2$ scenarios are shown in Figure 9.8. The natural zones estimated under the CCCM and GISS scenarios are similar for the territory north of 45° latitude and east of 100°-106° longitude, where a cool, moist desert is estimated, and to the west, where the region is estimated to be a cool steppe. (In Figure 9.8, the cool steppe zone is not distinguished from the desert zone for the CCCM and GISS scenarios.) Under the GFD3 scenario the Holdridge model suggested that the cool steppe is merged into one zone with the cool desert scrub; on the whole, it looks much more like the base scenario than the maps formed by the other two models. The results generated using the UK89 scenario are the closest to the results of the baseline scenario (Figure 9.7).

Figure 9.7. Holdridge Life Zones Generated from Historical Climate Data (1961-1990)

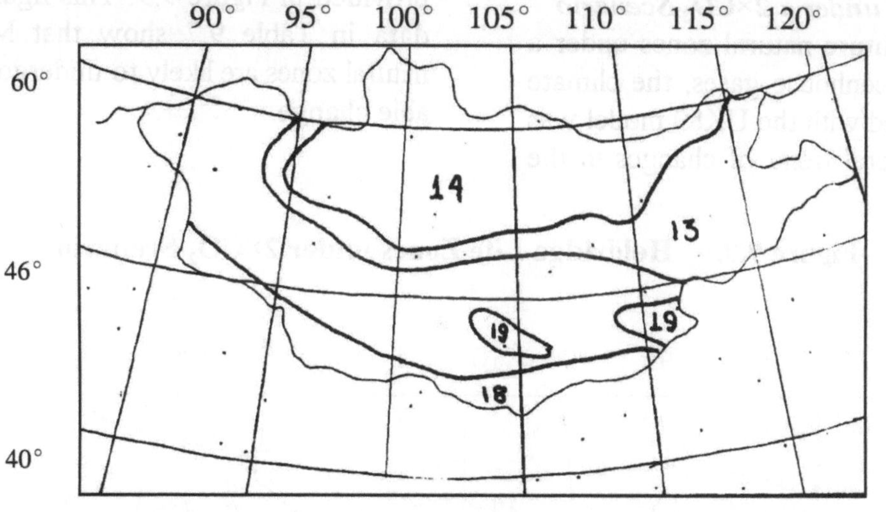

Regions: 12 - cool temperate desert 13 - cool temperate desert scrub 14 - cool temperate steppe
18 - warm temperate desert 19 - warm temperate desert scrub

Figure 9.8. Holdridge Life Zones under 1×CO_2 Scenarios

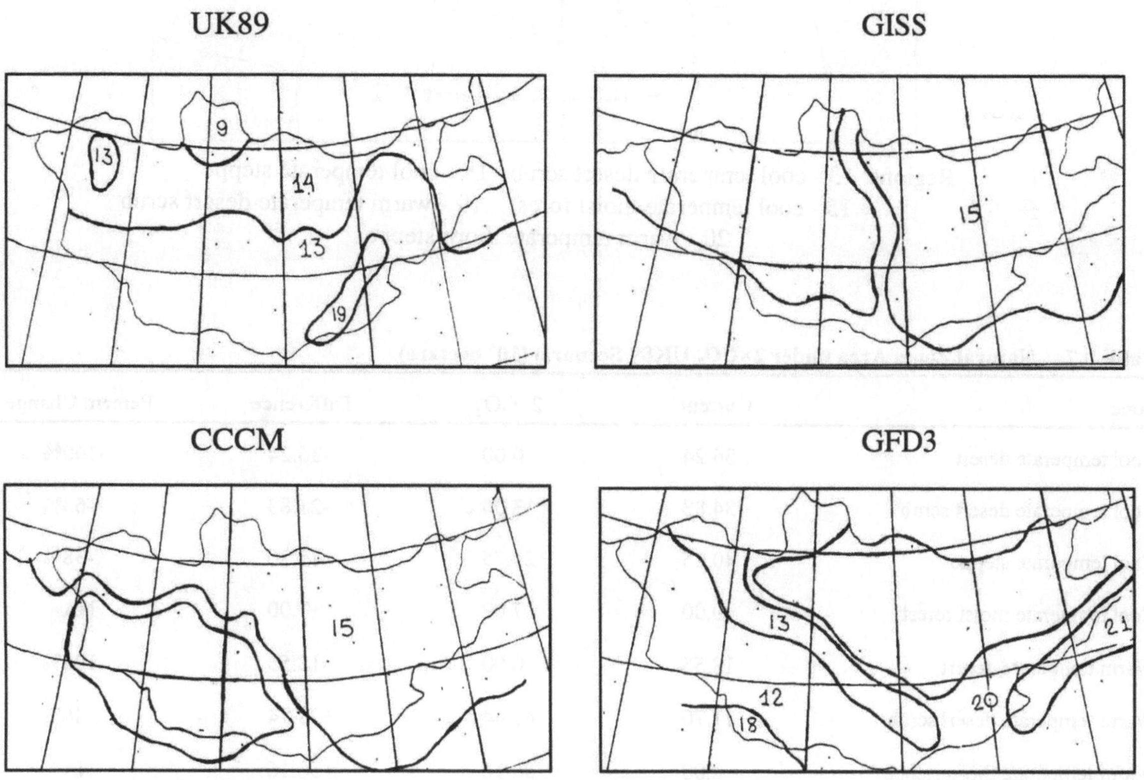

Regions: 9 - boreal moist forest 12 - cool temperate desert 13 - cool temperate desert scrub
14 - cool temperate steppe 15 - cool temperate moist forest 18 - warm temperate desert
19 - warm temperate desert scrub 20 - warm temperate thorn steppe 21 - warm temperate dry forest

Results

Natural Zones under a 2×CO_2 Scenario

In estimating future natural zones under a doubling of greenhouse gases, the climate scenario obtained with the UK89 model was used. The general trend of changes in the natural zones of Central Asia under the greenhouse gas doubling scenario are provided in Figure 9.9. This figure and the data in Table 9.7 show that Mongolia's natural zones are likely to undergo considerable change.

Figure 9.9. Holdridge Life Zones under 2×CO_2 Scenario

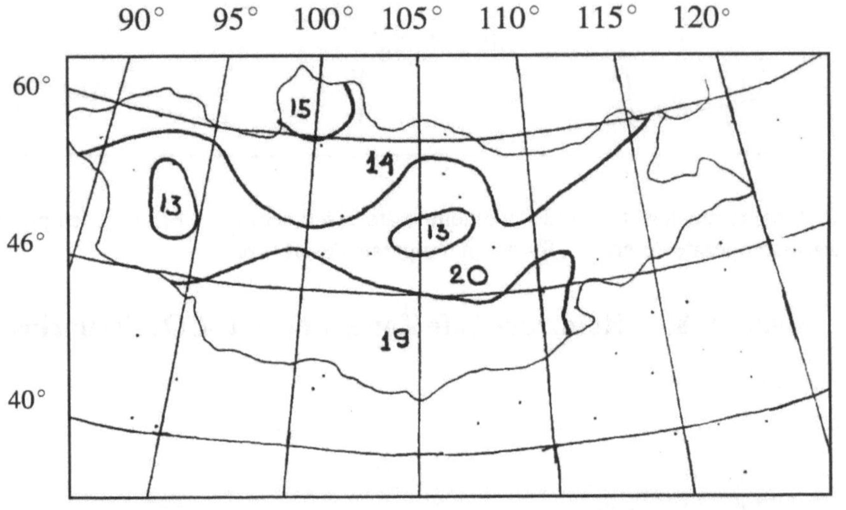

Regions: 13 - cool temperate desert scrub 14 - cool temperate steppe
15 - cool temperate moist forest 19 - warm temperate desert scrub
20 - warm temperate thorn steppe

Table 9.7. Natural Zones Area under 2×CO_2 UK89 Scenario (10^6 hectare)

Zone	Current	2×CO_2	Difference	Percent Change
Cool temperate desert	36.24	0.00	-36.24	-100%
Cool temperate desert scrub	34.83	13.00	-21.83	-63%
Cool temperate steppe	40.05	24.75	-15.30	-38%
Cool temperate moist forest	0.00	7.00	+7.00	NA
Warm temperate desert	18.55	0.00	-18.55	-100%
Warm temperate desert scrub	11.70	40.44	+28.74	+246%
Warm temperate thorn steppe	0.00	56.16	+56.16	NA

Discussion and Conclusions

Two conclusions were drawn from this study. First, it is possible to simulate Mongolia's natural zones using the Holdridge model. Second, the results of the UK89 model show that the characteristics of the Mongolian Gobi desert and steppe zones would be altered in the region of the Khangai Mountains under a 2×CO_2 scenario. This latter conclusion is a preliminary one.

Water Resources Sector

Methods Used

Selection of Study Sites

Three river basins were selected with a range of climatic and geographic conditions: the Bogd, Bulgan, and Khalkhgol river basins (Figure 9.2). The selection criteria included basin size and varying climate and basin characteristics (Tsegmed et al., 1969).

Bogd River. This river is a tributary of the Zavkhan River and was the smallest catchment area modeled. The Bogd River Basin covers an area of 1,610 km^2 and is located within a mountain range with many rivers and permanent snow that can affect the formation of spring floods. This basin has the largest discharge of the three basins studied. Most of its runoff occurs as summer floods during June to August. The average monthly precipitation in the basin area ranges from 2 to 90 mm; ice formation begins during the end of November.

Bulgan River. This river is located in the northwestern part of the country and originates from the highest mountain range in Mongolia, the Mongol Altai. The area can be divided into two main watersheds: the upper portion of the river in the mountainous area; and the lower portion, which is characterized by lowlands with sandy soil. The basin covers an area of 3,128 km^2 and experiences high temperature variation (the maximum temperature is +17.8°C and the minimum temperature is -27.5°C). The monthly mean precipitation ranges from 2 to 105 mm, and 80% of the annual precipitation occurs between June and August.

Khalkhgol River. This river basin is located in the eastern part of the country and covers an area of 17,120 km^2 (43% of this basin is within the boundaries of Mongolia). Most of the mean monthly precipitation occurs during the three-month summer, and the remaining portion of the year receives about 10% of the total annual precipitation. This river is covered by ice for approximately six months out of the year. The thickness of the ice cover on the Khalkhgol is about 1.0 m to 1.5 m. The runoff from floods due to snowmelt does not exceed the summer season runoff.

Hydrological Assessment Methodologies

To assess the vulnerability of various parts of the country to water resources, the Basin Conceptual Model (BCM) was used (refer to text box). This model was selected because it is better suited for Mongolia's climate than some standard models like CLIRUN and WATBAL (Benioff et al., in press). The CLIRUN model did not account for the cold temperatures of Mongolia and the WATBAL model did not produce good results.

The BCM is a monthly balance model that uses annual monthly mean values of precipitation, temperature, potential evapotranspiration, and runoff (Yates and

Basin Conceptual Model (BCM)

The Basin Conceptual Model (BCM) is a monthly balance model that uses annual monthly mean values of precipitation, temperature, potential evapotranspiration, and runoff (Yates and Strzepek, 1994; Grishanin, 1979).

Model input data and state variables consist of:

Peffi	=	effective precipitation at time i (mm)
Pmi	=	measured precipitation at time i (mm)
Evi	=	actual evapotranspiration at time i (mm)
PETi	=	potential evapotranspiration at time i (mm)
Si	=	active basin storage in month i (mm)
Ri	=	basin runoff in month i (mm)
Ai	=	accumulation at time step i (mm)
αi	=	accumulation index for i ($0 < \alpha < 1$)
Ti	=	temperature at time interval (°C)

The model parameters consist of:

Ts	=	solid snow threshold, completely frozen (-3°C)
Ti	=	upper temperature threshold, liquid above this value (+3°C)
Ke	=	evapotranspiration parameter
Kg	=	active basin storage parameter
Kw	=	winter basin runoff parameter
Ks	=	inverse of the maximum river basin storage capacity (l/mm)

Storage S_{t-1} is written as:

$$S_{t-1} = Peffi - Si - Evi - Ri \qquad (1)$$

effective precipitation, Peffi, is given by:

$$Peffi = \alpha i(A_{i-1} = Pmi) \qquad (2)$$

where

$$\begin{aligned} &\alpha = 0 && \text{for } Ti \le Ts \\ &\alpha = 1 && \text{for } Ti \ge T_1 \\ &\alpha = (Ti - Ts)/(T_1 - Ts) && \text{for } Ts < Ti < T_1 \end{aligned} \qquad (3)$$

snow accumulation, Ai, is written as:

$$Ai = (1 - \alpha)(A_{i-1} + Pmi) \qquad (4a)$$

evapotranspiration, Evi, is given as:

$$Evi = PETi(1 - \exp(- KeS_{i-1}) \qquad (5)$$

Runoff, Ri, is given in the winter season; winter runoff conditions prevail when

$A_{i-1} > 0$ and $Ti < T_i$) by:
$Ri = KgS_{i-1} + KwPeff$ (6a)

for the summer season, runoff is given by:

$Ri = KgS_{i-1} + (Pi - Ii)^2/P - 4Ii$ for $(Pi - Ii) > 0$ (7a)
$Ri = KgS_{i-1}$ for $(Pi - Ii) \leq 0$

where Ii is the monthly total initial abstraction:

$Ii = 0.2\ (1/k = S_{i-1})$ for $(1/Ls - S_{i-1}) > 0$
$Ii = 0$ for $(1/Ks - S_{i-1}) \leq 0$

The model described above did not produce satisfactory results. Therefore, we made some modifications in the findings for snow accumulation and the estimation of basin runoff for both seasons. Thus, equations 4a, 6a, and 7a became the following:

snow accumulation is written as:

$Ai = (1 - \alpha)(A_{i-1} + Pmi)$ for $Ti \geq Ts$
$Ai = (1 - \alpha)(A_{i-1} = SfPmi)$ for $Ti < Ts$ (4b)

runoff, Ri, is given in the winter season:

$Ri = Sf(KgS_{t-1} + KwPeffi)$ for $Ti < Ts$
$Ri = KgS_{i-1} + KwPeff$ for $Ti \geq Ts$ (6b)

for the summer season, runoff is given by:

$Ri = KgS_{i-1} + (Pi-Ii)/P - K_bIi$ for $(Pi-Ii) > 0$ (7b)
$Ri = KgS_{i-1}$ for $(Pi-Ii) \leq 0$

where

K_b = coefficient of surface runoff
Sf = coefficient of snow melting.

Strzepek, 1994; Grishanin, 1979). It uses the previous month's storage to compute infiltration, evapotranspiration, and runoff. It contains six parameters, two of which are the upper and lower temperature bounds on the freezing and melting processes.

This assessment was carried out using primary data collected by the Hydrometeorological Institute and the Water Policy Institute. The scenarios selected give uniform, annual changes in temperature and precipitation, within a range of plausible future climates.

The three study sites that were selected for testing the BCM are intended to show the range of the model's applicability by including a basin in a climate region that is dominated by snowfall and colder temperatures. A split sample test was used on all three basins to evaluate the hydrologic model. For this test, the historical record was broken into two segments, one for calibration and the other for validation. Because of the short data record for these rivers, the first 10 years (1970-1980) were used for calibration and the remaining 10 years (1980-1990) for validation.

Since the focus of this work is on determining the model's applicability for assessing the impacts of climate change on a river basin, incremental scenarios (i.e., usually put in the framework of a sensitivity analysis by applying a range of potential climates) were chosen for their simplicity. In this approach, incremental scenarios are cast as a set of plausible future climates. These scenarios will enable the generation of tables that will lend insight into the sensitivity of the basins and the model to climate variations. The scenarios chosen give uniform, annual changes in temperature and precipitation in various combinations, with the expectation that they will cover a range of plausible future climates.

Calibration and Validation of the BCM

The annual precipitation, temperature, and discharge records seem to indicate that the basins may be sensitive to small variations in precipitation during the summer. This may be explained by the region's arid climate. Also, representations of the melting rate produce a significantly different runoff regime when model parameters are changed. Figure 9.10 plots the mean monthly observed and modeled discharges for the basins. The model appears to consistently under-predict the winter runoff and also tends to under-predict the runoff in the transition period when most of the runoff forms (from May to September). It appears that the BCM definitely follows the historical runoff trends. An example of the analysis of historical trend is shown in Figure 9.11.

The calibration and validation values used in the BCM for the Bogd, Bulgan, and Khalkhgol rivers are shown in Table 9.8. The correlation coefficients and average error are given in Table 9.9.

Table 9.8. Calibration and Validation Values used in the BCM

Parameter	Bogd	Bulgan	Khalkhgol
Evapotranspiration parameter, Ke	0.012	0.009	0.0023
Active basin storage parameter, Kg	0.60	0.03	0.04
Winter basin runoff parameter, Kw	0.2	0.1	0.8
Inverse of the maximum river basin storage capacity (1/mm), Ks	0.0050	0.0024	0.0016
Coefficient of surface runoff, K_b	0.99	0.10	0.05
Coefficient of snow melting, S_f	0.20	0.40	0.01
Solid snow threshold, Ts	-4	-1	-3
Upper temperature threshold, T_1	0	+1	0

Figure 9.10. Observed Discharge Compared to BCM Discharge for Bogd, Bulgan, and Khalkhgol River Basins

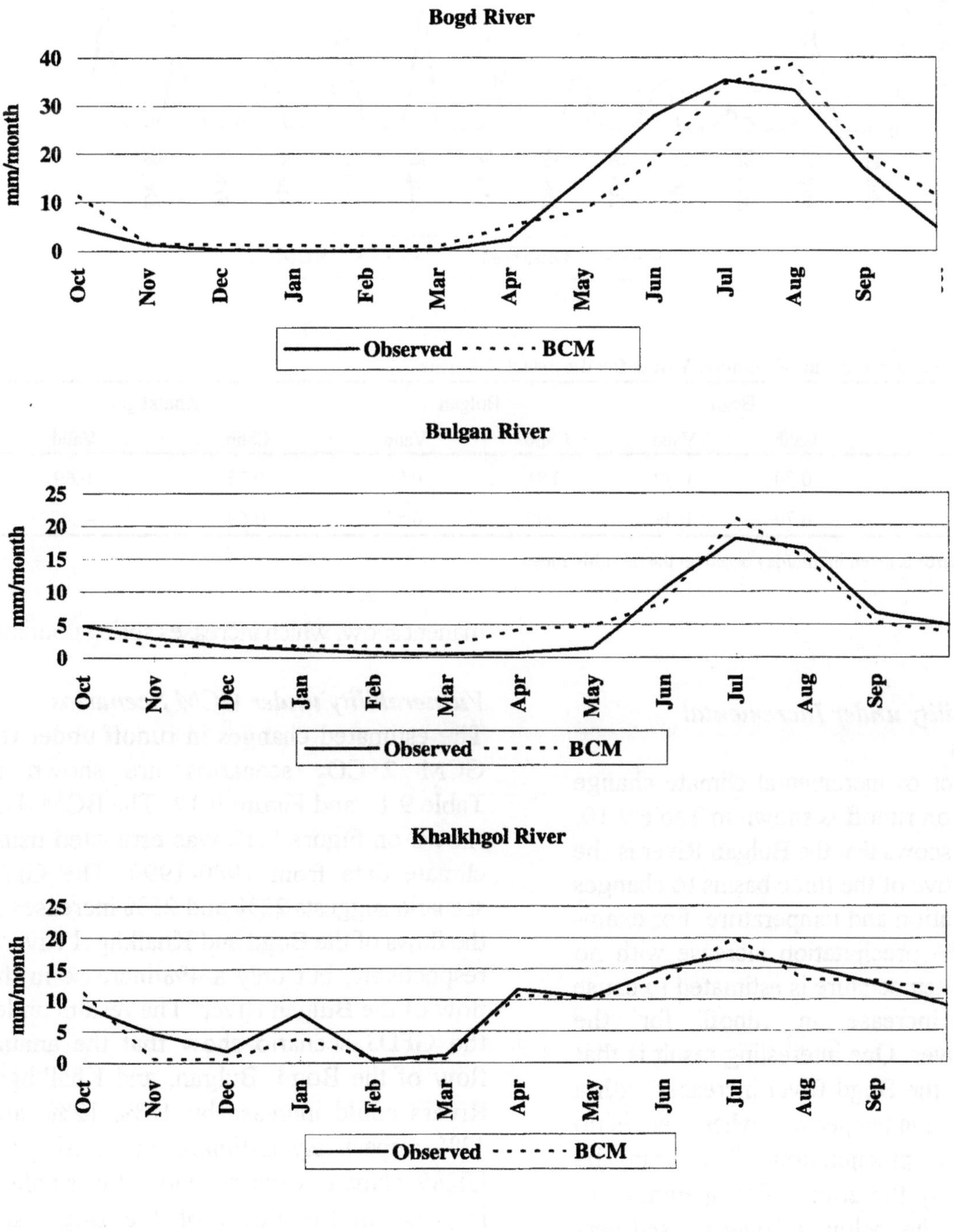

Figure 9.11. Historical Trend of Monthly Mean Discharge Compared to BCM Estimated Discharge for the Bogd River

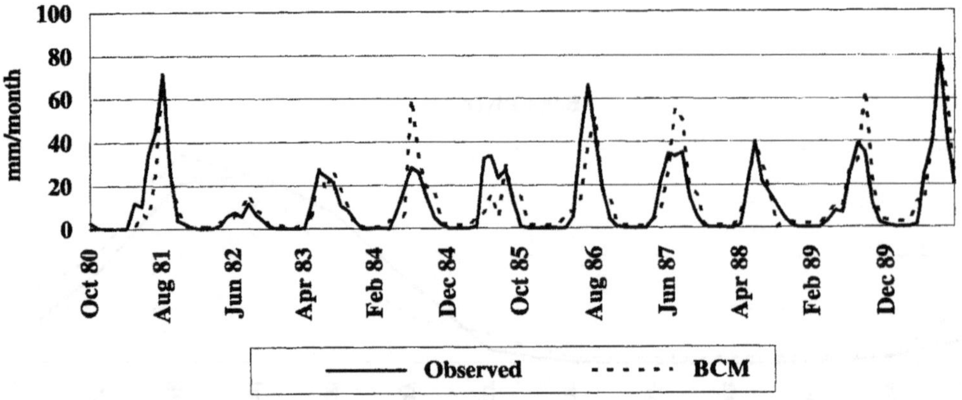

Table 9.9. Calibration and Validation Values for the Selected Basins

	Bogd		Bulgan		Khalkhgol	
	Calib.	Valid.	Calib.	Valid.	Calib.	Valid.
Correl	0.70	0.79	0.81	0.85	0.75	0.69
Avg. err [1]	0.79	0.49	0.00	0.55	0.69	-0.47

[1] Average error is given in mm/day based on the monthly mean.

Results

Vulnerability under Incremental Scenarios

The impact of incremental climate change scenarios on runoff is shown in Table 9.10. This table shows that the Bulgan River is the most sensitive of the three basins to changes in precipitation and temperature. For example, a 20% precipitation increase with no increase in temperature is estimated to cause a 79% increase in runoff for the Bulgan River. One interesting result is that runoff for the Bogd River increases with a 2°C increase in temperature when there is no change in precipitation. This can be explained by the effect of temperature increases on the melting of river ice and permanent snow, which increases spring flooding.

Vulnerability under GCM Scenarios

The estimated changes in runoff under the GCM 2×CO_2 scenarios are shown in Table 9.10 and Figure 9.12. The BCM data shown on Figure 9.12 was estimated using climate data from 1980-1990. The GISS scenario suggests 22% and 25% increases in the flows of the Bogd and Khalkhgol Rivers, respectively, but only a 4% increase in the flow of the Bulgan River. The results under the GFD3 scenario show that the annual flow of the Bogd, Bulgan, and Khalkhgol Rivers could increase by 18%, 12%, and 13%, respectively. Estimates made using the UK89 climate scenario show the smallest increase in the flows of the Bogd and

Table 9.10. Annual Percent Change in Runoff for the Selected Basins under Incremental Scenarios[1]

Bogd	P + 0%	P + 10%	P + 20%	P - 10%	P - 20%
T = 0	0%	+18%	+32%	-8%	-21%
T + 2	+3%	+16%	+30%	-10%	-23%
T + 4	+2%	+13%	+29%	-11%	-24%
Bulgan					
T = 0	0%	+42%	+79%	-20%	-42%
T + 2	-2%	+29%	+63%	-28%	-47%
T + 4	-13%	+15%	+47%	-36%	-53%
Khalkhgol					
T = 0	0%	+3%	+19%	-23%	-33%
T + 2	-16%	-3%	+12%	-27%	-36%
T + 4	-20%	-8%	+6%	-31%	-39%

[1] P = precipitation, T = temperature.

Table 9.11. 2×CO_2 GCM Estimates of Percent Change in River Flow

GCM Scenario	River Basin Bogd	 Bulgan	 Khalkhgol
GISS	+22.4%	+3.6%	+25.3%
UK89	+10.2%	-0.3%	+6.5%
CCCM	+21.0%	+26.3%	+15.4%
GFD3	+18.4%	+11.6%	+12.6%

Khalkhgol Rivers; this scenario also suggests a small decrease in the flow of the Bulgan River. The CCCM scenario results show the largest increase in the flow of all three rivers.

Discussion and Conclusions

This study was designed to provide a preliminary investigation of Mongolia's water resources with a regional disaggregation. In particular, the study was designed to examine the water availability balance in various regions under climate change scenarios. Because some of the models (e.g., CLIRUN, WATBAL) were not suitable for Mongolia's climate, the BCM was used with the addition of some model parameters. The model performed well under calibration and validation in all basins.

The results show that all three river basins are more sensitive to precipitation change than temperature change. But the snowmelt-dominated basins show less sensitivity to precipitation change in the winter when the rivers are covered by thick ice; these basins are also quite sensitive to temperature threshold values, where changes in temperature parameters produced significantly different values.

The strong seasonal variation in runoff in all three basins points to the need to incorporate seasonal parameters within the model. The results of the GCM scenarios also show that runoff might be increased by at least 3% to 25% for the selected basins.

Figure 9.12. Comparison of Estimated Mean Monthly Runoff for 2×CO_2 Scenarios and Observed Data

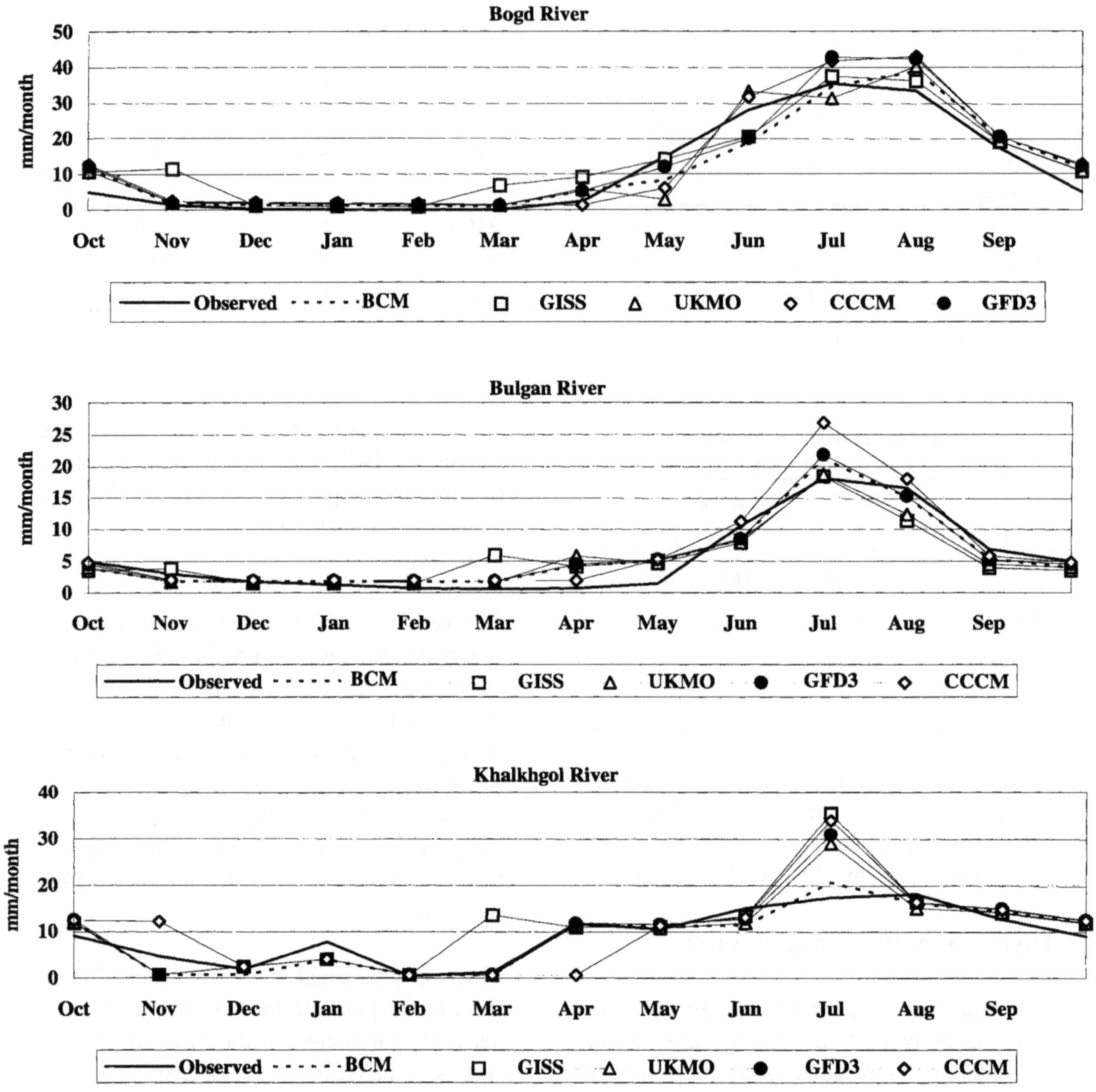

This study only considered hydrologic vulnerability. We did not look at water demand, supply, and economic impacts because there were no initial data that could be used for this study.

Summary Conclusions

The results of the vulnerability assessments show that climate change will have a significant negative impact on Mongolia's agricultural, forestry, and water resources sectors. These findings suggest that rising temperatures and reduced precipitation in some regions will increase the amount of arid and semi-arid areas, resulting in increased yields of forage crops and reduced or unchanged yields of grain and seed crops. Higher temperatures can also strain the interaction between plants and animals. This may produce changes in the composition of species, although such ecosystem changes are not anticipated to be dramatic. Plants and animals will not optimally adjust to their altered ecosystems for a span of years.

The fact that temperatures during the growing season (June to August) have been strongly reduced during the last 50-60 years in Mongolia (despite increases in annual mean temperatures) requires special consideration. Temperature reductions during the summer months are expected to cause a shortened growing season. The severity and frequency of drought may increase in the Gobi desert area with climate change. This could lead to a greater use of drought-tolerant crops, increased irrigation or changes in agricultural activity. A warmer climate may also lead to some expansion of agricultural lands, but Mongolia's soil types would limit additional expansion.

Results from the Holdridge model show that Mongolia's vegetation cover may be seriously altered. The study results suggest that some elements of subtropical desert may be observed in the current southwestern and southern desert. But these are questionable findings and need further evaluation.

The study results also show that water resources in selected river basins are more vulnerable to precipitation changes than to temperature changes.

It should be noted that the vulnerability assessments for the three sectors were conducted separately. It is clear that any changes in any one sector will affect the others. Also, social and economic analyses were not conducted because of the lack of necessary data.

References

Benioff, R., S. Guill, and J. Lee (eds.) (In press). *Vulnerability and Adaptation Assessments: An International Guidebook.* Dordrecht, The Netherlands: Kluwer Academic Publishers.

Dagvadorj, D. 1993. "Numerical Assessment of Climate Change Impacts on Grain Yield in Mongolia." *The First PRC-Mongolia Workshop on Climate Change in Arid and Semi-Arid Region over the Central Asia.* May 8-11, 1993, Beijing. pp. 104-108.

Dagvadorj, D., R. Mijiddorj, and L. Natsagdorj. 1994. "Climate Change and Variability Studies in Mongolia." In: *Papers in Meteorology. Annual Scientific Journal of Hydrometeorological Res., Ins. No. 17.* Ulaanbaatar. pp. 3-10.

Grishanin, K.B. 1979. *Dynamics of Stream Flows.* Russia: Leningrad, Hydrometeoizdat. pp. 85-125.

IPCC. 1990. *Climate Change: The IPCC Impacts Assessment.* Canberra: Australian Government Publishing Service. 268 pp.

Mijiddorj, R. 1994. "Air Temperature Changes over the Territory of Mongolia." In: *Papers in Meteorology, Annual Scientific Journal of Hydrometeorological Res., Ins. No. 17.* Ulaanbaatar. pp. 11-17.

National Development Conception. 1994. *National Development Board.* Ulaanbaatar.

National Standard of Mongolia. UST 41-93: Main Climate Data of Mongolia. 1993. *National Board of Standard Meteorology.* Ulaanbaatar.

Natsagdorj, L. and A. Namkhai. 1993. The latest climate changes in Mongolia. *The First PRC-Mongolia Workshop on Climate Change in Arid and Semi-Arid Region over the Central Asia.* May 8-11, 1993. Beijing. pp. 51-54.

Parry, M. 1990. *Climate Change and World Agriculture.* London: Earthscan Publications.

Parry, M. and Z. Jiachen. 1991. "The Potential Effect of Climate Changes on Agriculture." In: *Proceedings of the Second World Climate Conference: Climate Change: Science, Impacts and Policy.* Cambridge University Press.

Tsegmed et al. 1969. *Mongol Orhyn Phizic Gazar Zyi.* Ulsiin Hevlel, Mongolia: Institute of Geography, Mongolian Academy of Sciences. pp. 140-160.

Yates D. and K. Strzepek. 1994. *Comparison of Models for Climate Change Assessment of River Basin Runoff.* Laxenburg, Austria: IIASA.

Chapter 10

Vulnerability and Adaptation Assessments for Sri Lanka

Upali Weerakkody

Department of Geography, University of Ruhuna, Matara

SUMMARY. Beach erosion induced by sea level rise is a critical problem in Sri Lanka. Because accelerated sea level rise can be expected in the near future as a result of global warming, the country's beaches may be further eroded, putting pressure on its coastal settlements, commercial activities, and land use systems. Rising sea levels would inundate adjacent areas of wetlands and lowlands, pushing them farther inland. The southwestern coast, the most densely populated part of the island, may be among the areas most severely affected by rising sea levels, even though many parts of the coastline are already protected by groin fields and revetments. This chapter assesses the degree of vulnerability expected in the coming decades, in terms of the loss of land, the loss of sand spits and river mouth barriers, and the inundation of land adjacent to wetlands and lowlands under scenarios of 0.3 m, 1.0 m, and 2.0 m sea level rise. The results show that the beaches of southwestern Sri Lanka would be lost under current protection levels. The endangered areas could include wetlands, where cultivated paddies would be destroyed by inundation and saltwater intrusion. The existing policies on coastal protection are reviewed in light of the findings of the vulnerability assessment, and further recommendations are made.

INTRODUCTION

Sri Lanka, an island in the Indian Ocean (Figure 10.1), is affected by global atmospheric and large-scale oceanic changes as a result of southwestern and northeastern monsoon systems. In 1982, El Niño caused some US$150 million in damages from drought in Sri Lanka and southern India (UCAR, 1994).

The country's coastline stretches some 1,585 km and contains a variety of marine landforms and coastal vegetation. In the past hundred years, the beaches created by sea level recession have gradually eroded (Swan, 1974) and reshaped the entire coastline.

The 5.0 m sea level rise during the Mid-Holocene period submerged a land area approximately 1 km to 7 km wide. The lowlands were converted into bay beaches during the Mid-Holocene sea level rise of 5.0 m. When the Mid-Holocene sea level regressed to 2.0 m above the present sea

Note: This chapter does not contain a detailed description of the assessment methodologies that were recommended by the U.S. Country Studies Program. Refer to Chapter 2 for an overview of these methodologies.

J. B. Smith et al. (eds.), Vulnerability and Adaptation to Climate Change, 207–224.

Figure 10.1. Map of Sri Lanka

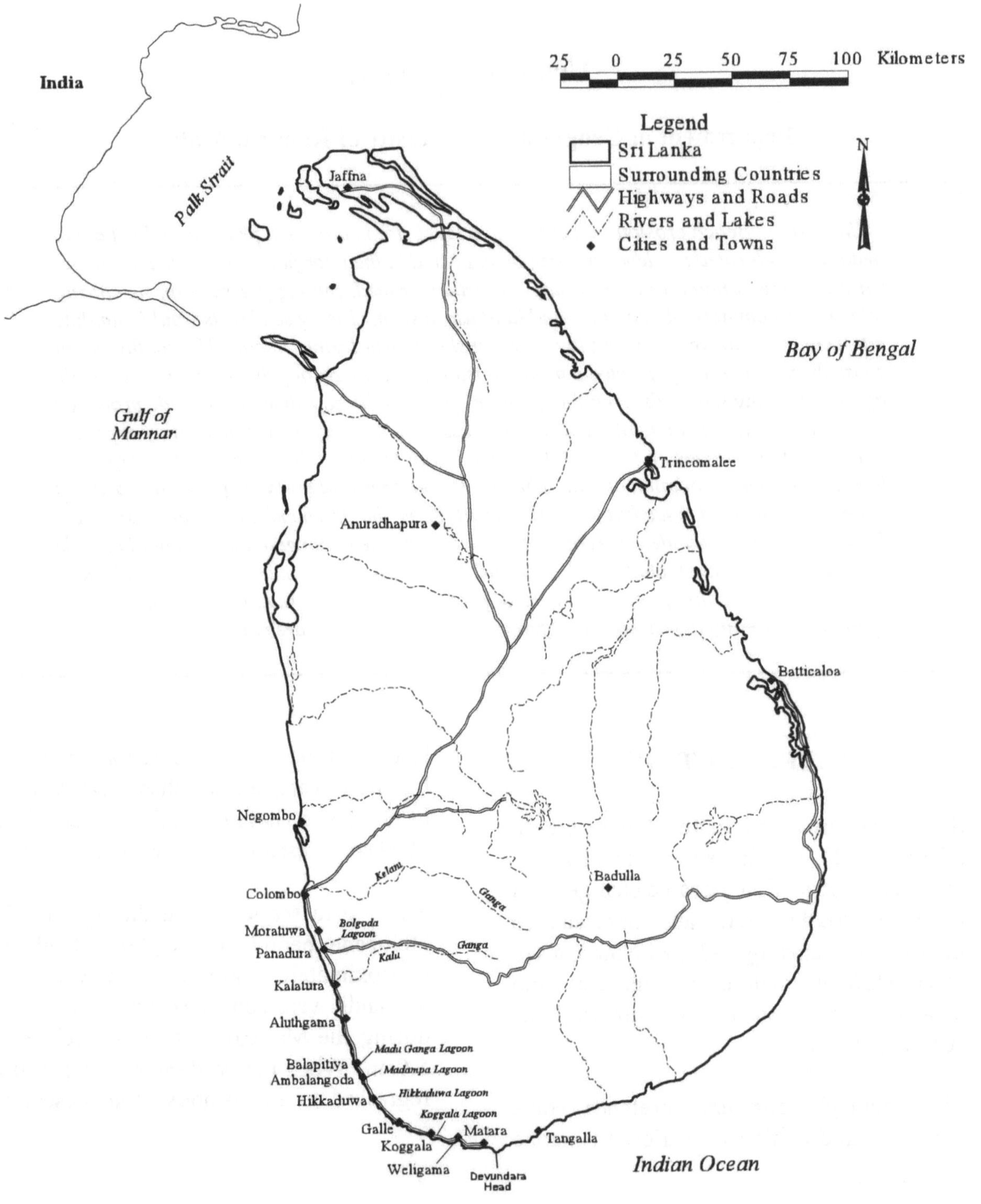

level, depositional features in the bay systems arose. Bays with depositional landforms were ultimately converted into lagoons when sea level dropped to its present level (Weerakkody, 1992a).

Tide gauge records dating from 1906 show that the sea level has increased slightly with tidal cycles in the past decades in response to increasing global temperatures. The changes in Sri Lanka's sea level indicate some tidal cycles (Swan, 1974), and the fluctuations of tides in February between 1901 and 1964 show that the sea level has risen 12 cm. There is also much evidence from meteorological records of increased temperatures in Sri Lanka over the last century in response to global atmospheric changes.

In the past century, cultivated littoral lands have been abandoned because of the intrusion of saline water. Hundreds of settlements have been washed away (Swan, 1974; see also John Fraser's Map of 1840, Department of Ceylon Archives of Sri Lanka). Some tens of kilometers of railroads and highways have also been washed away by sea waves.

Because of the island's location, the sea level rise of the past decades has created many environmental problems such as beach erosion, salinity intrusion into upstream estuaries and lagoons, and decreases in water supply to metropolitan, rural, and nature conservation areas (CEA, 1991). For example, water supply to both metropolitan and rural areas depends entirely on rivers and reservoirs. The amount of water available for drinking, industrial, and irrigation purposes is governed by the velocity of river discharge. The stagnation of water in rivers and lagoons during the dry season, in association with rising sea levels, has caused the intrusion of saline water into upstream estuaries and lagoons. This has also led to reduced water quality in rivers, which adversely affects water supply to metropolitan and rural areas and to irrigated paddy lands. Further inundation of the wetlands and lowlands will severely affect these lands in the future.

The most severe problem associated with sea level rise is related to beach erosion. All of the country's coasts, except the south coast, which receives sediment from the eroding beaches of southwestern Sri Lanka, are eroding more rapidly than they are accreting. Figure 10.2 shows the percentages of Sri Lanka's coasts that are subject to erosion and accretion. These data indicate that Sri Lanka has experienced severe effects from the sea level rise that took place during the last century.

The objective of this assessment is to examine potential land loss from both beach erosion and inundation caused by accelerated sea level rise, using sea level rise scenarios of 0.3 m, 1.0 m, and 2.0 m. Current adaptation policies to mitigate land loss and inundation, and the effects on land use and commercial activities are reviewed. Adaptation options to address the impact of saltwater intrusion of water resources are also briefly outlined. This chapter addresses only the coastal sector; the final vulnerability and adaptation assessment for Sri Lanka will also address the effects of climate change on the human health sector.

Figure 10.2. Coastal Erosion and Accretion in Sri Lanka

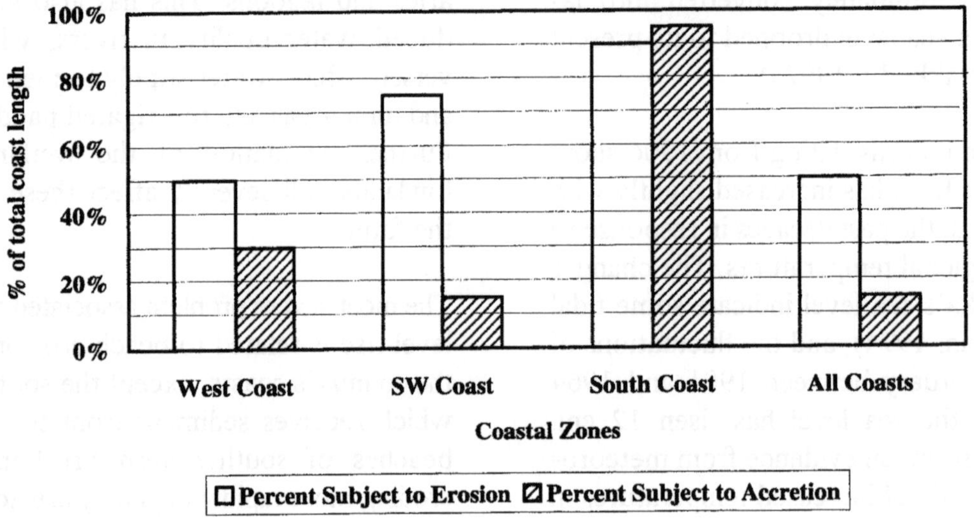

Source: Department of Coast Conservation, 1990.

Table 10.1. Scenarios of Accelerated Sea Level Rise for Sri Lanka (m)

	IPCC		Based on Observed Data for Colombo Harbor	
Year	0.3 m	1.0 m	0.3 m	1.0 m
2000	0.030	0.090	0.020	0.059
2010	0.060	0.180	0.039	0.118
2020	0.090	0.270	0.059	0.176
2030	0.120	0.360	0.078	0.235
2040	0.150	0.450	0.098	0.294
2050	0.180	0.550	0.118	0.359
2060	0.210	0.640	0.137	0.418
2070	0.240	0.730	0.157	0.477
2080	0.270	0.820	0.176	0.536
2090	0.300	0.910	0.196	0.595
2100	0.330	1.000	0.213	0.653

Scenarios

The scenarios provided by the Intergovernmental Panel on Climate Change (IPCC) and data collected from the tide gauge for Colombo harbor were used in this study to assess future sea level rise in Sri Lanka as a result of climate change (Table 10.1). Even though a 2.0 m sea level rise scenario is beyond the IPCC projections, it was used in this study as a sensitivity analysis. The 2.0 m scenario is useful for a country like Sri Lanka, which has a coastal zone management plan for various aspects of the littoral environment.

Coastal Resources Sector

Methods Used

Selection of Study Sites

As previous studies on beach erosion and recent geomorphology and land use surveys reveal, the southwest coast is the most vulnerable of Sri Lanka's coastal areas to sea level rise. In this area, erosion over the long term has been estimated to be three times greater than the accumulation of beaches, and the coast has receded hundreds of meters during the last seven decades.

Wind-generated waves and swells resulting from the southwest monsoon are the main factors governing the erosion process along the southwestern coast. The waves and swells are modified by morphometric features and landforms when they reach the shore. Significant modifications of waves are closely associated with headlands, reefs, beach rocks, rock outcrops, islets, etc. In addition, the orientation of the coastline governs the erosion associated with the waves generated by the southwestern monsoon. This is illustrated by the engineering structures constructed along the southwestern coast in the last several decades. The structures are mainly concentrated in coastline areas that abruptly change spatial orientation. These facts suggest that the southwestern coast should be given high priority among the coastal zones of Sri Lanka.

The Hikkaduwa area is one of the most degraded parts of the southwest coast; during the last century, this area has lost valuable wide sandy beaches. Hundreds of tourist hotels and popular recreational beaches as well as abundant coral reefs are found along this coast.
Since an expected sea level rise will immediately affect the southwest coast of Sri Lanka, the coastal stretch from Devundara Head to Negombo was selected for the present study (Figure 10.1). The study area forms an arc of 225 km and faces the southwest monsoon, which brings heavy seas from May to August. The beaches are eroded during the monsoon season and the materials taken from the beaches are transported by longshore currents toward the northern and southeastern coasts. The eroded beaches are sometimes redeposited during the intermonsoon period (October to March) in which calm seas prevail. However, they may be eroded again during the next monsoon, when waves may damage road fronts or highways, homestead gardens, tourist hotels, and other coastal resources.

Method of Coastal Classification

The classification of coastal stretches of the study site is based on practical application (Weerakkody, 1993a). The geomorphologic units, morphometry, and refracted waves of the coast were surveyed using 1:10,000 black and white panchromatic aerial photographs from 1987. The coast was divided into systems and subsystems using criteria such as concavity, beach plan curvature, indentedness, and morphologic units. Based on the identified systems and subsystems, areas subject to beach erosion and inundation were studied.

Method Used to Estimate Land Loss

Many scientists have studied beach erosion in Sri Lanka over the last three decades. Some studies have identified the causes of erosion without taking future sea level rise into account (Zeper, 1960; Swan, 1974; Gerritsen, 1974; Weerakkody, 1991). Therefore, the findings of these scientists are insufficient for the purposes of this study.

Contour lines of each system or subsystem in the southwestern coastal zone were drawn based on spot heights obtained from measurements of beach profiles. The estimates of vulnerability to sea level rise presented in this chapter are rough because the measurements necessary for drawing contour lines have not been completed. Using these contour lines and the scenarios of sea level rise,

Table 10.2. Categories and Subcategories of Beaches and Coasts

Beach/Coast Category	Subcategory	Landform/Feature
Sandy beach	—	Berm Cusps Terraces
Barrier beach	Barred by coral reef	Continuous reef Reef patches
	Barred by beachrock	Beachrock on beach Beachrock on shore
Cliff coast	High — over 15 m Medium — 8-15 m Low — less than 8 m	—
Coast with dune	High dunes Low dunes	Foredune Seif dune
Rocky coast	High degree of relief Low degree of relief	Homogenous rock Heterogenic rock
Estuaries	Broad	Riverine Lagoonal
	Narrow	Riverine Lagoonal
Bay beaches	Protected by two headlands	Bay head Bay mouth
	Protected by single headland	Bay head Bay mouth
Pocket beaches	Protected by two headlands Protected by single headland	Protected pocket Nonheadland pocket
Artificial beaches	Sandy beach with rock outcrops Sandy beach without rock outcrops	—
Coast with artificial harbor	Fisheries Commercial	—

the areas of inundation were calculated for each coastal system or subsystem.

Because Sri Lanka's financial resources for videomapping are limited, the area was surveyed using a checklist containing 10 natural and artificial beach and coast categories and 19 subcategories (Table 10.2). To assess vulnerability to the sea level rise scenarios of 0.3 m, 1.0 m, and 2.0 m, the coastal zone was divided into 1 km stretches and details of the land use, structures, and geomorphological features of each coastal stretch were measured. If the stretch consisted of homogeneous landforms, only one beach profile was made. If the hinterland of a coastal stretch consisted of lowland areas, marshlands, valleys, or hollows, more than one beach profile was taken. The same process was applied to obtain 1.0 m and 2.0 m contours.

Methods Used for Economic Analysis

In separate field work, a socioeconomic survey was conducted for each 1 km stretch of the selected coastal areas. It consisted of a questionnaire on type of land use, value of land, important human activities, and structures built along the coast. This survey formed the basis for the economic and social data used in the study. However, because the quantitative survey has not yet been completed, the results discussed here should be viewed as preliminary.

In addition, base maps were compiled using 1974 aerial photographs at a scale of 1:25,000. Because the scale of the photographs was not large enough to plot all necessary details of the coast, the data were complied using the checklist. In all, 225 checklists were filled out, each representing a 1 km stretch of the study area's coast. Each checklist indicates whether a given land use or structure is observed or not. By tabulating the information included in each checklist, a table was compiled which shows how many structures or how many acres of land in specific uses lie in the study area. The number of tourist hotels, for example, was obtained by tabulating the number of tourist hotels in each kilometer of the coastal zone.

Adaptation Alternatives and Evaluation Methods

In this study four adaptation alternatives were evaluated:

- retreat
- selective protection
- full protection
- reclamation/expansion seaward.

Results

Landforms and Their Classification

The study area consists mainly of a straight coastline with sandy beaches, and many former barriers can be observed in the hinterland. Because of the study area's near absence of hard rocks and sea cliffs, its natural protection against sea-induced erosion is minimal. However, scattered rock outcrops may be observed at several shore and offshore locations. Although such outcrops are thought to protect the coast, headward erosion is taking place within two rock outcrops that behave as a sluiceway (Weerakkody, 1992a). Within the study site, coral reefs are in successive stages of development from east to west (Mergner and Sheer, 1973). If a protective coral reef is destroyed by human activities or natural processes, a sluiceway is formed; thus, headward erosion of the immediate beach occurs. Many sluiceways have formed along the destroyed coral reefs or rock outcrops along the southwestern coast.

The short-term effect of rising sea level, however, is governed by the occurrence of headlands on the coast. A few headlands along the southwestern coast protect beaches against erosion. Therefore, from a geomorphologic point of view, an analysis of headland efficiency is valuable in determining which coastal segments of the study area are vulnerable to accelerated sea level rise. Swan (1965) measured the efficiency of headland protection along the coast from Uswetakeyyawa to Tangalla (within the present study area). He categorized the headlands along these coasts as follows:

- efficient, stable, resistant to erosion
- declining efficiency due to retreat
- temporarily efficient with small headlands (rock knobs) that might become isolated during storms
- inefficient, with poorly defined or without headlands

- defunct (headland cut off from mainland or destroyed).

Weerakkody (1992a) categorized the coast by these five types of headland efficiency (Table 10.3).

Some characteristics of headland interference with littoral drift can be derived from this table. The results can be summarized as follows:

- The Uswetakeyyawa to Balapitiya area, which is partly situated in the zone of decreasing littoral drift mentioned by Gerritsen (1974) consists of 50% Type A (efficient) headlands. Thirteen of these 14 headlands are situated between Kalutara and Balapitiya (Swan, 1965). It can thus be deduced that sediment is captured and transported northward by the longshore currents from the Balapitiya coastal tract to Kalutara.

- The Balapitiya to Dickwella area has only a few efficient headlands, as compared to the Uswetakeyyawa to Balapitiya area. Nine (12%) Type A headlands are not sufficient to interfere with the littoral drift on this stretch of coast.

- The Dickwella to Tangalla area consists of 15 headlands, 7 of which are efficient (Type A) headlands. This means that in this area, the littoral drift can be efficiently controlled by the headlands. The distribution of efficient, stable, resistant headlands in this area begins at Dickwella Bay, which is situated beyond the study area. Accordingly, short-term protection against sea-induced erosion can be expected along the coast from Balapitiya to Devundara (or Dickwella). This coastal stretch includes some 104 km, which account for approximately half of the study area. The other half of the study area, stretching from Balapitiya to Negombo, is less affected by rising sea level than the Uswetakeyyawa to Balapitiya area.

Table 10.3. Headland Efficiency in Three Coastal Tracts from Uswetakeyyawa to Tangalla

		Uswetakeyyawa to Balapitiya		Balapitiya to Dickwella		Dickwella to Tangalla	
Type	Efficiency	Number of Headlands	Percent of Headlands	Number of Headlands	Percent of Headlands	Number of Headlands	Percent of Headlands
A	Efficient: stable headland, resistant to abrasion	14	50%	9	12%	7	47%
B	Declining efficiency due to retreat	3	12%	19	26%	6	40%
C	Temporarily efficient	2	7%	16	23%	0	0%
D	Inefficient: poorly defined or without headland	6	21%	22	30%	2	13%
E	Defunct: headland cut off or destroyed	3	11%	8	9%	0	0%
Total		28	100%	74	100%	15	100%

Table 10.4. Landforms and Geomorphological Features along the Southwest Coast

Landform and Geomorphological Feature	Number[1]	Percent[2]
Coral reef/beachrock	67	12%
Headland	43	8%
Sea cliff	35	6%
Rock outcrop	83	15%
Islet	12	2%
Bay mouth beach	38	7%
Estuary and outfall	29	5%
Sand spit	26	5%
Straight coast	176	32%
Lagoonal beach	23	4%
Artificial harbors	24	4%

[1] Number of km in which the landform or geomorphological feature was observed.
[2] Percent of km in which the landform was observed.

These differences should be taken into account when formulating policies and implementing adaptation alternatives. Table 10.4 describes the most dominant landforms and geomorphological features of the southwestern coast.

The bay mouth beaches, which represent 7% of the study area, are least affected by rising sea level. Another 8% of the area is protected by headlands, but the efficiency of protection depends on the factors discussed above. No natural protection can be expected along the coast stretching from Balapitiya to Devundara Head.

The areas most vulnerable to rising sea level are the sand spits, straight coasts, and islets. They comprise some 39% of the study area. Artificial harbors will also experience severe effects from sea level rise.

However, there will be compensation between the materials removed by wave action as a result of rising sea levels and the materials transported by estuaries and lagoonal outfalls. Therefore, lagoonal beaches and beaches adjacent to estuaries and lagoonal outfalls will be relatively more protected than other types of beaches.

If any spot height was found to be far from the coastline or boundary of a coastal zone, it was classified as a "highly vulnerable area" to sea level rise. If a well-established sea cliff or highland area made up of hard rock (granite, gneiss, etc.) was identified, it was classified as a relatively "stable area." The remaining areas were classified as "moderately vulnerable" to sea level rise. This simple classification provides a rough overview of the effects of sea level rise on the study area's coast. Table 10.5 identifies the degree of vulnerability for stretches of the southwest coast.

Land Loss Estimates

Because the simplified classification is not sufficient to obtain a detailed picture of sea level rise impacts, the area of land loss for each kilometer stretch of the coast was calculated using 0.3 m, 1.0 m, and 2.0 m contour lines.

Erosion. Table 10.6 shows the estimated land loss due to beach and cliff retreat induced by erosion.

Inundation. The areas of coastal lowland inundation, especially areas adjacent to marshlands, lagoons, and estuaries, are somewhat larger than those lost through erosion. Table 10.7 shows the inundation of areas adjacent to the marshlands, lagoons, and estuaries.

Table 10.5. Classification of the Southwest Coast on the Basis of Vulnerability to Sea Level Rise

Highly Vulnerable[1]	Moderately Vulnerable[2]	Relatively Stable[3]
Estuary of the Nilwala	Meddewatta	Dondra Head-Meddewatta
Kapparatota-Ahangama	Kotowegoda	Unawatuna-Galle
Gintota estuary	Matara-Pulliena	Galle-Gintota
Narigama-Hikkaduwa	Bambarenda-Pemuyana	Gintota-Narigama
Hikkaduwa-Scenigama	Mirissa-Kapparatota	Ambalangoda
Scenigama-Ambalangoda	Ahangama-Koggala	Galkanda-Induruwa
Bentota-Aluthgama	Koggala-Unawatuna	Aluthgama-Beruwala
Colombo harbor-Palliyawatta	Balapitiya-Godakanda	Kalamula-Moratuwa
Palliyawatta-Kapungoda	Induruwa-Bentota	Colombo harbor
Kapungoda-Duwa	Beruwala-Maggona	
	Maggona-Kalamulla	
	Moratwa-Colombo	

1 Spot height far from the coastline or boundary of a coastal zone.
2 All areas that are not defined as highly vulnerable or relatively stable.
3 Well-established sea cliff or highland areas made up of hard rock.

Table 10.6. Land Loss of the Southwest Coast due to Erosion[1]

Sea Level Rise	Land Loss (km^2)
0.3 m	6.0
1.0 m	11.5
2.0 m	23.5

1 Assuming no protection.

Sand Spits and River Mouth Barriers. Under the 2.0 m sea level rise scenario, sand spits could either disappear or develop. The sand spits and barriers at river mouths may be destroyed by newly introduced erosion as a result of accelerated sea level rise. Alternatively, when sea level rises, the sediment brought by rivers and lagoons into the sea may be barricaded and pile up in front of river mouths or outfalls, as Sri Lanka's coasts are experiencing today. [However, this may be due to the equilibrium profile of Johnson (1919), which has been criticized.]

Table 10.8 shows estimates of the length of sand spits and river mouth barriers along the southwestern coast that would be lost if beach erosion progresses.

Table 10.7. Inundation of Lowlands Adjacent to Marshlands, Lagoons, and Estuaries on the Southwest Coast (km^2)

	Sea Level Scenario		
	0.3 m	1.0 m	2.0 m
Adjacent to Marshlands			
Goyyapana	0.0	1.0	6.5
Galle-Magalle	0.5	2.0	4.0
Akurala	0.0	4.0	11.0
Ahungalla	0.0	0.5	2.0
Induruwa	0.0	0.5	4.0
Bentara Ganga	4.0	6.5	12.0
Palliyawatta-Pinwatta	1.5	2.0	4.5
Nugape-Bopitiya	4.5	12.0	17.0
Endigastenna-Elenegoda	3.0	4.5	8.0
Subtotal	9.5	33.0	69.0
Adjacent to Lagoons			
Guarantuwa	0.0	2.5	4.0
Ratgama	3.0	5.5	8.0
Hikkaduwa Ganga	2.5	3.3	6.0
Madampe	3.5	7.0	15.1
Randombe Lake	0.0	1.0	5.0
Madu Ganga	1.0	2.5	4.0
Panadura Ganga	6.0	8.0	13.0
Baire Lake	0.0	1.0	1.0
Negombo Lagoon	12.0	23.0	34.0
Subtotal	28.0	53.8	90.1
Adjacent to Estuaries			
Gin Ganga	1.0	3.5	5.0
Kelani Ganga	2.0	4.5	9.0
Subtotal	3.0	8.0	14.0
Total	40.5	94.8	173.1

Table 10.8. Length of Sand Spits/River Mouth Barriers Lost due to Erosion

Sand Spit/River Mouth Barrier	Length (km)
River mouth barrier of Nilwala River at Matara	3.15
Small sand spit at the outfall of Koggala lagoon	0.25
River mouth barrier of Gin Ganga	1.00
Balapitiya	1.00
River mouth barrier of Bentara Ganga	2.50
River mouth barrier of Kaluganga	3.50
River mouth barrier of Panadura Ganga	7.00
River mouth barrier of Kelani Ganga	0.50
Sand spit stretched from Duwa to Pamunugama	14.00
Barriers developed along the Negambo lagoonal outfall	1.00

The sediment carried by rivers and lagoons is the main source of sediments that feed the beaches of Sri Lanka's coasts. The erosion of sea cliffs may also produce some sediment load to the beaches, but there are few sea cliffs along the southwestern coast. The coast from Devundara to Ambalangoda, which covers half of the study area, is fed by sediment transported into the Madampa, Hikkaduwa Canga, Ratgama, Gin Ganga, and Koggala lagoons, and the Polwatu and Nilwala rivers. However, the sediment load they carry is not sufficient to feed the entire study area and prevent beach erosion (Weerakkody, 1993b). Table 10.9 shows the feeding distance to beaches by river and lagoon.

If runoff brought by the southwestern monsoon increases due to global warming, all of the rivers in the study area will discharge high water loads to the Indian Ocean. This may result in an excess load of sediment being transported to the river mouths, giving rise to the formation of river mouth barriers in front of stream outlets (in contrast to the decrease in river mouth barriers via the processes induced by accelerated sea level rise).

Table 10.9. Beach Feeding Capacity of Drainage Systems between Ambalangoda and Devundara

Drainage	Feeding Distance[1] (m) Toward North/Northwest[2]	Toward South/East[3]	Basin Area (m^2)	Discharge (acre/feet)
Madampa Lagoon	340	640	23	137
Hikkaduwa Lagoon	1,800	480	na	na
Ratgama Lagoon	3,520	3,280	4	23
Gin Ganga	3,480	480	360	1,150
Osangoda Lagoon	520	1,200	na	na
Gluganga Lagoon	1,080	120	na	na
Koggala Lagoon	4,560	880	25	137
Midigama Ela	880	560	na	na
Polwatu Ganga	4,000	0	91	255
Garanduwa Lagoon	680	600	na	na
Nilwala Ganga	720	2,720	375	1,210

1 Feeding distance is defined as the distance from the mouth of the drainage to the point where the beach width in front of the drainage takes half of that width.

2 The orientation changes as northwest/north at the Galle beach.

3 The orientation changes as south/southeast at the Galle beach.

More construction in the densely populated study area will increase sand mining in river beds. Thus, any benefit gained by increased precipitation to feed the beaches via excess sediment load is likely to be minimized by sand mining activities in both river beds and beaches.

Socioeconomic and Land Use Activities

The southwestern coast from Devundara to Negombo is the most densely populated area of Sri Lanka. It holds the Greater Colombo area in the north, as well as small cities and suburbs in the coastal zone, creating high pressures on coastal resources.

Because socioeconomic activities are complex along the coastal zone, they were categorized in 10 major types. Fishing, the area's dominant economic activity, was categorized as shore area fishing, offshore fishing, ma-del (nearshore fishing in which a large net and two canoes are used), lagoon fishing, and prawn and crab fishing. Well-developed commercial fishing management systems are not employed in the study area, and the fish are sold in local markets. Other important socioeconomic activities include dry fish processing, coral mining, sand mining, coconut based industries, and tourism. Fishing in the shore area (using traditional crafts and powered boats) and coconut industries are the most widely distributed economic activities along the coast.

Although hundreds of land use activities were observed along the southwestern coast, only the five most common types were analyzed for this study: (1) commercial; (2) mining, industrial, and fishing; (3) services; (4) agricultural; and (5) residential. Each group consists of several major types of land use activities, as shown in Table 10.10.

The distribution of land uses along the southwestern coast is not even. Some areas such as Hikkaduwa, Negombo, Mt. Lavinia, Wellawatta, Colombo, Ratmalana, and Dehiwala have dense tourist hotels and other commercial activities. The coasts from Polhena to Habaraduwa, from Ratgama to Akurala, from Randombe to Panadura, and from Kelaniya to Negombo are under mainly subsistence or commercial agriculture land uses.

One of the most beautiful beaches on the southwestern coast is the Hikkaduwa area; its land remains in traditional use patterns compared to the Colombo metropolitan area. The impacts of sea level on the Hikkaduwa area will be several because the nation's most attractive tourist centers are situated along this coast. This area's coral reefs should be protected along with its immediate beaches. In addition, the lowlands adjacent to the wetlands and lagoons should be protected to conserve the area's natural resources. The placement of huge boulders to protect this area against sea-induced erosion has partially destroyed the natural beauty of the existing beaches. This coastal zone's sandy beaches could be completely eroded, with the exception of areas protected by revetments. Rising sea levels may also raise the base levels of lagoons, and cause the migration or partial destruction of coral reefs, and the collapse of highways.

The study area's land use patterns do not entirely reflect its socioeconomic activities. The most destructive human activity that directly affects erosion is mining. Sand mining activities along the Sri Lankan coast are prohibited, but illegal sand mining is practiced in certain areas where enforcement is weak. Coral mining in the study area is of

Table 10.10. Major Categories of Land Use on the Southwest Coast (km with each type of land use)

Commercial			Mining, Industrial, and Fishing			Services			Agriculture				Residential		
H	TC	NTC	M	IN	FI	BRS	R	SS	HG	CA	SA	LI	PH	TH	PG
81	61	122	193	98	162	55	96	135	185	148	201	26	112	153	140

H = Hotel, TC = Tourist center, NTC = Nontourist center, M = Mining, IN = Industrial site, FI = Fisheries centers, BRS = Bus/railway station, R = Road, SS = Service station, HG = Homestead garden, CA = Commercial agriculture, SA = Subsistence agriculture, LI = Livestock, PH = Permanent house, TH = Temporary house, PG = Playground.

three types: the mining of living reefs, the mining of raised reefs in the hinterlands, and the collection of coral on beaches.

Coral mining has destroyed reefs and has severely affected the dissipation of wave energy on beaches, leading to erosion (Weerakkody, 1992a; 1993a). Mining activities in the hinterland contribute to the inundation of lowlands; hundreds of mine holes filled with brackish water provide the conditions necessary for the intrusion of sea water.

Some areas like Akurala and Wellameda have already been protected by the Department of Coast Conservation, which built a huge revetment to overcome these problems.

Two other economic activities that are endangered by sea level rise are ma-del fishing (which concentrates on shore and near-offshore areas) and transportation. Since canoes are kept on beaches and nets are prepared on them, the loss of beaches will adversely affect fishing communities. The greatest impediment is that access to the beaches has been restricted by revetments and groins. In the future, traditional fishing activities may vanish entirely from the southwestern coast due to beach erosion and the construction of revetments along the coastline.

The highway and railroad from Colombo to Matara were constructed parallel to the coastline; therefore, a 1.0 m rise in sea level will severely damage many parts of them. The amount of damages and the cost of protection work for roads and railways will be calculated in a future study.

Evaluation of Adaptation Policies

Several environmental studies have been launched by Sri Lanka during the last decade to assess the nation's vulnerability to sea level rise and to evaluate adaptation alternatives. The Regional Study on Global Environmental Issues carried out under the Climate Change in Asia Program is one of the most in-depth of these (CEA, 1991). Two strategies are suggested to protect wetlands:

- allow the wetlands to function as they are, so that sea level rise will produce natural changes in the ecosystem
- enact protective measures for each wetland depending on its economic and ecologic importance.

This assessment shows that wetlands will be expanded by inundation, which will cause many environmental problems for the inhabitants in those vulnerable areas. Therefore, wetlands should be allowed to function as they are, but measures should also be taken to protect the communities living around them. This is especially true for the Negombo lagoonal area and the Kalutara and Bentara estuarine areas.

The Department of Coast Conservation gains experience every year in conducting protection work at sites on the southwestern coast. For example, at the time this research was being conducted, sand bags were being laid along the Telwatta beach at Malawenna. This temporary protection was undertaken to protect the Colombo-Galle road.

Adaptation Alternatives

In providing protection, relief, and vital utilities to coastal communities threatened by sea level rise, the two strategies for protecting wetlands may be divided into four alternatives: retreat, selective protection, full protection, and reclamation/expansion seaward.

Retreat

This policy has been practiced in Sri Lanka for the past six decades. Houses and tourist hotels constructed along the coastal zone are severely damaged by waves, especially during the southwestern monsoon, which generates heavy seas and stormy weather in the study area. Damaged houses are evacuated if temporary mitigation measures such as laying sand bags and building masonry walls are not effective. Evacuation has been very common in the study area during the last 50 years. However, if the property is very valuable, the owners often try to build structures to protect their homes. It is necessary to receive approval from the Department of Coast Conservation to build any structure that protects against erosion.

The policy of retreat should not be applied to areas that are densely populated and clustered with commercial activities. Since the highway and railway line that connect Colombo and Devundara (and eventually, further south) run parallel to the coastline, any destructive event generated by sea level rise should be taken into consideration immediately. Once the proposed highway connecting Colombo and the southern cities is built, the importance of protecting sea-induced erosion of railroads and highways may be reduced.

Selective Protection

Low coastal lands and lowlands adjacent to wetlands are the most vulnerable areas to the immediate effects of sea level rise. Such areas should be allowed to migrate landward, especially wetlands. The demarcation and definition of the coast, estuaries, and other coastal lands in the Coast Conservation Act of 1981 support the evacuation of lands that should be preserved for their natural value. Selected areas of such lands should be protected, depending on their ecological status, fauna and flora, cultural values, and interconnectivity with other ecosystems.

Full Protection

Full protection implies taking action to prevent erosion and inundation at all costs (i.e., holding the present shoreline). Some steps have already been taken to prevent erosion along the southwestern coast of Sri Lanka. Some 37,900 m of revetments and 3,400 m of groins have been constructed to prevent erosion. In addition, the policy of preservation at all costs is being practiced in

Sri Lanka under the Coastal Zone Management Plan (Department of Coast Conservation, 1990). The following elements of the plan deal with controlling erosion.

- using appropriate erosion control techniques that are cost-effective and socially and environmentally acceptable
- directing new development away from eroding shorelines
- requiring that new construction in erosion-prone areas be set back from the shore at a sufficient distance so that it will not be threatened by erosion for some time
- regulating human activities that contribute to accelerated erosion such as coral and sand mining
- protecting and, where possible, restoring the natural features that inhibit erosion.

Reclamation/Expansion Seaward

As revealed by the present geomorphologic survey, the most vulnerable areas of the southwestern coast are sand spits, river mouth barriers, and straight sandy beaches. To protect some sand spits, the Negombo beaches received artificial beach nourishment along a 7 km stretch of coast at a cost of US$3.75 million.

Artificial nourishment may be a feasible alternative to the policy of preservation at all costs. Full protection is difficult to maintain in the case of river mouth barriers and sand spits because they cannot be preserved using a groin or revetment. However, financial constraints must be overcome before other such programs can be undertaken in Sri Lanka.

Lessons Learned from Historical Practices to Protect the Coast

Sri Lanka's coast has been protected for more than 200 years. Groin fields in the Bay of Galle indicate that people struggled against sea-induced erosion as early as the 17th century. At the same time, however, they destroyed valuable landforms such as sandbanks and beach rocks, which protect against erosion (Weerakkody, 1987).

Since the 1970s the nation's eroding coasts have been protected by various government departments. The Department of Railways, for example, laid revetments or groins to protect against sea incursion on rail lines, while the Department of Irrigation constructs groins to protect river mouths against the formation of sand bars and to prevent the intrusion of saline water and the inundation of lowland catchment areas.

These activities have led to the construction of hundreds of engineering structures along the coast without any prior assessment (Gerritsen, 1974). The structures were built for such purposes as protecting against sea-induced erosion, constructing harbors, creating barriers to saline water intrusion, and preventing the development of barriers to river mouths. Engineering structures have been built along the coastline from Matara to Ambalangoda, and these structures have sometimes introduced erosion to coastal segments (Weerakkody, 1990). This practice occurred before the Department of Coast Conservation was established in Sri Lanka.

After having studied the distribution of all engineering structures built up to 1987, Weerakkody (1990) concluded that the changing orientation of the coastline in relation to wind direction is the major factor governing coastal erosion in Sri Lanka.

The establishment of the Department of Coast Conservation is the greatest step in achieving organization of coastal management activities in Sri Lanka. At present, the department is the main body governing the implementation of coastal protection work, and planning and management of the coastal zone. A Master Plan that covers construction work, protection work, legal aspects, development, and planning is now being implemented by the department.

Following the policies adopted by the department, huge revetments were placed along some endangered segments of the southwestern coast. Revetments of 4,873 m and 2,970 m along the west and south coasts, respectively, have been built very recently, at US$600 per meter of revetment. In the near future, another revetment will be built along the rest of the coastline. However, there is no evidence that rising sea level was seriously taken into account in designing and constructing the revetments (tidal changes were the primary consideration).

Furthermore, protection and maintenance of the hundreds of revetments, groins, masonry walls, etc., already built along the coast will cost millions of dollars if sea level rise accelerates.

Retreat from areas attacked by waves has been practiced in several areas of Sri Lanka during the past decades. For example, the slums north of the Hikkaduwa Ganga outfall were evacuated during several years of sea-induced erosion. In a similar manner, hundreds of houses damaged by sea waves along the coast of Akurala have been evacuated in the last two decades. In the future, this policy will be maintained along some coastal segments. Thus, retreating from some areas for a short period of time is essential. However, no studies or programs have been carried out to educate people on this matter.

Discussion and Conclusions

Some steps are being taken to conserve the coast and recommendations for the future are being developed. These include (1) prohibiting coral and sand mining, which will help conserve beaches; (2) educating the public on the hazards of coastal erosion and accelerated sea level rise; and (3) reserving some coastal lands for nature and delimiting the coastal zone. This has been stipulated by the Coast Conservation Act of 1981; however, enforcement is weak due to the absence of qualified staff. As a result of sea level rise, salinity intrusion and flooding will render some of the low-lying agricultural areas unfit for cultivation. Large tracts of these lands have been identified by this study.

In addition, the following adaptation alternatives are proposed to adapt to the decline in water resources that is likely to be caused by sea level rise. The following strategies are proposed to ensure the optimum use of water resources:

- introduce low-cost miniature reservoirs for local irrigation and rehabilitate small tanks in the dry zone

- recycle water used for domestic, industrial, and recreational purposes
- improve water use practices and techniques
- ensure sound watershed management
- grow crops that consume less water
- reallocate water to meet growing demands
- increase the availability of water by utilizing additional groundwater
- reforest and revegetate hillsides to trap runoff, reduce erosion, replenish groundwater supplies, and increase wood supplies
- increase the productivity of existing irrigated lands through crop diversification and higher cropping intensities.

References

CEA (Central Environmental Authority of Sri Lanka). 1991. *The Greenhouse Effect and its Impact on the SAARC Region: Sri Lanka Country Report.* Colombo: Ministry of Environment and Parliamentary Affairs.

Department of Coast Conservation. 1990. *Coastal Zone Management Plan.* Sri Lanka Department of Coastal Conservation, Colombo.

Gerritsen, F. 1974. *Coastal Engineering in Sri Lanka.* Report of UN Mission. Honolulu: University of Hawaii.

Johnson, W.D. 1919. *Shore Processes and Shoreline Development.* New York: John Wiley & Sons, Inc.

Mergner, H. and G. Sheer. 1973. "Physiographic Zonation and the Ecological Conditions of Some South Indian and Ceylon Coral Reefs." *Proc. 2nd International Coral Reef Symposium*, pp. 3-30. Brisbane.

Swan, B. 1965. "Coast Erosion Principles and a Classification of the Southwestern Ceylon's Beaches on the Basis of Their Erosional Stability." *Ceylon Geographer.* vol. 19(1-40), pp. 1-16.

Swan, B. 1974. *Coastal Erosion Hazard of SW Sri Lanka.* University of New England, Australia. Unpublished Report.

UCAR (University Corporation for Atmospheric Research). 1994. *El Niño and Climate Prediction Reports to the Nation.* Boulder, Colorado, USA.

Weerakkody, U. 1987. "Use of old document in mapping coastal changes." *ITC Journal*, 1987-2, pp. 139-144.

Weerakkody, U. 1990. "Engineering Structures of the Southwestern Coast of Sri Lanka." Paper presented to the Symposium on Ecology and Landscape Management of Sri Lanka, March 12-26, 1990, Taj Samudra Hotel, Colombo.

Weerakkody, U. 1991. *Changing Coastline of Southwestern Sri Lanka.* Colombo: Natural Resources, Energy and Science Authority of Sri Lanka (NARESA).

Weerakkody, U. 1992a. *Geomorphological Evolution of the Southwestern Coast of Sri Lanka*. Unpublished report. Colombo: NARESA.

Weerakkody, U. 1992b. "The Holocene Coasts of Sri Lanka." *The Geographical Journal*. vol. 158, no. 3, November 1992, pp. 300-306.

Weerakkody, U. 1993a. *Spatial and Temporal Diversity of Coral and Algal Communities of Southwestern Sri Lanka with Special Emphasis on Destructive Human Activities.* Unpublished report submitted World Wildlife Fund, Washington, DC.

Weerakkody, U. 1993b. "Beach Feeding Capacity of Rivers and Lagoons of the Southwest Coast of Sri Lanka." *Journal of National Science Council of Sri Lanka*. vol. 21(1), pp. 111-123.

Zeper, J. 1960. *Sea Erosion Studies and Recommendations on Coast Protection in Ceylon.* The Hague: Netherland Bureau for International Technical Assistance. Ministry of Foreign Affairs, The Hague, The Netherlands.

Chapter 11

Vulnerability and Adaptation Assessments for the Czech Republic

Bedřich Moldan,[1] Jaroslava Kalvová,[1] Ivana Nemešová[2]
Josef Hladný,[3] Rudolf Brázdil,[4] Jaroslav Rožnovský,[5] Bohuslav Vinš[6]

[1] Charles University, Prague
[2] Institute of Atmospheric Physics, Czech Academy of Sciences, Prague
[3] Czech Hydrometeorological Institute, Prague
[4] Masaryk University, Brno
[5] Mendel University of Agriculture and Forestry, Brno
[6] National Forest Committee, Prague

SUMMARY. This chapter summarizes the results of the Czech Republic's assessments of vulnerability and adaptation to climate change. This assessment has two principal parts: detailed analysis of possible scenarios of future climate, and analysis of vulnerability and adaptation for agriculture, forestry, water resources, and health. The findings are summarized here for all sectors except the health sector, where the assessment found that the potential impact of climate change was insignificant (Kveton and Kazmarová, 1995). For the agriculture sector, a decrease in precipitation could lead to a change in winter wheat yields of -3% to +16% depending on the climate scenario. A number of adaptation measures are proposed, including improved soil management, integrated pest control, and careful water use. The potential impact on forests could be large, especially because there are already several problems (e.g., prevalence of Norway spruce monoculture) in this sector that would be exacerbated. The main adaptation mechanisms proposed are increased species diversity and sustainable forest management. The impact on water resources was estimated by modeling four catchments. The result under most of the scenarios was, at a minimum, a one-third decrease in runoff.

INTRODUCTION

The Czech Republic is a landlocked country in the middle of Europe, with four neighbors: Poland to the north, Slovakia to the east, Austria to the south, and Germany to the west (Figure 11.1). The surface area is 78,864 km². It is composed of the historical territories of Bohemia, Moravia, and a part of Silesia. It was established on January 1, 1993, when the former Czechoslovakia was split.

Note: This chapter does not contain a detailed description of the assessment methodologies that were recommended by the U.S. Country Studies Program. Refer to Chapter 2 for an overview of these methodologies.

J. B. Smith et al. (eds.), Vulnerability and Adaptation to Climate Change, 225–247.

Figure 11.1. Map of the Czech Republic

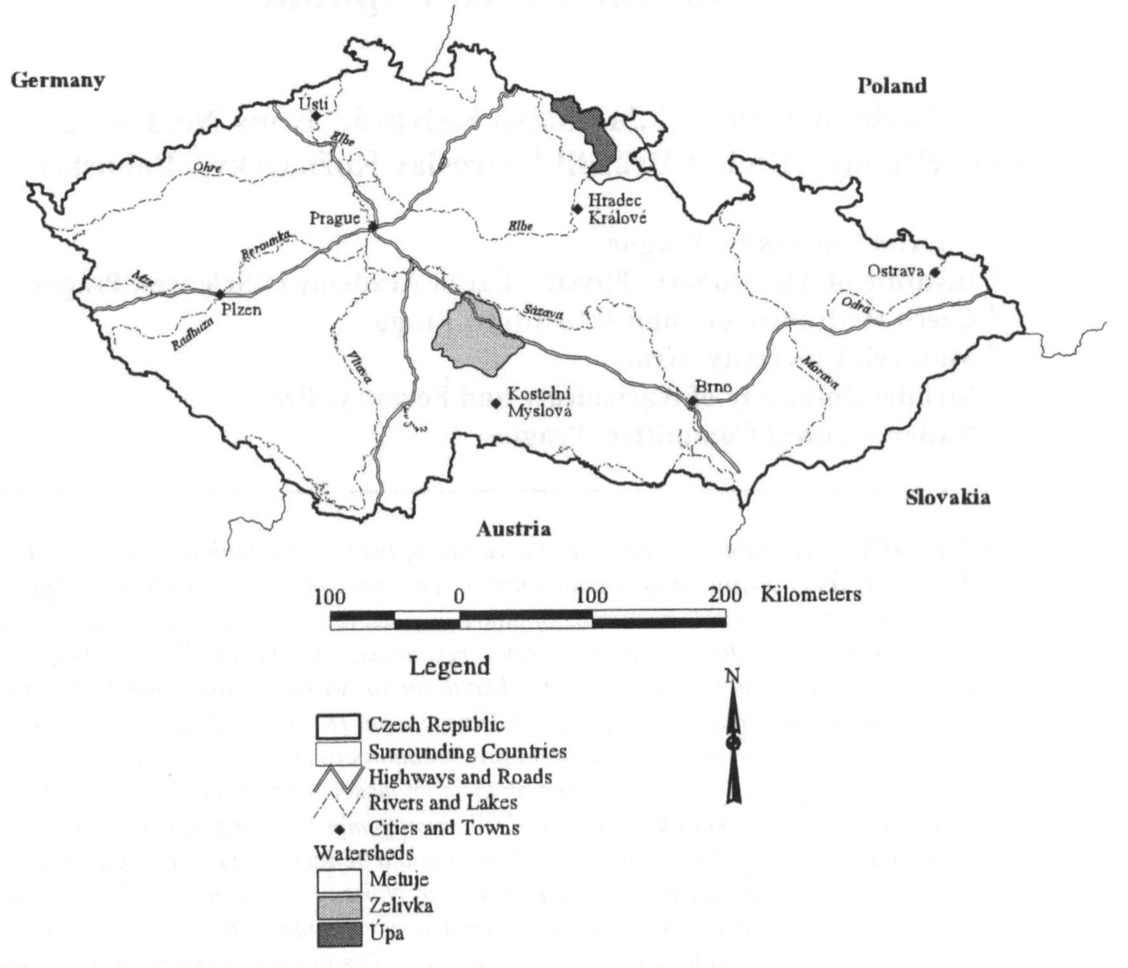

The Czech Republic has 10,362,000 inhabitants. The capital, Prague, has a population of 1.2 million. There are seven towns with more than 100,000 inhabitants; together they constitute 23% of the population. The demographic projections used in this study assume that any changes in the total population will not exceed 0.2% by 2000.

The Czech Republic has an Atlantic-continental moderate climate. It is a northern mild zone, with a typical rotation of four seasons and variable weather over the year. Mean annual temperature ranges between 1.0° and 9.4°C (between 8.8° and 18.5°C in the summer and between -6.8° and 0.2°C in the winter). Its complicated orography also affects the precipitation regime. The annual distribution of precipitation exhibits a continental pattern. The long-term average of precipitation is 686 mm (this is the mean of the annual values measured at 92 measuring sites for 1961-1990). The driest regions exhibit an annual average precipitation of 410 mm, and the highest annual average of 1,500 mm occurs on mountaintops. The snow cover is not stable, and the winters in the lower areas are usually without a permanent snow cover.

There are three major European river basins in the Czech Republic. The western part of the country is drained by the Elbe River to the North Sea, Central and South Moravia are drained by the Danube River (to the Black Sea), and North Moravia (Silesia) is drained by the Odra River to the Baltic Sea. Most of the country's rivers originate in the Czech Republic. The total stream flow in 1993 amounted to 10.1 million m^3, out of which 2.2 million m^3 (22%) were utilized.

The economy of the Czech Republic is undergoing transformation. The process of transformation started in 1990, and one of the first effects was reduced production. Most of the inefficient production ceased, as did exports to the former socialist countries.

The country has a long tradition of industrial activity. In 1990, about 47.7% of the economically active population of the Czech Republic was employed in industry and construction. Energy generation, iron and steel, machinery, glass and ceramics, cement and construction materials, chemicals, textiles, and food industries rank among the most important sectors. Industry is the largest energy consumer; in 1990 it consumed 48% of available energy.

In 1992, approximately 55% of the land in the Czech Republic was agricultural, and 33% was forested. In 1990, when the transformation of property and legal relationships began, the contribution of agriculture to the GDP dropped. The area of arable soil for cultivation of certain products has declined significantly. The use of pesticides and fertilizers has also been reduced. The production of rapeseed oil plants has increased, because the oil can be used as a renewable source of energy.

The current species composition of forests in the Czech Republic is a result of an intensive exploitation of monocultures for production of wood. A high proportion of monocultures adversely affects the stability of forest ecosystems. The forests in the country have also been heavily affected by SO_2 and NO_x emissions. This emissions load reduces the resistance of trees to wind, snow, insect parasites, and parasitic fungi; today more than 60% of the forest exhibits some level of damage.

The study of vulnerability and adaptation to climate change was started within the activities of the National Climate Program (NKP) of the Czech Republic and Slovak Federative Republic. It was officially started on January 1, 1991, as the response of the Czech Republic to the World Climate Program of the World Meteorological Organization (WMO). The NKP is a nongovernmental association of several institutes, including the Academy of Sciences, Charles University in Prague, Masaryk University in Brno, the Czech Hydrometeorological Institute, and others. In 1993, the NKP published a study entitled "Risk of Climate Change and its Mitigation" (Moldan, 1993).

The NKP is also responsible for the Czech Republic's study under the U.S. Country Studies Program. The Czech study has five elements that cover emissions inventory, assessment of vulnerability and adaptation, and proposals for mitigation measures at different levels. Element 2 of the study is devoted to the analysis of vulnerability and adaptation, and its results are summarized in this chapter.

Element 2 of the Czech Republic Country Study focused on two issues: climatology and the assessment of vulnerability and adaptation for the relevant sectors: agriculture, forestry, water resources, and health. These sectors were selected on the basis of a preliminary study that examined several sectors, including other sectors such as industry, building and construction, and transportation, which were not significantly affected by climate change. During the vulnerability and adaptation assessment, it became evident that the potential impact of climate change on health would not be significant, and thus that sector is not addressed here. This chapter presents the work on scenario development and, in addition, the methods used for the assessments of the effects on agriculture, forestry, and water resources are summarized.

SCENARIOS

To construct scenarios of estimated climatic change in the Czech Republic, the outputs of four general circulation models (GCMs) were used: the Canadian Climate Centre Model (CCCM), the Geophysical Fluid Dynamics Laboratory (GFDL) equilibrium (GFD3) and 1% transient (GF01) models, and the Goddard Institute for Space Studies (GISS) model (Benioff et al., in press). Two of these models (GISS and CCCM) were selected as the appropriate GCM scenarios for the Czech Republic. In addition, an incremental scenario was developed.

Baseline Climate Scenario

In selecting a baseline period, preference was given to 1961-1990 over 1951-1980. The data on temperature and precipitation from the later period are, in contrast to the data from the 1950s, of sufficient quantity and quality and also available on computer media. It might be a disadvantage to include the 1980s in the reference period, but according to Kalvová et al. (1995), this period does not cause serious limitations in the interpretation of scenarios for the Czech Republic.

The climatological database used for this study included monthly mean temperatures from 57 climatological stations, monthly totals of precipitation from 92 stations, and daily records of both (as well as solar radiation) from 16 stations situated in two areas selected as study sites in the agriculture sector analysis. The Czech Hydrometeorological Institute was the main source of the data used.

GCM Scenarios

To select suitable GCMs, the annual and seasonal values of the $1\times CO_2$ climate simulations in the area between 35° to 65°N and 10°W to 30°E were compared with CLIM data. We also assessed the ability of the GCMs to simulate annual cycles of temperature, precipitation, and solar radiation in several grid points near or in the Czech Republic. Comparisons were performed both for individual grid points and for averaged values from several grid points.

The temperature and precipitation averages in the Czech Republic were characterized by area mean values. Annual fluctuations in temperature and precipitation were based on monthly data. Monthly precipitation totals were expressed as a percentage of the annual precipitation total.

A comparison of the annual temperature and precipitation with $1\times CO_2$ scenarios shows that, for temperature, the best results were provided by the GISS and CCCM models. The remaining two models, and especially GFD3, have a rather high annual range of temperature, and prominent kinks appear in the curves of the annual fluctuations of temperature (this is also true for solar radiation with GFD3).

For precipitation, the GISS, CCCM, and GFD3 simulations were close to the observed data. The GF01 model moves the "alpine" precipitation maximum in Central Europe farther eastward and underestimates summer precipitation. Because the GISS model was relatively suitable for both temperature and precipitation, it was selected to generate a climate change scenario in the Czech Republic. The CCCM was selected as a second model to generate a climate change scenario because it performs quite well for both temperature and annual solar radiation. Another reason for using the CCCM scenario was its lower sensitivity to a $2\times CO_2$ climate condition as reflected in global temperatures (see Kalvová et al., 1995).

The first scenario (Table 11.1) is based on the GISS model, and the second (Table 11.2) is based on the CCCM. GISS and CCCM are used as upper and lower bounds, respectively, of the estimated changes. However, the range is rather small.

The values in Tables 11.1 and 11.2 apply to the whole Czech Republic. Because of the small size of the country in terms of the GCM resolutions and the uncertainties of estimates of temperature and precipitation changes on a regional scale, the gradient of changes across the Czech Republic territory was not taken into account.

Table 11.1. Adjustment Statistics for the GISS Scenario

	Jan	Feb	Mar	Apr	May	Jun	Jul	Aug	Sep	Oct	Nov	Dec	Annual
Temperature (°C)	+6.2	+5.7	+4.9	+4.0	+3.2	+2.7	+2.6	+2.7	+3.2	+4.1	+5.0	+ 5.8	+4.2
Precipitation Ratio	1.18	1.40	1.20	1.41	1.20	1.04	1.10	1.09	1.02	1.19	1.19	1.08	1.16

Table 11.2. Adjustment Statistics for the CCCM Scenario

	Jan	Feb	Mar	Apr	May	Jun	Jul	Aug	Sep	Oct	Nov	Dec	Annual
Temperature (°C)	+3.3	+4.3	+3.5	+2.4	+2.4	+2.8	+3.4	+2.8	+3.4	+3.1	+3.3	+2.2	+3.1
Precipitation Ratio	1.32	1.28	0.96	1.35	0.83	0.90	0.82	1.06	0.95	1.08	1.30	1.35	1.07

An example of the comparison of the GISS and CCCM 2×CO_2 scenarios to observed monthly mean temperatures and precipitation is shown in Figure 11.2. According to the GISS scenario, the new equilibrium 2×CO_2 climate is characterized by a decline in the annual range of temperature and an increase of the annual precipitation totals. The higher winter warming (approximately 41%) and lower warming in summer (when compared with the annual values) in this scenario correspond with our previous estimates (Kalvová, 1993; Kalvová and Brázdil, 1993; Kalvová and Nemešová, 1994). The CCCM scenario generates a dangerous combination in the summer of temperature increases and a 10% precipitation reduction.

For the impact assessments, the values in Tables 11.1 and 11.2 were assigned to 2075, and the values for the 2030 and 2010 were linearly interpolated.

Incremental and Other Scenarios

The incremental scenarios are based on a combination of annual mean temperature changes of 2 °C, 3 °C, and 4 °C and changes in annual precipitation totals of -5%, 0%, +5%, and 10%. In selecting values for this scenario, the results of preceding studies (analogues, correlation techniques) were also taken into account (see Kalvová et al., 1995).

Vulnerability can depend on how the prescribed changes in annual means are distributed among individual months. To take this into account, two or three variants of annual distributions were prescribed for each change in the annual mean (for example, constant increase in temperature and in precipitation in all months or higher warming in winter compared to summer). It was therefore possible to combine the changes in the annual distribution of temperature with several variants of the annual fluctuations of precipitation. Experts from the respective vulnerability sectors selected combinations that, according to their experience, could express the sensitivity of the sector best.

In addition, regional scenarios for the Šumava and Džbán regions were developed for the forestry sector assessment. For these regional scenarios, detailed spatial temperature and precipitation patterns were developed based on meteorological data, statistical approaches, and the GISS and CCCM scenarios.

Secondary Scenario Characteristics

In addition to the primary scenario characteristics (temperature, precipitation), secondary characteristics (Viner and Hulme, 1994) were used in assessing the potential impacts of climate change on the Czech Republic. These were mainly characteristics based on daily mean temperatures and daily extreme temperatures (sums of daily temperatures above threshold values, e.g., 8 °C, 10 °C, number of summer days, frost days, and so on). When only monthly values were available, estimates of changes in these characteristics were obtained using regression models (Kalvová, 1993; Kalvová and Brázdil, 1993). To incorporate changes in the frequency of extreme events in the scenarios we need a sufficient number of high quality meteorological daily data sets.

For the time being, we have been processing daily data necessary for estimates of impacts of the climatic change on agriculture via the

Figure 11.2. Comparison of the GISS and CCCM 2×CO_2 Scenarios to Observed Monthly Mean Temperature and Precipitation (2075)

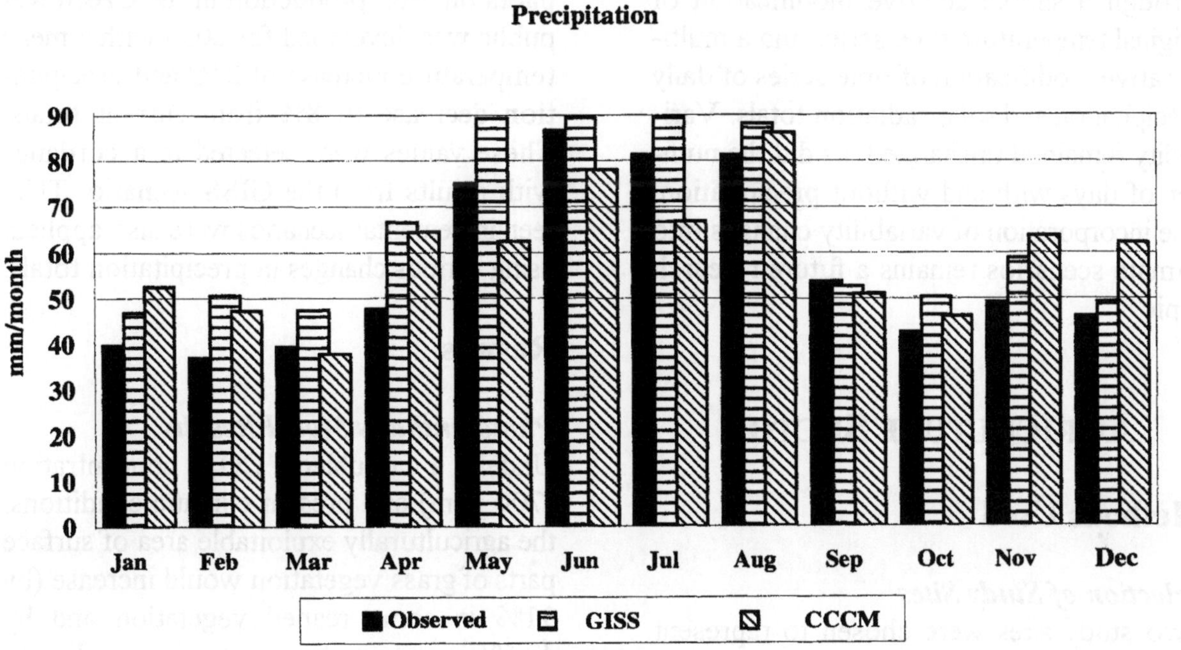

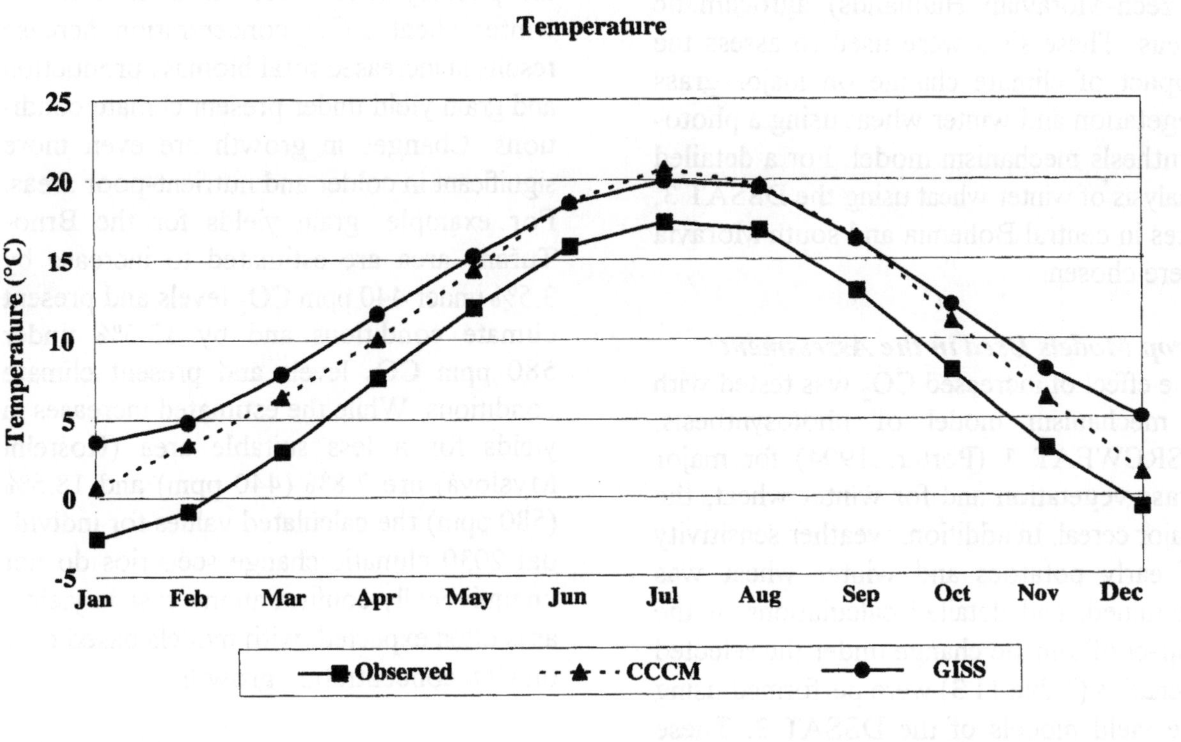

Decision Support System for Agrotechnology Transfer version 3 (DSSAT 3) (Benioff et al., in press). Time series of daily values for a 2×CO_2 climate were obtained through a simple additive modification of original temperature time series and a multiplicative modification of time series of daily precipitation and solar radiation totals. Variability remained unchanged, as did the number of days with and without precipitation. The incorporation of variability changes into climate scenarios remains a future research topic.

AGRICULTURE SECTOR

Methods Used

Selection of Study Sites

Two study sites were chosen to represent warm (south Moravia) and moderately warm (Czech-Moravian Highlands) agroclimatic areas. These sites were used to assess the impact of climate change on major grass vegetation and winter wheat using a photosynthesis mechanism model. For a detailed analysis of winter wheat using the DSSAT 3, sites in central Bohemia and south Moravia were chosen.

Crop Models Used in the Assessment

The effect of increased CO_2 was tested with a mechanism model of photosynthesis, ASRCWEAT 1 (Porter, 1994) for major grass vegetation and for winter wheat, the major cereal. In addition, weather sensitivity of early potatoes and winter wheat was examined, and detailed calculations of the impact of climatic change under the selected scenarios (Table 11.3) were performed using the yield models of the DSSAT 3. These models correspond, within a certain degree of tolerance, to cultivation conditions in the Czech Republic (Table 11.4). An incremental scenario of possible climate change impacts on crop production in the Czech Republic was developed for 2030 with a mean temperature increase of 2°C and precipitation decrease of 8% from current totals. These values were selected in accordance with results from the GISS scenario. Thirteen incremental scenarios were also applied, using various changes in precipitation totals.

Results

Photosynthesis Model Results

Under a doubled CO_2 concentration (700 ppm) and present climatic conditions, the agriculturally exploitable area of surface parts of grass vegetation would increase (by 31% in some reaped vegetation and by 42.5% in unreaped vegetation) according to the photosynthesis mechanism model. For winter wheat, a CO_2 concentration increase results in increased total biomass production and grain yield under present climate conditions. Changes in growth are even more significant in colder and nutrient-poor areas. For example, grain yields for the Brno-Tuřany area are estimated to increase by 3.5% under 440 ppm CO_2 levels and present climate conditions and by 12.5% under 580 ppm CO_2 levels and present climate conditions. While the estimated increases in yields for a less suitable area (Kostelní Myslová) are 7.8% (440 ppm) and 18.5% (580 ppm) the calculated values for individual 2030 climatic change scenarios do not unequivocally confirm an increase in yields, as is often expected, with models based only on CO_2 concentration growth.

Table 11.3. Adjustment Statistics for the Agriculture Assessment

Scenario	Change[1] of	Jan	Feb	Mar	Apr	May	Jun	Jul	Aug	Sep	Oct	Nov	Dec
1 Incremental	Tmp (°C)	2.9	2.7	2.3	1.9	1.5	1.3	1.2	1.3	1.5	1.9	2.4	2.7
	Pcp Ratio	0.95	0.95	0.95	0.95	0.95	0.95	0.95	0.95	0.95	0.95	0.95	0.95
2 Incremental	Tmp (°C)	2.9	2.7	2.3	1.9	1.5	1.3	1.2	1.3	1.5	1.9	2.4	2.7
	Pcp Ratio	1.00	1.00	1.00	1.00	1.00	1.00	1.00	1.00	1.00	1.00	1.00	1.00
3 Incremental	Tmp (°C)	2.9	2.7	2.3	1.9	1.5	1.3	1.2	1.3	1.5	1.9	2.4	2.7
	Pcp Ratio	1.05	1.05	1.05	1.05	1.05	1.05	1.05	1.05	1.05	1.05	1.05	1.05
4 Incremental	Tmp (°C)	2.9	2.7	2.3	1.9	1.5	1.3	1.2	1.3	1.5	1.9	2.4	2.7
	Pcp Ratio	1.05	1.26	1.25	1.13	0.96	0.82	0.71	0.81	0.95	1.10	0.98	0.99
5 GISS	Tmp (°C)	2.9	2.7	2.3	1.9	1.5	1.3	1.2	1.3	1.5	1.9	2.4	2.7
	Pcp Ratio	1.09	1.19	1.09	1.19	1.09	1.02	1.05	1.04	1.01	1.09	1.09	1.04
6 Incremental	Tmp (°C)	2.9	2.7	2.3	1.9	1.5	1.3	1.2	1.3	1.5	1.9	2.4	2.7
	Pcp Ratio	1.32	1.28	0.96	1.35	0.83	0.90	0.82	1.06	0.95	1.08	1.30	1.35
7 CCCM	Tmp (°C)	1.5	2.0	1.7	1.1	1.1	1.3	1.6	1.3	1.6	1.5	1.6	1.0
	Pcp Ratio	1.15	1.13	0.98	1.16	0.92	0.95	0.92	1.03	0.98	1.04	1.14	1.16

Month columns: Jan–Dec.

[1] Tmp denotes temperature changes in °C; Pcp denotes precipitation changes as ratios.

Table 11.4. Actual and Calculated Yields of Winter Wheat for 1975-1990 (t/ha)

	Actual Yield	Modeled Yield under Different Levels[1] of Fertilization: 0 kg	30 kg	60 kg	90 kg	120 kg	150 kg	180 kg	210 kg
Mean	5.89	3.671	4.331	4.875	5.173	5.283	5.324	5.323	5.324
Max.	8.65	4.803	5.486	5.834	6.172	6.290	6.337	6.339	6.339
Min.	4.49	2.511	3.451	4.202	4.365	4.437	4.447	4.455	4.459

[1] Levels of fertilization in kg nitrogen per hectare.

When temperature and precipitation increases are taken into account (using scenario 5 in Table 11.3), the relevant changes in yield are -0.6% to 18.5% (440 ppm) and 37.5% to 9.8% (580 ppm), respectively. The respective changes will to a large extent depend on meteorological conditions during the short grain-creation period.

DSSAT 3 Results

Crop production will also be considerably influenced by expected changes in temperature and precipitation. The expected increase in temperatures is estimated to extend the warmest areas to the Dyje-Svratka and lower-Moravian basins, the Bohumín region, and parts of central and northwest Bohemia. By 2030 (again, using scenario 5 in Table 11.3), a 20-30 day prolongation of the frostless period is expected, changing the beginning of the vegetation period to early March and its end to late October in many areas; today it is generally limited to between April and September.

Higher temperatures will also undoubtedly influence the growth and development of plants to allow earlier germination, and thus the maturing or harvest periods could move up by at least 10 to 14 days. However, earlier growth could increase the danger of damage from frosts in the late winter. The expected temperature change would support heat-loving plant strains (e.g., half-ploughed maize, early vines). On the other hand, there is a serious danger of stress from more frequent high temperatures.

According to Seljaninov's moisture index, without a considerable precipitation increase and with the expected evapotranspiration increase, greater parts of south and central Moravia, central and northwest Bohemia, and low and middle Labe and Vltava drainage areas will be subject to droughts, which could reduce crop yields in the country's most productive areas. The effect of increased temperatures could be offset to some extent by increased precipitation, as was estimated in several climate change scenarios. However, it is difficult to assess the distribution of this precipitation throughout the year, and, according to the simulations, the increase in precipitation could also increase soil erosion in May, June, and September.

Weather sensitivity of early potatoes and winter wheat was examined using a ratio of yields. For potatoes, the ideal temperature is 17°C to 25°C, their growth stops when temperatures are above 30°C, and above 40°C their cellular structure is damaged. With winter wheat, higher temperatures during hibernation may make the wheat susceptible to damage from rapid temperature drops or thick snow cover.

The correlation coefficient between actual winter wheat yields and yields simulated by the DSSAT 3 model in Zabcice in 1975-1990 reaches values of approximately 0.6, depending on the selected degree of fertilization.

The climate change scenarios in Table 11.3 and the DSSAT 3 were used to simulate growth and development characteristics of winter wheat for the Hradec Králové study site. These results revealed that, in comparison with current conditions, scenarios 1-7 show a reduction in the flowering period of 7 to 17 days, a reduction in the period of physiological maturity of 8 to 17 days, and increases in grain yields (16%), leaf area, biomass, and straw. In warm and dry vegetation periods, grain yields were less than in normal conditions (but only up to 3%). Equally, in the scenario combining temperatures and precipitation according to the CCCM model (Table 11.3, scenario 7), the yield characteristics show a decline compared to current conditions.

It is also important to recognize that in the Czech Republic, fertilization has been reduced by more than 50% compared with the 1980s. A reduced level of fertilization with mineral fertilizers, mainly nitrogen, would result in a reduction in yield, and in many instances of as much as 50%.

Agricultural Pests and Diseases

Conditions for the development and impacts of agricultural pests and diseases may be changed considerably. Critical phases of their development depend on temperature. For example, the Colorado beetle requires a temperature of 15.2°C for the first larva phase development and 29.7°C for the fourth larva phase development. If warming occurs, some years may produce conditions that may allow a second summer generation as a result of a hastened individual phases. Further, an increased occurrence of viral diseases is predictable in large areas, mainly because peach aphid may occur as early as May, when the potato plant is more susceptible to viral infections. Higher temperatures and current precipitation levels are ideal conditions for disease and pest development, and increasing numbers of generations of mature individuals. Similarly, a more frequent occurrence of fungal diseases, and potato and hops molds, is expected. Maintaining high levels of crop production will require a higher level of chemical protection, which will increase the cost of crop production.

Impact of Climate Change on Soil Conditions

This study did not include specific estimates of the impact of climate change on soil conditions; however, climate change can affect soil conditions, especially soils negatively affected by cultivation practices. This mainly concerns the condition of the arable land subsurface, reduction of soil retention ability, and soil microbial activity. Higher temperatures and reduced precipitation will change the soil humidity regimes, extend areas of soil threatened by the humidity stress, and reduce soil microbial activity. Because soil humidity affects plowing resistance, climate change will affect cultivation practices. High temperatures and low precipitation will lower soil humidity, especially in August and September, when sowing beds are usually prepared. This will require a new type of cultivation system to keep costs low.

Adaptation Alternatives

Given the potential level of climate change impacts on agriculture, adaptation measures will become critical. The effects of climate change may be more critical than any economic development issues facing this sector (Rožnovský, 1993). All the proposed adaptation measures presented here are based on contemporary scientific knowledge.

In agriculture, proposed adaptation measures will have to address the following:

Food production in corresponding quality and quantity. A change in the composition of cultivated crop and utility animal species will be necessary to adapt to estimated changes in climate. It will be important to introduce some species and adapt existing ones to the changed conditions.

Multicropping. Multicropping should be practiced and monocultures discouraged. That is, in terms of the stability of the ecosystem, it would be efficient to grow four to five crops on one farm. In dry areas, it will be necessary to focus on crops with shorter vegetation periods and lower humidity requirements.

Reduction of soil humidity loss. Appropriate systems of cultivation guaranteeing suitable conditions for soil life and plant growth must be adopted.

Erosion control. New cultivation procedures must not increase the intensity of erosion processes. The analyses show that erosion might increase under climate change. Erosion from wind could increase due to increased temperatures and/or decreased precipitation, particularly in the warm and dry areas of southern Moravia. Any proposals for long-term and high-cost erosion control measures (e.g., terraces, windbreaks, erosion control reservoirs) must consider potential climatic change impacts.

Potential of crop production with irrigation. The use of irrigation will have to be considered. Microirrigation appears to be suitable. Under climate change conditions, agricultural production may not be possible in the most fertile and warm localities without irrigation.

Plant protection through integrated procedures. Climate change could bring increased disease pressures, and some pest species will have several generations in a single season. Crop-spraying at the beginning of the season will have to be performed substantially earlier than today.

Discussion and Conclusions

In order to ensure the stability of the ecosystem under conditions of climate change, it will be important to avoid monoculture agriculture practices. In addition, greater attention should be focused on the selection and verification of suitable crop varieties and to implementation of a variety of preferred agriculture management practices, including new methods for fertilization, use of soil hillage systems, weed control, measure to control wind erosion, irrigation, and integrated pest management.

FORESTRY SECTOR

Methods Used

Selection of Study Sites

To create a model for the whole country, the ISÚ Biography Register, based on *cadastral* areas of the Czech Republic (12,950 units averaging 6 km^2), was used (Kopecká and Buček, 1994). On the regional level, two study sites were chosen: Šumava, a mountainous area in southwestern Bohemia, and Džbán, an upland area in central Bohemia.

The Šumava National Park stretches to the mountains of Šumava in southwest Bohemia, the slopes of which gradually change from hills and ridges to a highland. It reaches altitudes of 500 to 1,350 m above sea level with a prevalence of spruce and beech forests. The Šumava model territory was selected as a representative mountain area with a high forest percentage, a prevailing representation of Norway spruce (*Picea abies* L. Karst; the main forestry tree species in the Czech Republic), and an expected change in species dominance. It was determined that the shift in forest vegetation zones at this site was suitable for the examination of forestry adaptation measures.

The Džbán territory in the northwestern part of central Bohemia is formed by an upland plain with a mild inclination, altitudes of between 300 m and 500 m above sea level. It is situated in a rain shadow and it is the driest area of the Czech Republic. About 20% of the area is forested; 35% is Norway spruce and another 35% is Scotch pine

(*Pinus sylvestris* L.). Before human influence, these tree species were not represented. It is an area with a high degree of vulnerability to climate change and an urgent need for adaptation measures.

Forest Ecosystem Models Used

Three fundamental methodological approaches were applied:

- analyses of forest ecosystems via inventory and monitoring databases, including a synthesis of development trends
- simulation of changes in site conditions and reactions of the forest ecosystems based on climatic scenarios in 2010 and 2030
- analyses and syntheses of estimates of forest ecosystem vulnerability to climate change and proposals of preventative forest management measures.

Using meteorological data, regional climate scenarios of spatial temperature and precipitation patterns were developed for the Sŭmava and Džbán regions. The ecological forest network in the Czech Republic is divided into vertically defined forest vegetation zones (see Table 11.5) based on the relationship between climate and the ecosystem expressed by the natural composition (Plíva, 1985; Henzlík, 1994). Soil properties are used for horizontal segmentation. Detailed forest types (zone, soil, and biocenosis) are joined into forest site types (zone and soil), edaphic categories (similar soil properties), and so-called ecological rows (broader frames).

The methodology employs records of climatic data from 1961-1990 and outputs of GISS and CCCM climatic models linearly interpolated for 2010 and 2030 (Černý and Hadaš, 1994). To estimate shifts in forest vegetation zones on the basis of a GISS scenario (Kalvová et al., 1995) (0.98°C increase for 2010 and 1.86°C for 2030 and precipitation ratios of 1.038 for 2010 and 1.075 for 2030), a nondynamic correlative model based on a relation between contemporary climatic conditions and vegetation types was applied. For change estimation, a scenario based on the annual mean temperature and a scenario based on the annual

Table 11.5. Forest Vegetation Zone Characteristics

Zone	Altitude (m ASL)	Average Annual Temperature (°C)	Annual Rainfall (mm)	Vegetation Period (days)
1 Oak	<350	>8.0	<600	>165
2 Beech — oak	350-400	7.5-8.0	600-650	160-165
3 Oak — beech	400-550	6.5-7.5	650-700	150-160
4 Beech	550-600	5.5-6.0	700-800	140-150
5 Fir — beech	600-700	6.0-6.5	800-900	130-140
6 Spruce — beech	700-900	4.5-5.5	900-1,050	115-130
7 Beech — spruce	900-1,050	4.0-4.5	1,050-1,200	100-115
8 Spruce	1,050-1,350	2.5-4.0	1,200-1,500	60-100
9 Dwarf pine	>1,350	<2.5	>1,500	<60

mean of precipitation divided by the annual mean of temperature were applied. In addition, regional scenarios for Šumava and Džbán were developed based on meteorological data and the GISS and CCCM scenarios.

Adaptation Alternatives and Evaluation Methods

At this preliminary stage in our evaluation, our method has been to examine alternate adaptation measures formulated in proposals to the forest administration. These measures were based on expert assessments and were developed according to the reliability of the estimates and the feasibility of implementation within realistic time horizons (Kubík and Vinš, 1993).

Results

The forest vegetation zone shift, according to the scenario using the ratio of annual mean precipitation to annual mean temperature, is more moderate than the shift estimated by the GISS and CCCM scenarios, and it approaches realistic conditions of vegetation reactions to the expected climatic change. Under the GISS scenario it is estimated that the forest surface with the climatic conditions of zones 1-3 would increase from the present share of 21% to 47% in 2030. Forest areas with climatic conditions of zones 4-8 would be reduced (Kopecká and Buček, 1994).

According to the regional climate scenario for the Šumava site, a gradual temperature increase by 2010 will be accompanied by increased precipitation. This will result in a shift in forest vegetation zones toward higher elevations together with the elimination of zone 8 and the lower part of zone 7, and enlargement of zones 6, 5, and 4. By 2030, the model shows a stronger tendency for increased precipitation than for higher temperatures. This would result in a gradual regeneration of zone 7, a slight decrease in the area of zone 6, and a more substantial reduction of zones 5 and 4. At higher altitudes, the decline in Norway spruce will encourage the establishment of European beech (*Fagus sylvatica* L.) (Černý and Hadaš, 1994).

In the Džbán site, two forest vegetation zones are distinguished: zone 2 in its eastern part (65%) and zone 3 (35%) in the western part of the study site. According to the regional scenario, the forest vegetation zones will be joined, and only the lower half of zone 2 will remain. The Norway spruce and probably the European beech are the most endangered tree species in the area.

Economic Analyses

Economic analyses of the impact of global climate change were not performed because current models of climate change are limited in their usefulness for ecosystem response simulation in the very differentiated forest sites in the Czech Republic. The cost of measures to address ecological destabilization and the effects of climate change on forest management in lower forest vegetation zones (forest regeneration and afforestation, soft management methods and increased stand tending, changes in usable biomass, etc.) is estimated to be hundreds of thousands of dollars a year (Henzlík, 1994).

In the Czech Republic, the additional costs for maintaining the natural resource uses of forests have been analyzed (Kreèmer and Matìjíèek, 1993; Kreèmer, 1994a). The analysis covers water management, soil protection, recreation, and preservation (Kreèmer, 1994b). Previous research implies that these are appropriate measures in the Czech Republic regardless of the possible extent of global climate change. The additional costs were calculated to be approximately US$45.4 million (1995 dollars) to protect the environmental uses of the forested sites (Kreèmer and Matìjíèek, 1993).

The currently poor condition and excessive density of forest roads as well as the poor condition of structures in watersheds present a threat to the cultural landscape that would be intensified by climate change.

Adaptation Alternatives

Adaptation to climate change in the Czech Republic will to a large extent include measures implemented to address other threats to the forests. Eurytope and pioneer species (e.g., birch, aspen) are better able to adapt to environment changes than stenotope climax species (e.g., silver fir, European beech). Therefore, the ratio of deciduous tree species should be increased, particularly in the lower forest vegetation zones. The objective of maintaining and establishing mixed stands is directly related to the optimal exploitation of growth conditions and the risks associated with potential climate change. Restoring the biological diversity of the forests, revitalizing forest stands through effective forest management, and applying sustainable forest management principles are critical to forestry policy. A realistic estimate of the optimal deciduous species ratio increase would be 10% over the next four to five decades.

It is unlikely that the gradual mean temperature change would significantly increase forest stand mortality. Greater losses may be caused by more frequent anomalies in weather, particularly by sudden temperature changes toward extreme values in winter and early-spring seasons, episodic and regional low precipitation in critical parts of vegetation periods, and increased frequency of extreme gales, heavy rains, and snow or ice.

Forest stands are already increasingly endangered by previously unrecognized harmful factors and by the increased intensity of other factors that were once controllable. The damage from these factors in many cases is likely to be exacerbated by climate change. Examples of such factors include pests and insects as well as anthropogenic factors. The engraver beetle and a number of leaf-eating insects pose a threat to the forests that is likely to increase under conditions of climate change. The growing aggressiveness and virulence of various fungi is also significant, particularly those participating in tracheomycoses, and by honey fungus (*Armillaria melea*). Damage will increase under conditions of climate change because of the slow adaptation of forest tree species and their stands and the faster proliferation of various plant pests.

Long-term research on watersheds such as the Beskydy mountains indicates that anthropogenic factors may result in some very negative effects on hydrology and soil (Kreèmer and Jarabár, 1981). If forest stands decline because of such factors, they will be critically endangered under conditions of climate change. In addition, because climate change will increase the sensitivity of forest ecosystems to recreational use, preventive measures should be taken to protect forests used for recreation (including an emphasis on fire protection).

Production processes of forests will be influenced by climate change within the complex interactions in forest ecosystems. Neither actual changes in CO_2 concentration nor, for instance, nitrogen deposition will necessarily influence the biomass increment and total production in forest ecosystems (Marek et al., 1995).

Discussion and Conclusions

There are several specific features of the forestry sector in the Czech Republic:

- Within this relatively small country, there are highly varied natural conditions.
- The territory is situated on the boundary line of oceanic and continental climate effects.
- The forests have long been affected by different human activities, including fundamental changes in tree species composition over the last 200 years and intensive air pollution over the last five decades.
- There is a relatively high population density and considerable interest in non-wood-producing forest uses.
- Above all, there is a very valuable and extensive set of information about forest stands and site conditions (and their changes during the past five decades).

Adaptation options in anticipation of global climate change are to a large extent identical with measures to ensure stability and well-managed forest use without climate change. In the Czech Republic, these forestry measures are already urgent (Vinš, 1995).

The importance of impacts of the expected global climatic change on forests and forest management underlines the value of targeted and coordinated monitoring of forest stands and basic parameters of forest site conditions. These priorities are among others defined in Resolution H4 from the Ministerial Conference on Forest Protection in Europe — Helsinki 1993 (Poláček, 1993).

The climate change scenarios and mathematical models available do not currently address many bioclimatically important factors, and they do not describe the climatic regime to the degree necessary for analysis of the effects on live organisms. Forest science and research in the Czech Republic so far have provided neither critical values nor intervals which could define responses of forest ecosystems which could be used to apply dynamic models of forest ecosystems such as the forest gap model (Benioff et al., in press) or Forcyte. Research needs to be undertaken to:

- develop climatic and biometric parameters
- estimate decline in annual precipitation (e.g., drought periods) and forest stand mortality of various forest sites
- estimate warming and insect proliferation as well as the defensive potential of forest ecosystems.

Therefore, it was necessary to use only the forest vegetation zone characteristics of the typological classification system for the whole country to model the impacts of climate change on regional-scale forest ecosystem.

Public awareness and awareness within the forestry profession of the risks associated with climate change and its effects on forests are still insufficient. There has been no significant effort to change this situation through relevant educational campaigns. As the state forestry administration designs long-term solutions, education and consultancy to forest owners should be a primary objective. Education should also be aimed at leading political organizations to enhance their decision-making capabilities.

WATER RESOURCES SECTOR

Methods Used

Selection of Study Sites

In order to study the potential impacts of climate change on hydrological systems and water resources, four river basins were selected, representing the range of physiographic and hydrological conditions across the country:

- the Elbe Basin, which is the largest basin in the Czech Republic, with a drainage area of 50,762 km^2
- the Zelivka Basin, representing a hilly crystalline complex with a water supply reservoir serving Prague and with a drainage area of 1,189 km^2
- the Úpa Basin, drained by a mountainous stream with abundant precipitation, significant accumulations of water from snow, and with a drainage area of 461 km^2
- the Metuje Basin, representing a hydrological system with a significant groundwater component and a drainage area of 94 km^2.

The sensitivity of the storage capacities of water reservoirs to changes in hydrological conditions was studied at two sites, the Soutice Reservoir on the Zelivka River and the Česká Skalice Reservoir on the Úpa River.

Hydrological and Water Resources Methodologies

The hydrological systems of the four river basins under climate change were studied using the BILAN (Krejčová et al., 1994, Czech Republic) and CLIRUN (Kaczmarek, 1993; Poland; Benioff et al., in press) water balance models and, for comparison, the SACRAMENTO (Burnash and Ferral, 1980; USA) conceptual model. No significant differences in the runoff simulations using these three models were found. Of the two water balance models with monthly and annual steps (BILAN and CLIRUN), relatively better fitting results were obtained with the BILAN model. The study has, therefore, been based mainly on results from this model, with the SACRAMENTO model which uses daily steps complementing some of the results. In addition to the basin monthly areal precipitation series, the BILAN model requires as input monthly temperature and humidity data. The model simulates direct runoff, interflow, base flow, areal evaporation, soil moisture water content, and groundwater storage.

Incremental and GCM scenarios with alternatively uniform and nonuniform distributions of temperature and precipitation changes throughout the year were used for hydrological modeling runs. A temperature change of 2° or 4°C was estimated for individual months. These values were selected to coincide with the GISS climate scenarios. The distribution of precipitation changes was based on the long-term monthly averages from 1896 to 1992. The GISS, CCCM, and GFD3 $2\times CO_2$ model results were also used.

Simulated quantitative surface water and groundwater hydrological characteristics, estimates of economic development in the main sectors of national economy, and estimates of demographic features all served as background information for estimates of future water demand, potential future requirements for reservoir accumulation volumes, and the long-term water supply/demand balance. These socioeconomic factors were used to examine the sensitivity of the vulnerability and adaptation assessment for the Czech Republic. More than 50% of the current drinking water demand in the Czech Republic is met by water reservoirs. Therefore, the study also considered quantitative aspects of climate change impacts on their operation.

Climatological, hydrological, and water management data and information for the study were derived from databases of the Czech Hydrometeorological Institute and the T.G.M. Water Research Institute in Prague. The socioeconomic data were produced by the Czech Statistical Office.

Adaptation Alternatives and Evaluation Methods

It is not yet possible to estimate the potential changes in the runoff in particular regions of the country, or the relevant water management measures to be taken, because of the high variability and uncertainty of the GCM precipitation scenarios. Under these circumstances, the method used to assess adaptation alternatives was to formulate a strategy that the Czech Republic should adopt to reduce adverse effects of climate change on water resources.

Results

Results of the Hydrologic Vulnerability Assessment

Changes in runoff differ according to the climate scenario and hydrological model applied.

Using incremental and GCM scenarios, we found that serious effects could be expected in basins with currently low total precipitation. Basins with high precipitation appear to be relatively less sensitive. Significant changes could therefore be expected in basins with medium and low runoff. These conclusions can be used in the classification of basins in the Czech Republic. Table 11.6 illustrates the range of potential changes in annual runoff for the selected study sites.

For the most critical incremental scenario (+4°C and -5% precipitation), the base flow would decrease by about one third and even more during periods of low flow. It can reasonably be assumed that with a 2°C to 4°C warming, an increase of precipitation by even 5% would not ensure that groundwater runoff would be maintained at the original level. With decreasing precipitation, the groundwater runoff would decrease even in basins with large accumulations of groundwater. The effects could be critical, particularly during long or extreme droughts.

The impact of climate change appears to be more critical for those parameters that

Table 11.6. Climate Change Simulations of the Mean Annual Runoff[1]

Type of Scenario	Change in Temp. (°C)	Change in Precip. (%)	Change in Annual Runoff (%)	
			Low	High
GISS	+4.2	+16	-2	+3
GFD3	+4.0	+18	-8	-3
CCCM	+3.1	+7	-10	+2
Incremental	+2	-5	-33	-15
	+2	0	-23	-6
	+2	+5	-13	+5
	+2	+10	-2	+15
	+4	-5	-45	-25
	+4	0	-36	-16
	+4	+5	-26	-7
	+4	+10	-16	+3

[1] Results represent the minimum and maximum values obtained from all study sites.

characterize periods of low flow rather than those characterizing annual values. While for the most unfavorable incremental scenario (+4°C and -5% precipitation) the mean annual runoff would decline to about 60% of the current level, the low flow that happens 10% of the time would drop to about 40% of its current level.

Similarly, under the most adverse scenarios, reservoir accumulation volumes would need to be several orders of magnitude larger to ensure required outflow. Important effects on reservoir operation can be attributed to changes of temperature, because the necessary accumulation volume would increase due to increased temperatures even without changes in precipitation.

Results of Water Resources Supply and Demand Models

By 2075, the country's water supply and demand will depend not only on climate change but also on its socioeconomic development. No detailed conception of such long-term development of particular sectors has yet been prepared. Therefore, future water use has only been estimated in terms of incremental changes from current demand.

Estimates by 2075 in the most important categories of water demand indicate that drinking water, irrigation water, and water use by the power industry would remain at approximately the 1990 level under current climate conditions. This reflects population forecasts to the year 2020 from the Czech Statistical Board and expectations of structural transformation in the electrical energy sector. Water demand in other industries is estimated to increase.

Climate change is not expected to cause a change in drinking water demand; however, a 2°C increase in annual temperature is expected to increase irrigation water demand by 40% (assuming the same number of hectares are under cultivation). Surface water pollution should decrease in the relevant period, even under conditions of climate change, mainly because of continuing construction of wastewater treatment plants.

Table 11.7. Projected Demand for Water in 2075 in the Czech Republic

Water Use	Demand (m^3)
Drinking Water	1.0-1.3 billion
Irrigation	575 million[1]
Hydropower	924-1,201 million
Industry	652 million
Total	3.2-3.7 billion

1 Demand could increase by 40% to offset agricultural impacts of drier climate.

The estimates of future water demand are presented in Table 11.7.

Adaptation Alternatives

Until about 2010, mainly nonstructural adaptation measures are recommended in the Czech Republic. In the meantime, knowledge about climate change and its impacts are expected to become more detailed and would allow better-informed decisions on the nature and necessity of potentially costly structural measures and their implementation.

The following points have been proposed as a basic set of nonstructural measures aimed at reducing or eliminating the adverse effects of climate change on hydrological and water supply systems:

- gradual decrease of water consumption through a range of economic and administrative measures
- integrated management of groundwater and surface water resources to preserve water in aquifers and protect against pollution
- improved water supply and treatment efficiency, especially a program of wastewater treatment and use of treated water
- implementation of limits and notification of conditions for use of water during droughts, similar to those in the electricity and gas supply practice, including preparation of the relevant legislative tools and organizational arrangements
- improved real-time operation of water reservoirs.

Discussion and Conclusions

This assessment has demonstrated that even slight changes in climatological parameters could generate serious water management problems at sites where either increased demand or pollution may make water scarce, while at other sites even larger manifestations of implied changes may not cause problems. Therefore, it will be necessary to pay attention primarily to regions that currently have threatened water resources. Some of these regions have been examined in this study, e.g., the Morava River basin contains few rivers that could contribute to substantial reservoir accumulations, and currently monthly flows decline substantially during low flow periods.

It is generally assumed that measures to be taken in response to climate change in the water resources sector at the beginning of the next century might be very costly and time-consuming. Studies of climate change impacts on the water component of the environment should therefore continue, with a specific focus on impacts of extreme events such as floods and droughts, and determination of water budget changes in the hydrologic cycle.

SUMMARY CONCLUSIONS

The consequences of climate change will be important in the Czech Republic. This applies both to the seriousness of the purely physical changes of climate and to the impacts on various sectors of the society and the economy. The studies completed so far must be regarded only as preliminary work that should be further developed. The conclusions that we are able to make to date are neither sufficiently precise nor reliable. They reflect all the uncertainties that characterize research in this field. Among other things, a finer distribution of possible hydrometeorological phenomena (including extreme events) is urgently needed.

During our assessment we examined all potentially affected sectors of society and concluded that only three deserve close examination: agriculture, forestry, and water resources. The analysis of any sector and especially an assessment of its possible future development in the Czech Republic is a difficult task because the society is in a state of rapid political and economic change. There are neither government documents outlining economic policies nor official forecasts. Therefore it is almost impossible to use most of the sectoral models. Similarly, no comparative economic analysis of sectors was possible, and there was no serious effort made to integrate the analysis of the individual sectors.

Some intuitive remarks can be made. A relatively high degree of confidence may be attributed to both the vulnerability and adaptation assessments for the forestry sector. Even without analysis using sophisticated models, it is clear that climate change will enhance most of the problems that already exist. The effect of climate change on the agricultural sector is very complicated and in some cases controversial. The impacts are likely to be important, but because of the relatively short-term character of technology cycles, the danger of climatic impacts is diminished. The magnitude of the impact in the water resources sector is not clear. This study has shown that even slight changes in climate could generate serious water management problems at sites where either increased demand or pollution may make water scarce. However, there is a wide margin of uncertainty regarding future water needs. So far, no major change in existing policy on surface water reservoirs is proposed, but in light of the very probable decrease of runoff, the existing policy might be inadequate. Our knowledge to date is clearly insufficient. More research is needed based, above all, on further development of scenarios of changing climatic variables.

REFERENCES

Benioff, R., S. Guill, and J. Lee (eds.) (In press). *Vulnerability and Adaptation Assessments: An International Guidebook.* Dordrecht, The Netherlands: Kluwer Academic Publishers.

Černý, M. and P. Hadaš. 1994. Regional Scenario of the Forest Vegetation Zones Shift in the Forest Regions Džbán, Central-Bohemian Upland, and Šumava Mountains. Final Report. IFER, Davle, Czech Republic, 149 pp. (in Czech).

Henzlík, V. 1994. Lesprojekt's Forest Management Models. Adjustment of Models to Climate Change. In Proceedings of the Workshop, "Effects of Global Climate Change on Boreal and Temperate Forests." VÚLHM, Jíloviště-Strnady, Czech Republic.

Kaczmarek, Z. 1993. CLIRUN3 Identification and Water Balance Simulation. Institute of Geophysics, Polish Academy of Sciences, Warsaw.

Kalvová, J. 1993. The Czech Republic's Temperature Scenario. Report of the Project, "The Shift of the Forest Vegetation Zones in the CR According to the Regional Scenario of Climate Change." NKP CR, MZe CR, Prague, Czech Republic, 29 pp. (in Czech).

Kalvová, J. and R. Brázdil. 1993. Climate Change. In: *Risk of Climate Change and its Mitigation*. Moldan, B. (ed.), NKP CR, Prague, Czech Republic, pp. 48-91.

Kalvová, J. and I. Nemešová. 1994. A Temperature Scenario for the Czech Republic. In: *Contemporary Climatology*. Brázdil, R. and M. Kolář (eds.). IGU, UGI, Masaryk University, Brno, Czech Republic, pp. 293-299.

Kalvová, J. et al. 1995. The Czech Republic's Country Study. Final Report: Climate Change Scenario for the Czech Republic, NKP CR, Prague, Czech Republic.

Kopecká, V. and A. Buček. 1994. Use of the Register of Biogeography ISÚ for Spatial Evaluation of the Trends of the Vegetation Zones Change Caused by the Global Climate Change in the Czech Republic. Final Report. TERPLAN, Prague, Czech Republic, 52 pp. (in Czech).

Kreèmer, V. 1994a. Sustainable Development and Forest Management in the Czech Republic. Lesnictví — *Forestry*, 40, 1:48-54. (in Czech).

Kreèmer, V. 1994b. Preconditions for the Functionality of Produced Forest Infrastructure. In Proceedings of the Int. Symposium, "Czech Forest Policy in the Changed Socio-economical Conditions." Agricultural University Brno, Czech Republic, 128-130.

Kreèmer, V. and M. Jarabár. 1981. Forest and Water in Beskydy Mountains. Digest of Documents, Communications, and of the Czechoslovak Academy of Agriculture No. 10, Prague, Czech Republic, 84 pp. (in Czech).

Kreèmer, V. and J. Matìjíèek. 1993. Theoretical Groundwork and Calculations of Ecological Costs in the Forest Management. Ministry of Environment of the Czech Republic, Prague, Czech Republic, 56 pp. (in Czech).

Krejčová, K., L. Kašpárek, et al. 1994. Impact of Climate Change on Water Resources. Final Report in Czech. Vúv T.G.M., Prague.

Kubík, J. and B. Vinš (eds.), 1993. Sustainable Forest Management in the Czech Republic. Revue of Activities for the CSCE Seminar, Montreal. Ministry of Agriculture of the Czech Republic — Branch of Forestry, Prague, Czech Republic, 36 pp.

Kveton, V. and H. Kazmarová. 1995. The Czech Republic's Country Study. Final

Report for Human Health Sector. NKP CR, Prague, Czech Republic (in Czech).

Marek, M.V., J. Kalina, and M. Matoušková. 1995. Response of Photosynthetic Carbon Assimilation of Norway Spruce Exposed to Long-Term Elevation of CO_2 Concentration. *Photosynthetica*, 30:189-198.

Moldan, B. (ed.). 1993. Risk of Climate Change and its Mitigation. NKP CR, Prague, Czech Republic, 175 pp.

Plíva, K. 1985. Natural Forest Regions of the Czech Republic. ÚHÚL, Brandýs N.L., Czech Republic (in Czech).

Poláček, P. (ed.). 1993. Ministerial Conference on Forest Protection in Europe, Helsinki, June. Ministry of Agriculture of the Czech Republic — Branch of Forestry, Prague, Czech Republic, 33 pp. (in Czech).

Porter, J.R. 1994. A Model of Canopy Development in Winter Wheat. *J. Agric. Sci.* 102, 383-392.

Rožnovský, J. 1993. Agroclimatic Conditions and Phenological Evaluation from Point of View Possible Climate Change. Slovak Bioclimatological Society, 11, Bratislava, Slovak Republic, 52-55.

Viner, D. and M. Hulme. 1994. Climate Change Scenarios for Impact Studies in the UK: General Circulation Models, Scenario Construction, Methods, and Applications for Impact Assessment. A report prepared for the UK Department of the Environment Climate Change Impacts/Predictive Modeling LINK PECD 7/12/96, Climatic Research Unit University of East Anglia, Norwich, United Kingdom, 70 pp.

Vinš, B. (ed.). 1995. The Czech Republic's Country Study. Final Report for Forestry Sector. NKP CR, Prague, Czech Republic.

Vrkoč, F. 1994. Actual Problems of Structure of Crop Production. In: *Crop Production in Present Conditions. Economical and Ecological Aspects of Production of the Main Market Crops.* Prague, pp. 8-12 (in Czech).

Chapter 12

Vulnerability and Adaptation Assessments for Estonia

Are Kont,[1] Jaan-Mati Punning,[1] Jaak Jaagus,[2] Jaan Klõšeiko[1] Tiiu Koff,[1] Malle Mandre,[1] Hugo Roostalu,[3] Toomas Tamm[3]

[1] Institute of Ecology
[2] University of Tartu
[3] Estonian Agriculture University

SUMMARY. Because of its geography, wide coastal areas, water resources, forests, and wetlands, the environment of Estonia is sensitive to climate change and sea level rise. Therefore, the vulnerability and adaptation assessments focused on these sectors. This chapter gives preliminary results of our study. The CERES-Barley and SPUR2 models were used to assess crop productivity. A sea level rise of 1.0 m was used to evaluate the responses of the physical and natural systems in the low-lying coastal territories in the West Estonian Plain and the West Estonian Archipelago. To estimate the responses of forests to proposed climate change scenarios, two case study areas with different types of climate (continental and maritime) were selected and a modified forest gap model was used. In addition, an analysis was conducted using paleogeographical data from Estonia and neighboring areas. These data indicate that between 7,000 and 5,000 years ago (Holocene "climatic optimum") the temperatures in Estonia were similar to predicted future temperatures. Thus we can compare the state of ecosystems in the past with those predicted for future global climate change.

INTRODUCTION

Estonia is the northernmost Baltic country in Europe (between 57.30° and 59.49°N and 21.46° and 28.13°E). Its total area is 45,216 km². Over 1,500 islands and islets account for 9.2% of the territory. Estonia is a seaside country, and entirely within the drainage area of the Baltic Sea (Figure 12.1). Its coastline length, including the mainland and islands and islets, is 3,794 km. The coastline is characterized by numerous bays, peninsulas, and straits between islands. The coastal sea is shallow and full of shoals.

Situated in the northwestern part of the East European plain, Estonia is characterized by a flat topography. The average elevation is about 50 m, and the highest point is 318 m

Note: This chapter does not contain a detailed description of the assessment methodologies that were recommended by the U.S. Country Studies Program. Refer to Chapter 2 for an overview of these methodologies.

J. B. Smith et al. (eds.), Vulnerability and Adaptation to Climate Change, 249–275.

Figure 12.1. Map of Estonia

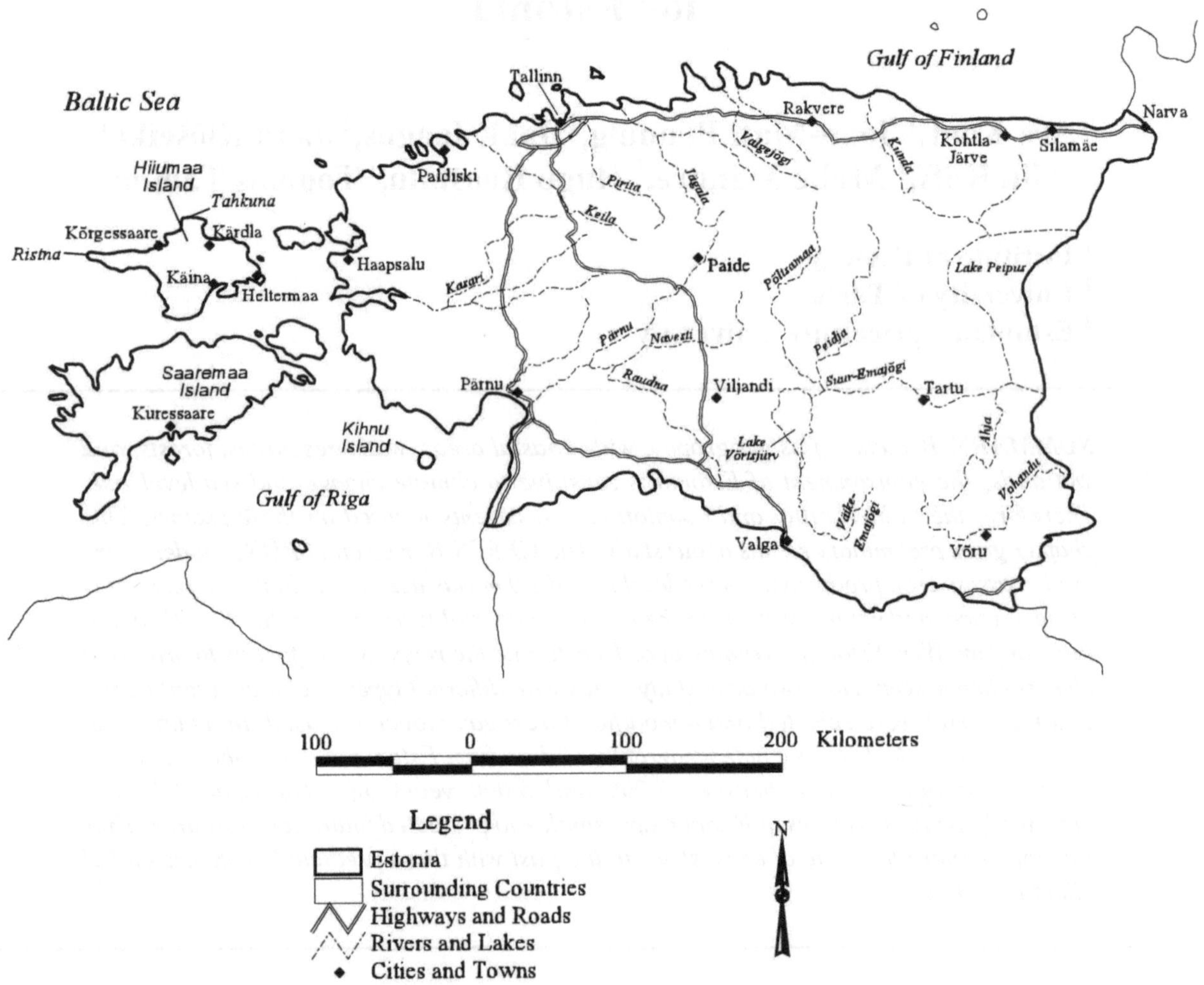

above sea level. The country can be divided into two regions: Lower Estonia and Upper Estonia. Upper Estonia comprises the more elevated areas in the central and southern parts, which were not flooded by the sea and ice lakes during early stages of the Holocene epoch. The soils in Upper Estonia are more fertile and the rural population is denser than in Lower Estonia.

Geologically, the country is located on the southern slope of the Baltic Shield, which is undergoing an isostatic land uplift; Estonia is being lifted 2.8 mm/year on the northwestern coast and 1.0 mm/year in the southwest. Only a comparatively small southeastern part of its territory is subsiding slightly. Isostatic land uplift could potentially reduce, but not alleviate, the impact of sea level rise on the Estonian coast.

Estonia belongs to the Atlantic continental region of the temperate zone; its climatic conditions can be briefly characterized as warm summers (17°C in July) and mild winters (-4.5°C in February). Since the annual amount of precipitation (600 mm/year) is approximately twice the evaporation, the climate is quite humid.

The forests of Estonia belong to a zone of mixed and coniferous forests with relatively favorable growth conditions and good diversity. The variability brought about by natural conditions (parent material of soil, relief, climatic differences) is increased by the fact that the majority of the forests of Estonia have been affected by human activities in various degrees and ways (cutting, drainage, fires etc.).

Today 47.6% of the country, approximately 2.02 million ha, is covered with forest (Statistical Office, 1995a). The total area of stands is 1.82 million ha, and there are 53,500 ha of young cultivated forests and forest nurseries. Forests with conifers as the dominant tree species make up 63% of the total forest and 66% of total forest yield; forests with deciduous trees as dominant species constitute 37% of the forest and 34% of the forest yield. Forestry and the forest industry have been and still are important contributions to the economy and employment of Estonia, constituting, together with agriculture, 11% of the labor force and 10% of GDP (Statistical Office, 1995b).

The peatland area of Estonia is approximately 10,000 km^2, or 22% of the territory (partly coinciding with forest areas).

Estonia is one of the smallest and least populated countries of Europe — its total population is 1,574,955 inhabitants (1990 Census). The population density is 35 inhabitants/km^2. About 71% live in urban areas, and 51% live in the five largest cities: Tallinn (484,400); Tartu (115,400); Narva (82,300); Kohtla-Järve (76,800); and Pärnu (54,200).

Although small in area, Estonia is relatively rich in natural resources, both mineral and biological, which have been the basis of the economy, and particularly of industry. The most important natural resource is oil shale. Other natural resources worth mentioning include phosphorite, dolomite, sand, gravel, peat, curative mud, and mineral water. The main industrial sector in Estonia is energy. Both the energy and chemical industries are based on oil shale.

Estonian agriculture specializes in livestock breeding, which provides 65% to 70% of agricultural production. When Estonia was part of the Soviet Union, agricultural production was based on state and collective farms. Animal husbandry was strongly dependent on imported fodder, and preference was given to pig husbandry and pork production. Most of the meat and dairy products were exported to the other republics of the Soviet Union.

Since the transition process began, the former state and collective farms have been replaced by private farms, which today number over 14,000. The end of state and collective farms, the development of private farms, the loss of previous markets, and high prices for fuel and fertilizers are the primary agents affecting the agricultural sector today. A decrease in farmed land (from 1.12 million ha in 1990 to 925.4 thousand ha in 1995) as well as in the number of cattle (1.9 times) is evidence of the changes in the sector. The projection for 1996 is for further decreases in sowed area, but yield and quality, and the main animal husbandry products, should increase from the 1990-1991 level.

The most important sectors for the vulnerability and adaptation assessment for Estonia are agriculture, forests, and coastal resources. Preliminary results for these sectors are included here. Many other sectors such as marine ecosystems, water resources, grasslands and livestock, human settlements are also important, and are integrated with the sectors of primary focus. The assessments of these sectors are still under development.

The main objective of the vulnerability and adaptation assessment for Estonia is to evaluate climate change and sea level rise impacts on natural systems and human activities and to evaluate possible adaptation options. Thus ecosystems that are most vulnerable to changing climate conditions were selected. Meteorological observation stations in different locations of Estonia were chosen for developing climate change scenarios.

Figure 12.2. Map of Estonia Study Sites

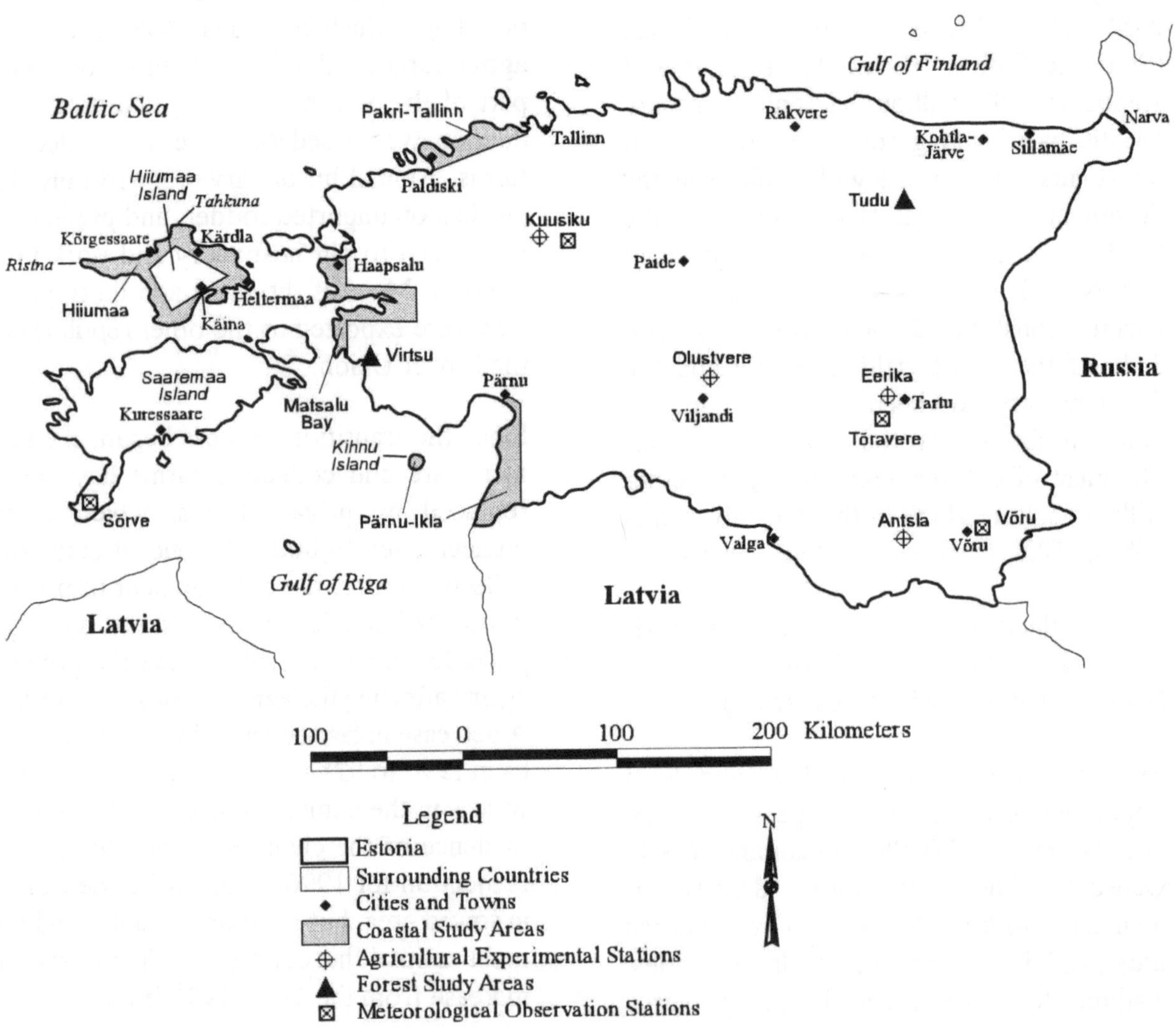

Four agricultural experiment stations were selected. Two (inland and coastal) study areas were selected in the forestry sector, and four study areas with different geomorphic shore types were selected to assess the effects of accelerated sea level rise on coastal resources (Figure 12.2).

SCENARIOS

Baseline Climate Scenario

The 30-year period (1951-1980) recommended for baseline climate scenarios is sufficiently long to provide a good representation of climate statistics. Although the 1980s may have been subject to some climate warming, the period 1961-1990 was chosen for developing baseline climate scenarios for Estonia because of concerns about the quality of data for Estonia from the 1950s. The later baseline period also allows the use of high quality solar radiation data measured at the Tõravere actinometrical station.

The baseline climate scenario was developed from spatial mean values of monthly temperature, precipitation, and solar radiation measured at meteorological stations in Estonia and averaged for 1961-1990. Mean temperature was calculated for 25 stations. Spatial mean totals of precipitation were found by interpolating average values into grid points using data recorded at more than 100 precipitation stations. We also have a technique that allows us to calculate spatial mean precipitation for smaller regions in Estonia, for example, for river basins. Solar radiation was measured only in Tõravere.

Despite the comparatively small territory, climate conditions across Estonia are very different. For instance, the monthly mean temperature in January varies from -2.5°C at Vilsandi, the westernmost island of Estonia in the Baltic Sea, to -7.6°C in eastern Estonia. Therefore, temperature was analyzed using not only spatial mean values but also the values measured at the three stations in different locations (Figure 12.2). Võru (58°N, 27°E) is situated in southeastern Estonia between two uplands; it has the most continental climate. Kuusiku (59°N, 25°E) lies in the plain in western Estonia, and Sõrve (58°N, 22°E) is located in the southwestern point of Saaremaa Island, with the most maritime climate.

The grid points where GCM results were calculated usually did not lie within the territory of Estonia, so it was necessary to apply spatial interpolation. In this study, GCM results (air temperature, precipitation) are interpolated for the center of Estonia, and also for the above-mentioned three stations as well as Tõravere (solar radiation). GCMs were chosen through comparisons between observed and calculated values.

Incremental Scenarios

Developing incremental scenarios is the simplest way to obtain climate change scenarios. They provide a wide range of potential regional climate changes and help identify sensitivities to changes in temperature and precipitation. For Estonia, increases in temperature (+2°C, +4°C, +6°C) were combined with no change, and with ±10% and ±20% changes in precipitation. As a result, 15 incremental climate change scenarios were developed.

The main disadvantage of using incremental scenarios is that they may not be physically plausible. Also, uniform climate changes throughout a year are not realistic. For instance, in higher latitudes (between 57°N and 60°N), including Estonia, estimated warming in winter should be much greater than in summer. Despite these concerns, incremental scenarios are feasible for Estonia, which has great variability in precipitation; warm and rainy years as well as hot and dry years have occurred.

GCM Climate Change Scenarios

Five GCM outputs were evaluated to determine which model provided the best results for Estonia: the Canadian Climate Centre Model (CCCM), the Goddard Institute for Space Studies (GISS) model, the United Kingdom Meteorological Office (UK89) model, and the Geophysical Fluid Dynamics Laboratory (GFDL) equilibrium (GFD3) and GFDL 1% transient (GF01) models (Benioff et al., in press). These outputs were combined with the baseline climate data to produce climate change scenarios that could serve as inputs for vulnerability assessments.

1×CO_2 Climate Scenarios

The GCM temperature outputs in 1×CO_2 scenarios are similar within the continental part of Estonia and a bit different in the West Estonian Archipelago. Figures 12.3 and 12.4 show the fluctuations of monthly mean temperatures at two stations for both the baseline scenario and the 1×CO_2 scenarios. UK89 output data are excluded from Figure 12.3 because the values were quite low; in the continental part of Estonia they are below -20°C in winter, and summer maximum values are generally less than +15°C. Such data are more typical of the climate of Siberia at the same latitudes.

The GF01 transient model underestimates temperature in winter and overestimates in summer. The rest of the GCMs also tend to underestimate temperature in winter. Only the CCCM output indicates higher temperatures in winter.

The best temperature results were obtained with the CCCM, especially during spring and autumn, when the other models underestimated temperature. CCCM overestimates winter and summer temperatures in mainland Estonia. The GISS model results are also satisfactory. They match observed data best for the winter, and underestimate temperatures for the spring and autumn.

GFD3 was the third suitable model for describing the annual fluctuations of temperature. This model underestimates temperature except for the summer. This model is better for western Estonia, where it does a good job of reflecting the influence of the Baltic Sea on temperature.

For precipitation, the deviation of GCM results from baseline data is substantially greater. Unfortunately, most of the GCMs highly overestimate precipitation during the cold half of the year, and underestimate it in the summer, particularly in July and August (Figure 12.5). Consequently, the models do not accurately describe the annual distribution of precipitation in Estonia.

Precipitation measurements are representative only for the small area around a gauge. Therefore, it is not reasonable to use data from single stations. GCM results were compared with spatial mean data for the

Figure 12.3. Baseline and 1×CO_2 Temperature Scenarios at Võru

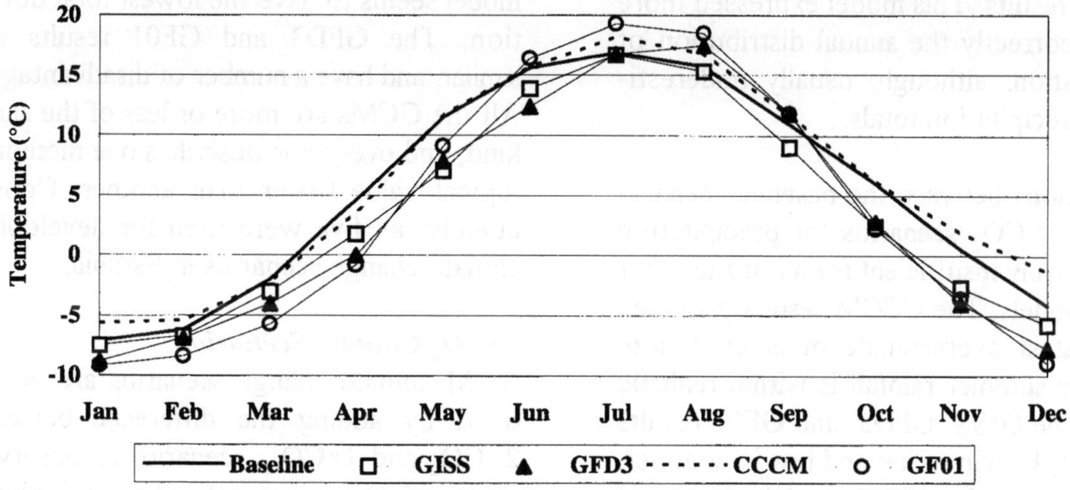

Figure 12.4. Baseline and 1×CO_2 Temperature Scenarios at Sõrve

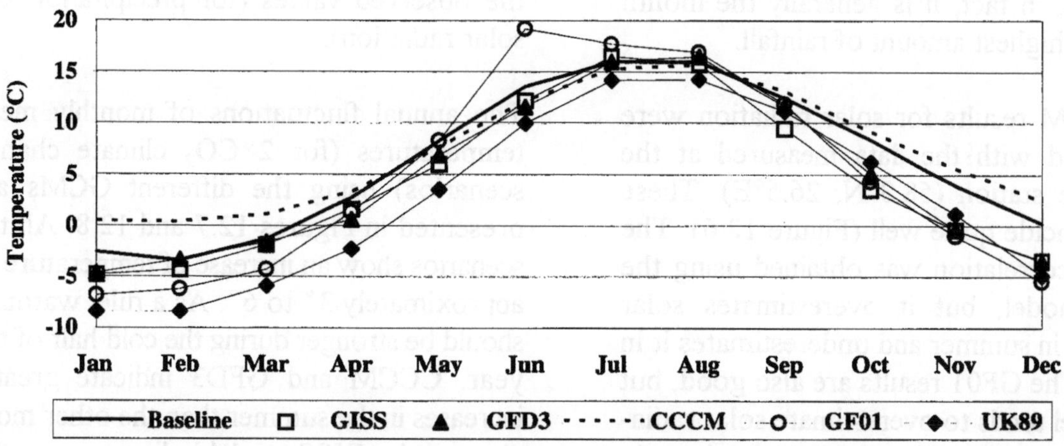

Figure 12.5. Baseline and 1×CO_2 Precipitation Scenarios for Estonia

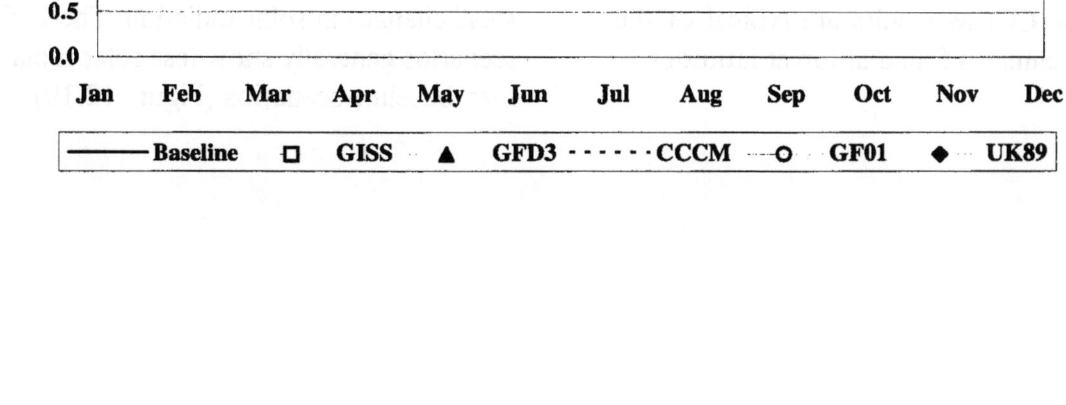

territory of Estonia. The UK89 model gave the best results. This model expressed more or less correctly the annual distribution of precipitation, although, usually underestimated precipitation totals.

Correlations between the baseline scenario and the $1\times CO_2$ scenarios for precipitation are uniformly insufficient for all of the other GCM models. The CCCM results provided the greatest overestimate of precipitation, and only summer rainfall is within realistic limits. The GISS, GFD3, and GF01 results are evenly high in winter and low in summer. According to the GFD3 model, the driest month in Estonia should be August, although, in fact, it is generally the month with the highest amount of rainfall.

The GCM results for solar radiation were compared with the data measured at the Tõravere station (58.3°N, 26.5°E). These data coincide quite well (Figure 12.6). The highest correlation was obtained using the GISS model, but it overestimates solar radiation in summer and underestimates it in winter. The GF01 results are also good, but the model tends to overestimate solar radiation in summer. The CCCM consistently overestimates radiation and the UK89 model strongly underestimates it.

Overall, the comparison of the GCM $1\times CO_2$ results with baseline scenarios indicates significant differences. No one GCM describes current climatic conditions in Estonia sufficiently. The UK89 model results seem to correspond to a much more severe climate usually found at higher latitudes. At the same time, the CCCM results are typical of the warmer climates found at lower latitudes.

Considering all three parameters, the GISS model seems to have the lowest total deviation. The GFD3 and GF01 results are similar, and have a number of disadvantages. All the GCMs are more or less of the same kind, and every one describes one meteorological value better than another. Consequently, all five were used for developing climate change scenarios in Estonia.

$2\times CO_2$ Climate Scenarios

GCM climate change scenarios are developed by adding the difference between $2\times CO_2$ and $1\times CO_2$ scenarios to observed values (for temperature) and by multiplying the ratios of $2\times CO_2$ and $1\times CO_2$ scenarios by the observed values (for precipitation and solar radiation).

The annual fluctuations of monthly mean temperatures (for $2\times CO_2$ climate change scenarios) using the different GCMs are presented in Figures 12.7 and 12.8. All the scenarios show an increase of temperature of approximately 3° to 6°. As a rule, warming should be stronger during the cold half of the year. CCCM and GFD3 indicate greater increases in the summer than the other models, and the GISS model indicates a smaller increase in summer than other models.

The climate change scenarios for precipitation are highly variable (Figure 12.9). In most cases, increased precipitation is expected. The UK89 and GISS models' results show the highest increase and CCCM and GFD3 the lowest.

The GCMs results indicate comparatively small changes in solar radiation. The $2\times CO_2$ scenarios generally show less solar radiation than baseline scenarios (Figure 12.10).

Figure 12.6. Baseline and 1×CO_2 Solar Radiation Scenarios at Tõravere

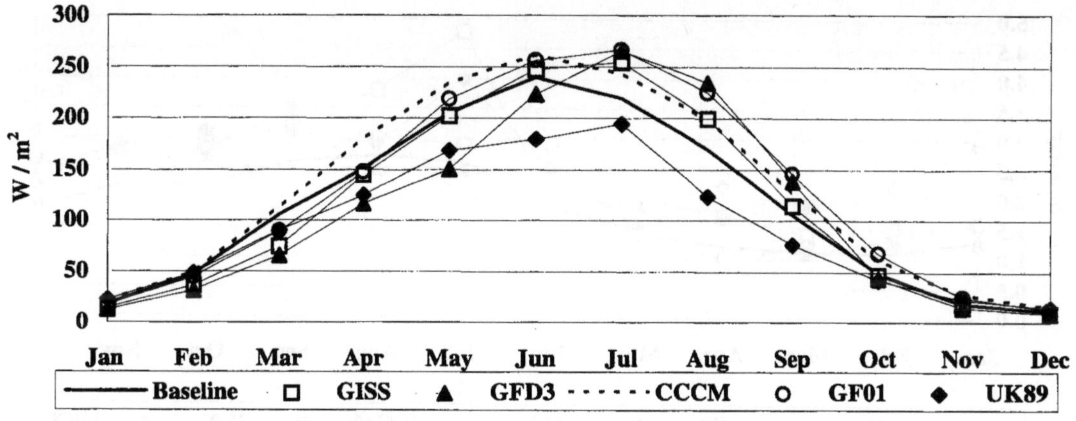

Figure 12.7. Baseline and 2×CO_2 Temperature Scenarios at Võru

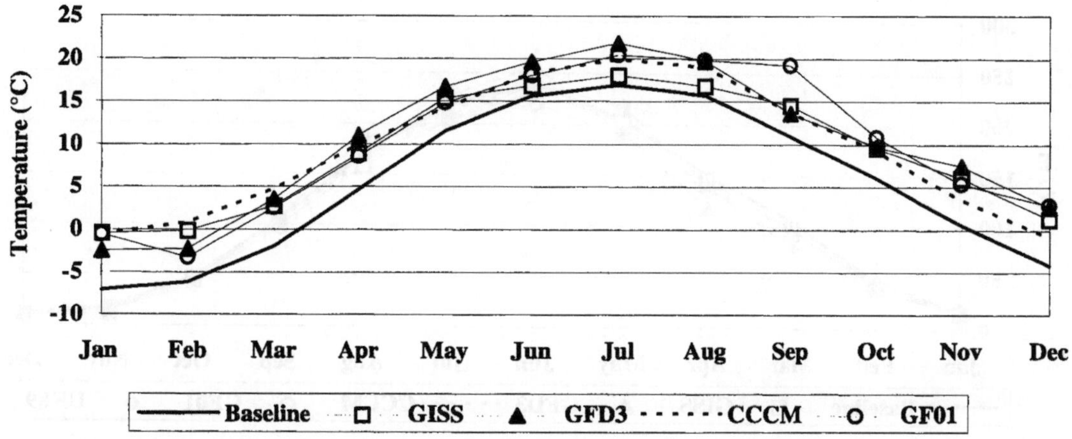

Figure 12.8. Baseline and 2×CO_2 Temperature Scenarios at Sõrve

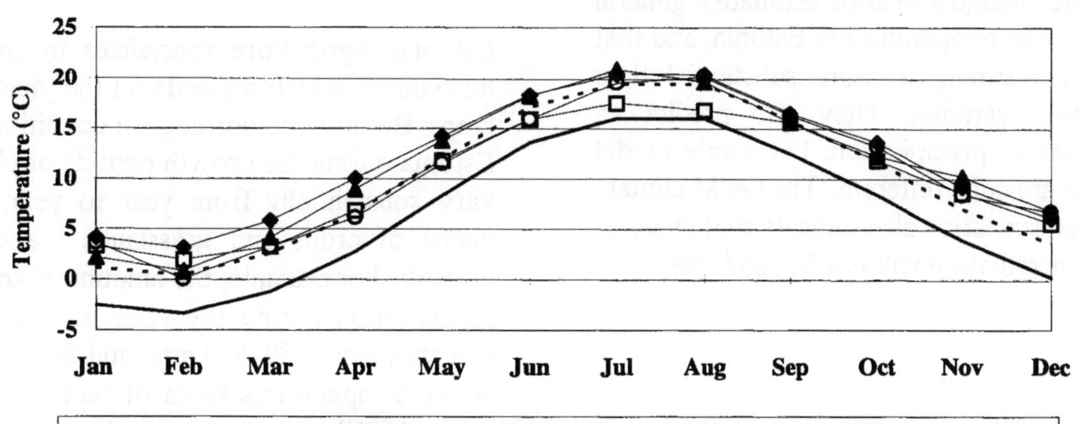

Figure 12.9. Baseline and 2×CO_2 Precipitation Scenarios for Estonia

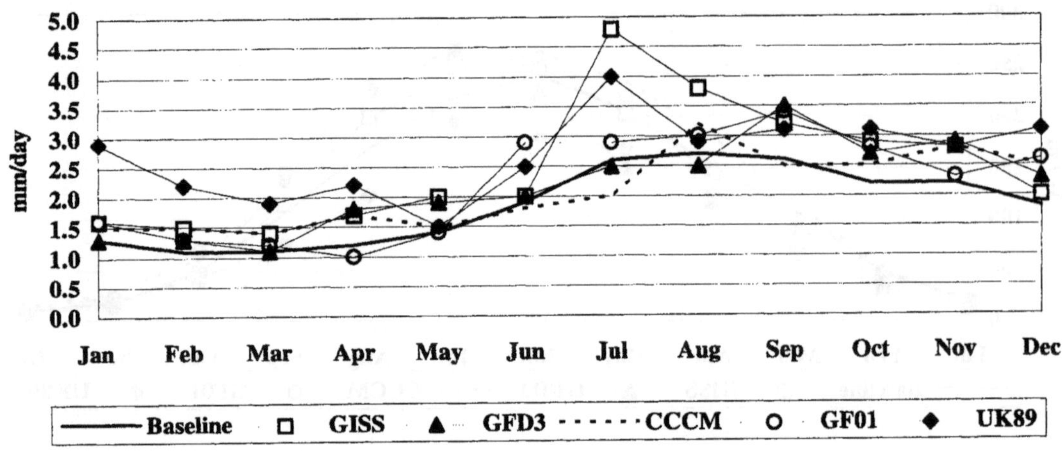

Figure 12.10. Baseline and 2×CO_2 Solar Radiation Scenarios for Tõravere

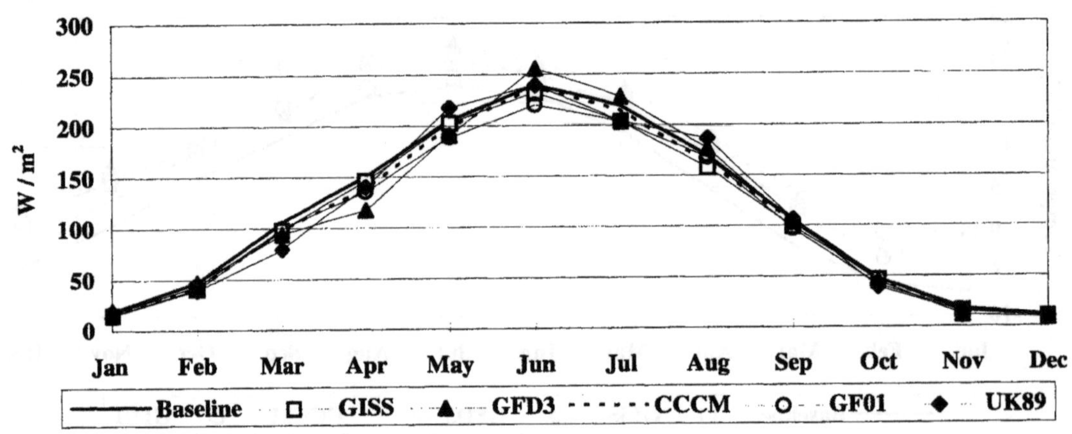

In conclusion, it can be said that the GCM climate change scenarios estimate a general increase in temperature in Estonia, and that winter warming is more substantial than summer warming. They also predict an increase in precipitation, but single model results are quite different. The GCM climate change scenarios also indicate that changes in solar radiation will not be significant.

AGRICULTURE SECTOR

Estonian agriculture specializes in animal husbandry, which depends on the yields of crops. Because meteorological conditions in Estonia during the growth periods of plants vary substantially from year to year, the yields of crops and grasslands are quite unstable. For example, the amount of annual precipitation during the vegetation period may vary from 150-500 mm, and the sum of active temperatures (sum of temperatures over +10°C) may vary as much as twofold.

Thus the yields of potatoes and cereals are variable in Estonia, differing sometimes as much as threefold (Table 12.1).

One important factor limiting the productivity of agricultural crops is the amount of water. A 10% deficiency in water supply decreases the yield of fertilizer-dependent cereal crops by 130-190 kg/ha.

Of the total arable land in Estonia, 55% is covered by grasslands and 38% by grain fields. The main cereal is barley (60% of the planted field devoted to cereal). Consequently, the vulnerability and adaptation assessment for the agriculture sector in Estonia has focused on the analysis of possible changes in yields of barley under different climate change scenarios. This assessment will also consider the impact of climate change on clover-timothy mixture yields.

Methods

Four case study areas — Eerika, Antsla, Olustvere, and Kuusiku (Figure 12.2) — were selected. All four areas have experimental stations where special long-term trials (1966-1994) have been carried out.

The CERES-Barley and SPUR2 models will be used for Estonia's crop productivity assessment.

Estonia is currently calibrating the models and verifying the data; therefore no specific yield results are available at this time.

To provide information for the adaptation analysis, the data on barley are based on 16 variants of fertilizing experiments.

Table 12.1. Probability of Obtaining Various Crop Yields under Current Climate

Crop (kg/ha)	10%	30%	50%	70%	90%
Barley	5,800	5,100	4,600	4,000	2,800
Oats	5,600	4,700	4,000	3,200	2,200
Summer Wheat	5,600	4,800	4,400	3,700	2,900
Rye	5,100	4,100	3,600	3,100	1,900
Winter Wheat	5,100	4,100	3,500	3,000	2,800
Potatoes	34,700	27,700	23,000	19,300	14,200
Grass/Clover Mixture	10,000	8,900	8,300	6,700	5,700

Preliminary Results

Despite the small size of Estonia, the soil and climate conditions are extremely variable and affect suitability for plant growth (Table 12.2).

The initial growth conditions for barley in central Estonia are 20% to 30% more favorable than in the northern, western, and southeastern parts of Estonia. Fertilizing can reduce these differences by 10% to 12%.

Herbage cultivation compared to cereals is less dependent on soil-climatic conditions. The most favorable sites for herbage are the till plains (Podzoluvisols) of central Estonia and the varved clay plains (based-saturated Gleysols) of west Estonia.

Earlier experiments using biophysical models for the productivity of various crops indicate that a temperature rise of 4°C would decrease barley yields by 20% to 30% (Roostalu et al., 1980). At the same time clover-timothy mixture yields would increase by 10%.

Does the Climate of the Future Have Analogues In the Past? Evidence from Estonia

Paleoecological approaches are widely used for studying environments in the past and for compiling predictions for the future (Chambers, 1993). The main principle of this approach is simple: on the basis of paleoecological data, the state of ecosystems in different time spans in the past can be used to assess the state of climatic conditions in the future.

Paleoecological Approaches

Many paleoecological models and approaches have been developed to reconstruct past environments using sequences of sediments such as peat, lacustrine deposits, and ice cores (Berglund, 1986). Naturally, the reconstruction of paleoecological conditions is highly complicated. The formal use of classical principles of actualism has often led to simplified and even wrong conclusions. The environmental variability of the ecosystems in space and time, their tolerance, and their inertia often aggravates the problem (Delcourt and Delcourt, 1988). When we look at human-induced changes (such as greenhouse gas warming) in the environment, the intensity of impact over time must also be considered (Punning, 1994).

Paleoecologists have often used intuitive methods to reconstruct paleoclimates from biological data. Multivariate statistical methods developed during the last two decades provide the most powerful approach. These methods are based on several hypotheses: (1) climate is the ultimate cause of changes in the paleobotanical data; (2) the ecology of the species considered has not varied between the period analyzed and the present time, and consequently the relationship between the species and the climate has been constant in time; and (3) modern observations contain all the information needed for interpreting fossil data (Birks, 1981).

Using pollen data and the transfer function method, Huntley and Prentice (1988) have reconstructed mean July temperatures across Europe 6,000 years ago. Palynological (science dealing with pollen) data from Estonia were well represented in this model. In addition, several other methods have been used to reconstruct paleoclimatic parameters. For example, Klimanov (1987) used an information statistical method based on correlation of subrecent pollen spectra with present climatic conditions. Variations in the stable isotopes of oxygen also offer certain possibilities for paleoclimatic reconstructions. The isotopic composition of lake lime is mainly determined by temperature, which has a direct influence on the kinetics of the equilibrium reaction between the carbonate ion and water, as well as by the isotopic composition of water, which in its turn is determined by the isotopic composition of atmospheric precipitation and the intensity of evaporation.

The main outlines of the dynamics of hydrothermic parameters in the northern and northeastern parts of Europe during the last 10,000 years have been reconstructed using results from investigations of glaciers and lake-marl isotopic-geochemical studies (Punning et al., 1984), determinations of macrofossil and pollen distributions in bog and lacustrine sediments (Klimanov et al., 1986; Koff, 1992), and the analysis of data published in many papers (Prentice, 1983; Punning and Raukas, 1985).

Climate

Generalized data permit the reconstruction of the main tendencies, and in a first approximation even to estimate absolute values of yearly temperatures from 10,000 years ago. The most widely used approach for the reconstruction of long-term climatic change based on palynological data uses the transfer function method (Huntley and Prentice, 1988). The generalized data show that in the beginning of this period the yearly temperatures were up to 4°C lower than the present ones, but about 9,000 years ago, all of northern Europe experienced a rapid increase in annual temperatures.

The time interval between 7,000 and 5,000 years ago is known as the Holocene "climatic optimum." During this time, the summer temperatures in the Northern Hemisphere were, depending on latitude, about 1° to 4°C higher than today. According to Huntley and Prentice (1988) the mean annual temperature gradient in Europe trends north-south, paralleling the gradient of annual insolation. Therefore the differences between reconstructed and present summer and winter temperatures have strong regional variations. It is estimated that in Estonia, the summer temperature was about 2.4°C higher during the climatic optimum than the current summer temperature.

According to current climate models, one of the effects of climate change due to increased greenhouse gas concentrations might be temperatures 2° to 4°C higher than today. Thus, we could consider the reconstructed state of ecological conditions during the Holocene climatic optimum as an analogue of the effects of a 2°C warming scenario. This assumes that the general atmospheric mechanisms at the time being considered were the same as today.

Reconstructed precipitation maps for 6,000 years ago show great variability. Guiot et al. (1993) show that in the region of mixed deciduous boreal forest, the precipitation anomalies compared with contemporary numbers were -100 mm to +100 mm. According to Jelovicheva et al. (1988), the anomaly was about +50 mm.

Forest Vegetation

In Estonia 6,000 years ago, the dominant vegetation formations were species-rich broad-leaved forests, growing on the Pandivere, Otepää, and Sakala uplands (Figure 12.11). Here, the dominant species was elm; however oak, lime, maple, ash, birch, and pine were also represented.

According to this analysis forests dominated by spruce were found only in southeastern Estonia. On the sandy soils of the coast were pine forests. The warmer and wetter climate increased the creation of marshes or fens. Large areas of the northeastern and central part of the country were covered by mire vegetation and swamp forests, where birch and alder dominated in the canopy. On the thin soils of the limestone plateaus in north and west Estonia, alvar forests of oak and pine covered wide areas. "Alvar" is a dry, calciphilous meadow on a limestone plateau; forests growing in these areas are called alvar forests. They are found only in Sweden and Estonia.

Figure 12.11 Schematic Vegetation Maps Comparing 6,000 Years Ago to Present

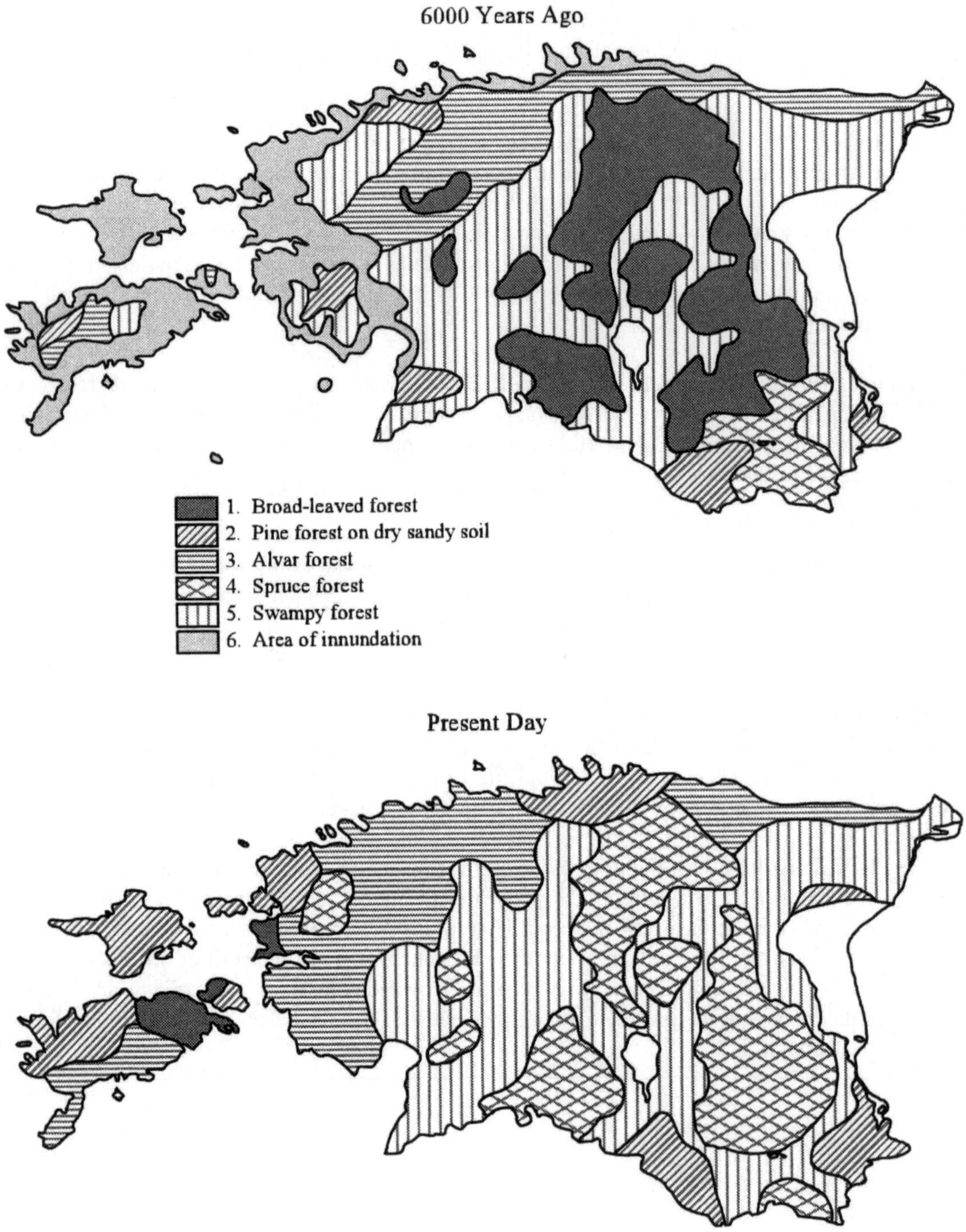

Comparison of the estimated structure of the Estonian forest in the warmest period of the Holocene with that of today (Laasimer, 1965; Valk and Eilart, 1974; Lõhmus, 1984; Kalda, 1995) indicates that the primary difference is a decrease in broad-leaved forests. Today they are found only in western Estonia. Human activities and changes in climate contributed to the change to species-poor forests. The dominance of spruce has been achieved by silvicultural methods. Hardwood forests of oak, elm, and lime are found only in small groves. Alvar forests have been preserved only locally, and today consist mainly of spruce and pine. The distribution of dry pine forests is related to the poorest sandy soils, and they usually develop as a result of forest fires (Kask and Laasimer, 1975). The swamp forests and mire vegetation have remained largely unchanged.

Sea Level Rise

The wide spectrum of coastal formations at different altitudes, remnants of subfossil fauna, and buried lagoon and lacustrine sediments demonstrate the complicated history of the Baltic Sea. Based on the results of geochronological, biostratigraphical, and geomorphological research, the eustatic sea level curve for the Baltic Sea has been reconstructed and used for characterizing the history of the Baltic Sea (Punning, 1993). The data obtained show that in the beginning of the second half of the Holocene, about 6,000 to 5,000 years ago, the ocean level reached its maximum height. The maximum value of the eustatic sea level was 2.0 m above today's level (Punning, 1987).

Depending on the neotectonic uplift of one or another region, the relative extent of sea level changes was and will be quite different. Using the formula:

$$dS = dE/dT,$$

where dS = flooded area, dE = rate of water sea level rise, and dT = rate of neotectonic uplift, it is possible to calculate the spatial extent of sea level rise.

In western Estonia we find ancient coastal formations of different ages tens of kilometers from the present shore. These formations, as well as lagoon and bog deposits covered by sediments, permit the reconstruction of the extent of the sea about 7,000 to 5,000 years ago (Kessel and Punning, 1984). Because of the high sea level and durable hydrological connection between the ocean and the Baltic Sea, the salinity of the Gulf of Finland reached 11-12 ppt (today it is 5-7 ppt) and such molluscs as *Littorina littorea, L. saxatilis,* and *Zippora membranacca* were widespread.

Discussion

Therefore, in principle, it is possible to find in the past analogous situations to those predicted as the result of greenhouse gas warming. The use of paleoecological reconstructions for these periods could provide very valuable information for understanding the state of the environment in the future. At the same time, there are many factors that will hinder the direct use of paleoecological data. In the interpretation of paleodata and comparison with future situations, one must take into account such complications as similarity of natural conditions, cause-consequence adequacy, stability of hierarchic correspondence of allogenic factors, and intensity in time.

Table 12.2. Suitability of Soil-Climatic Conditions for Barley and Herbage Cultivation (% in relation to Järvamaa District)

		Barley		Clover/Timothy Mixture in Field Crop Rotation	
		Without Fertilizers	With Fertilizers	Without Fertilizers	With Fertilizers
District	Region	$N_0P_0K_0$	$N_{90}P_{39}K_{75}$	$N_0P_0K_0$	$N_{90}P_{39}K_{75}$
Järvamaa	Central	100	100	100	100
Jôgevamaa	Central	96	98	100	104
Raplamaa	Central	86	91	94	94
Viljandimaa	Central	97	99	99	104
Tartumaa	Eastern	94	96	97	101
Ida-Virumaa	Northeastern	81	87	91	94
Lääne-Virumaa	Northern	91	95	91	92
Harjumaa	Northwestern	86	91	83	88
Pôlvamaa	Southeastern	84	91	93	97
Vôrumaa	Southeastern	79	86	92	97
Valgamaa	Southern	83	90	94	100
Pärnumaa	Southwestern	86	91	99	100
Hiiumaa	Western	76	88	85	88
Läänemaa	Western	82	89	91	92
Saaremaa	Western	75	87	82	86

FORESTRY SECTOR

Potential forest responses to global climate change were examined for the territory of Estonia. In Estonia, there are 87 native and over 500 introduced tree and shrub species (Karoles et al., 1994). The primary tree species are Scots pine (*Pinus sylvestris* L.) which account for 40% of standing volume, Norway spruce (*Picea abies* L. Karst.) which account for 26%, and birches (*Betula pendula* Roth., *B. pubescens* Ehrh.) which account for 26%. Generally, the computer simulations indicate that climate warming would increase the productivity of forest stands in the boreal zone (Kellomäki et al., 1988; Kellomäki and Kolström, 1992; Solomon and Leemans, 1990). Plant communities will respond to both increases in atmospheric CO_2 and associated climate change as a function of changing competitive advantages among the species (Solomon and Cramer, 1993). Climate warming may cause the extinction of or at least fewer prevailing genotypes and their gradual substitution by more adaptable genotypes. Furthermore, new plant communities may change the balance of nature. Forest responses to climate change are dependent on the type of forest under consideration.

Methods

Selection of Study Sites

Two study sites were selected to estimate the vulnerability of Estonian forests to climate change. Tudu (central Estonia) represents an inland site, and Virtsu (western Estonia) is located in the coastal zone (Figure 12.2). These sites were selected because they differ in both current climate conditions and forest type. Primary attention was given to the evaluation of changes in species composition and productivity under climate change.

Table 12.3. Basic Parameters Used to Define Species Attributes in the Forest Gap Model[1]

Species	Max. Observed Diameter (m)	Max. Observed Age (yr)	Max. Observed Height (m)	SEDmax[2]	Shade Tolerance[3]	Min PET Ratio[4]
Picea abies	1.5	350	47	50	1	0.94
Pinus sylvestris	1.7	450	40	40	3	1.20
Quercus robur	2.5	999	39	20	2	1.10
Betula pendula	0.7	150	31	100	2	0.98
Betula pubescens	0.6	120	25	100	2	0.97
Populus tremula	1.0	200	34	40	3	1.00
Alnus glutinosa	1.2	200	30	20	1	0.94
Alnus incana	0.5	70	20	40	1	0.98

1 Paal et al., 1989; Holdridge, 1967; Laas, 1967.
2 SEDmax — maximum yearly seedling establishment scaled to plot.
3 1 — the shade tolerant, 2 — intermediate, 3 — the shade intolerant; expert estimation.
4 Min PET Ratio - maximum potential evapotranspiration (PET) ratio allowing tree growth, as species specific bioclimatic index of drought tolerance. It is the quotient of PET and average annual precipitation (Holdridge, 1967).

Forest and Ecosystem Methodologies

The responses of Estonian forests to global climate change were evaluated using a modified version of the forest gap model (Botkin et al., 1972; Shugart, 1984). The Holdridge Life Zones Classification Model (Holdridge, 1967) was also used to determine potential evapotranspiration ratios for different species and different types of forest ecosystems.

The first step was to collect data for the forest gap model (Table 12.3). Four climate change scenarios for the two study areas were incorporated into the modeling framework (Table 12.4). The amount of water in soil available for plant growth in the model at a given PET ratio is dependent on field moisture capacity and wilting point. The forest gap model was calibrated using 4 to 13 tree species and extrapolated monthly temperature and precipitation data of the last 30 years for five sites in Estonia. Tree growth was estimated as an average of nine

Table 12.4. Incremental Climate Change Scenarios for Central and Coastal Estonia

Parameters	GDD[2]	PET Ratio[3]	AP[4]
Tudu (59° 12' N, 27° 00' E)			
Scenario 0 — Current Conditions[1]	2,257	0.72	670
Scenario 1 — +4°C	3,264	0.88	670
Scenario 2 — +4°C, -10% AP	3,264	0.98	604
Scenario 3 — +4°C, +10% AP	3,264	0.80	737
Virtsu (58° 36' N, 23° 30' E)			
Scenario 0 — Current Conditions[1]	2,504	0.82	620
Scenario 1 — +4°C	3,610	1.01	620
Scenario 2 — +4°C, -10% AP	3,610	1.12	560
Scenario 3 — +4°C, +10% AP	3,610	0.92	682

1 Data of J. Jaagus from Institute of Geography of Tartu University.
2 GDD — Annual Sum of Growing-Degree District Days = Yearly Sum of Daily Mean Temperatures above 0°C.
3 PET Ratio — Potential Evapotranspiration Ratio, mm/mm.
4 Ap — Annual Precipitation, mm.

sample plots with yearly timestep during 500 years.

Preliminary Results

Simulations of long-term forest development (500 years) indicate significant changes in the structure of forest communities under various climatic conditions for the two selected study areas (Figure 12.12). Competition, succession, and changes in species composition as a function of time were successfully reproduced by the forest gap model in the baseline scenario.

Scenario 0. During the first period of forest development, the pioneer community consists mainly of birch (*Betula* sp.). After 100 years, deciduous trees will be substituted by Norway spruce (*Bicea abies* L. Karst) in central Estonia or by pine-spruce mixed stands in the coastal area. These types of forest community development are characteristic of Estonia under present climate conditions.

Scenario 1. A 4°C increase in temperature brings structural changes to the forests. Scots pine (*Pinus sylvestris* L.) becomes the predominant species in central Estonia. During the first 100 years, high populations of birch (*Betula* sp.) and aspen (*Populus tremula*) are notable, but their share decreases as the forest ages. In coastal areas (Virtsu), the mass of Scots pine is estimated to reach its peak of more than 200 tons/ha in 40 to 50 years. At the same time, the masses of Norway spruce and birches are estimated to rapidly decrease. The total mass of a tree stand during the first 200 years may increase up to 15%.

Scenario 2. With increased temperature and decreased annual precipitation, drought-tolerant species such as Scots pine and oak (*Quercus robur*) will grow in central Estonia; only Scots pine will grow on the coast. This is the most drastic scenario for Estonian forests. In central Estonia, less precipitation and higher temperatures mean that the biomass will peak after 20 to 40 years and decrease after 200 years. In the coastal area, biomass peaking may be 30 to 40 years later, and the total biomass level will be lower than it is today.

Scenario 3. Increases in both temperature and precipitation favor intensive growth of deciduous trees during the first 120 to 170 years in both study areas. In later stages of stand development, deciduous tree species are replaced by Norway spruce. At the same time, Scots pine and few deciduous species will be characteristic in the coastal area.

Discussion and Conclusions

Disturbances in climate pattern need not be catastrophic to alter forest species composition. Scots pine seems to be rather tolerant to both temperature rise and changes in precipitation in both study areas. At the same time, Norway spruce and aspen are sensitive to climate change.

Spruce and aspen today have the highest average volumes of growing stock — 173 m³/ha and 212 m³/ha, respectively. Decreases as a result of climate change could hurt the Estonian timber industry. The results of our model calculations indicate that the competitive ability of Scots pine at higher temperatures is improving, particularly in the Virtsu study area. From

Figure 12.12. Alterations in the Forest Community Structures at Tudu and Virtsu

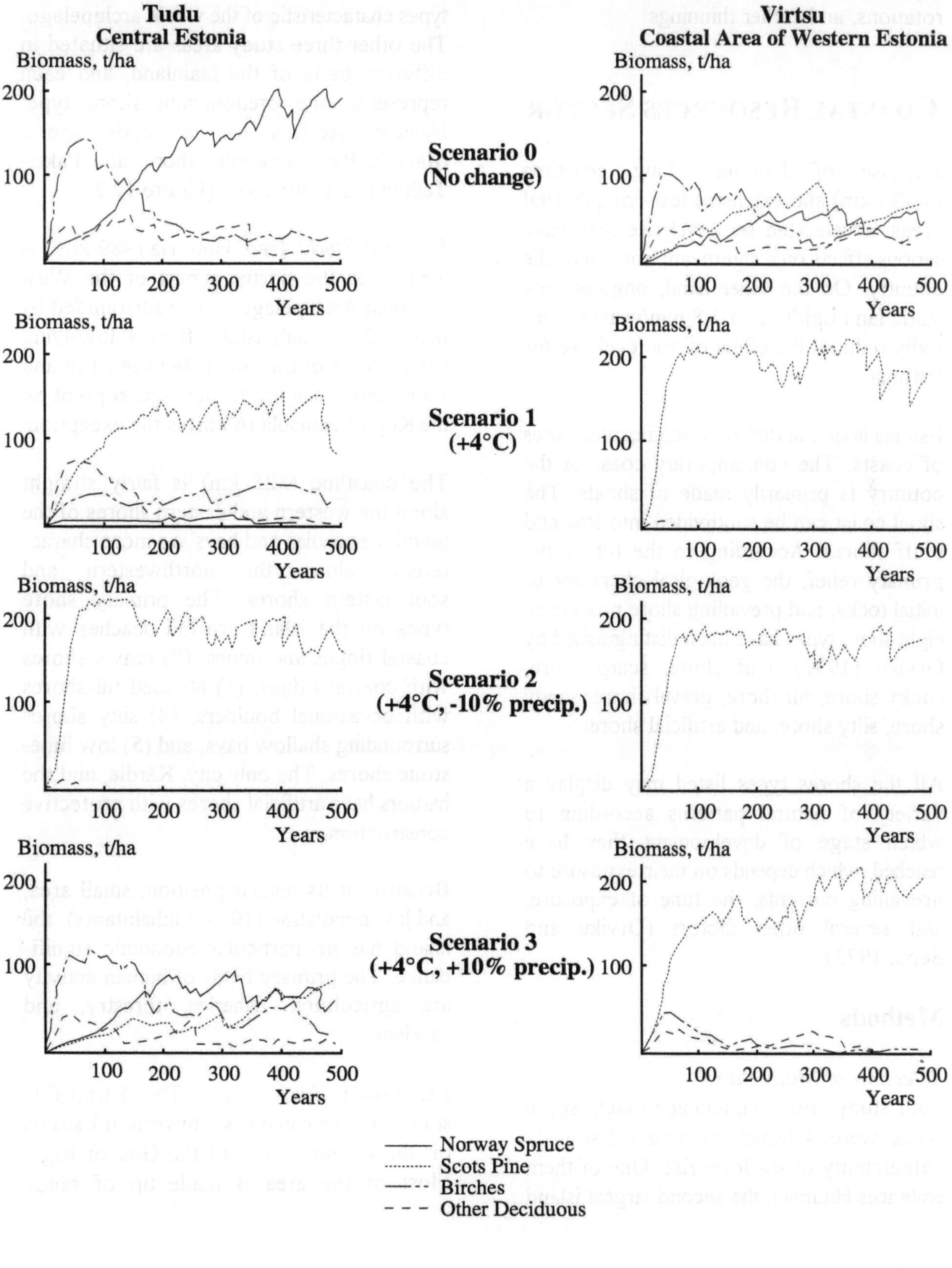

the point of view of forestry, this means higher productivity of forests, shortened rotations, and earlier thinnings.

COASTAL RESOURCES SECTOR

Because of Estonia's long coastline (3,794 km) and extensive low-lying coastal areas, accelerated sea level rise may have serious effects on a significant portion of the country. On the other hand, ongoing isostatic land uplift (1 to 2.8 mm/year) potentially reduces the effect of sea level rise for Estonia.

Estonia is rich in different geomorphic types of coasts. The contemporary coast of the country is primarily made of shoals. The shoal coast can be subdivided into low and bluff shores. According to the tilt of the primary relief, the geological character of initial rocks, and prevailing shore processes, eight shore types have been distinguished by Orviku (1992): cliff shore, scarp shore, rocky shore, till shore, gravel shore, sandy shore, silty shore, and artificial shore.

All the shores types listed may display a variety of natural patterns according to which stage of development they have reached, which depends on their exposure to prevailing currents, the time of exposure, and several other factors (Orviku and Sepp, 1972).

Methods

Selection of Study Sites

Four study areas containing all eight shore types were selected to assess Estonia's vulnerability to sea level rise. One of them embraces Hiiumaa, the second largest island of the West Estonian Archipelago, which represents a variety of geomorphic coast types characteristic of the whole archipelago. The other three study areas are situated in different parts of the mainland, and each represents one predominant shore type: Häädemeeste-Ikla has a sandy shore, Matsalu Bay has a silty shore, and Pakri-Tallinn has a cliff shore (Figure 12.2).

Hiiumaa Study Area. Hiiumaa (989 km^2) is located in the northern part of the West Estonian Archipelago, and is surrounded by nearly 200 small islets. It is a low-lying island; most of the land is between 1 m and 10 m above sea level. The highest point on the Kõpu Peninsula (63 m) is the exception.

The coastline (325 km) is fairly straight along the western and eastern shores of the island. Peninsulas and bays are more characteristic along the northwestern and southeastern shores. The primary shore types on the island are (1) beaches with coastal ridges and dunes, (2) gravel shores with coastal ridges, (3) abraded till shores with occasional boulders, (4) silty shores surrounding shallow bays, and (5) low limestone shores. The only city, Kärdla, and the harbors have artificial shores with protective construction.

Because of its insular position, small area, and low population (10,600 inhabitants), the island has no particular economic significance. The primary fields of human activity are agriculture, fisheries, forestry, and tourism.

Pärnu-Ikla Study Area. The Pärnu-Ikla study area is located in southwestern Estonia on the western coast of the Gulf of Riga. Most of the area is made up of sandy beaches with an extensive ridge of coastal formations covered by foredunes and dunes. The highest dunes of Estonia (40 m above sea level) are situated in its southern part. This is one of the most popular recreation areas in the country, with famous health resorts in Pärnu and numerous motels between Rannametsa and Ikla.

Sea Level Rise Methodology

Detailed measurements using large-scale (1:25,000) topographic maps and calculations according to the Bruun Rule were made along the coastline at 200 m intervals. There were two problems: (1) the Estonian coasts are too shoaly to apply the Bruun Rule without corrections, and (2) the overfill ratio was found quite approximately using the following coefficients:

- sand — 1.0
- silt and varved clays — 0.7
- till (loam full of shingle) — 0.4
- limestone — 0.1.

It is worth mentioning that till shores in Estonia are erodible to such an extent that fine-grained fractions are washed out. The remaining shingle ridges and boulder fields prove to be extremely resistant to further erosion.

Based on measurements, calculations, and field observations, new coastline positions were drawn on the topographic maps. Field observations were used to compare possible coastline changes obtained by the Bruun Rule to existing situations, and to make corrections where necessary. Two different coastline positions were drawn, depending on either normal weather conditions or potential storm surges. Both zones were subjected to inventories of land loss and of temporary damages.

The data on water-level fluctuations and storm surges were obtained from five meteorological and sea level observation stations in Hiiumaa (Table 12.5) and from two stations in Pärnu-Ikla (Table 12.6) whose data go back to the second half of the 1800s. Sea level fluctuations caused by tides in the Baltic Sea do not need to be taken into consideration in calculating possible coastline recession. This is due to the shallow

Table 12.5. Maximum and Minimum Sea Levels Obtained by Long-Term Observations in Hiiumaa (cm)[1]

	Month											
	Jan	Feb	Mar	Apr	May	Jun	Jul	Aug	Sep	Oct	Nov	Dec
Tahkuna												
max	144	135	135	126	104	104	112	118	126	138	136	153
min	0	-22	-10	-6	7	20	30	30	16	5	5	8
Ristna												
max	112	95	95	58	47	53	55	81	89	83	86	123
min	-48	-63	-68	-65	-50	-40	-25	-32	-29	-53	-56	-78
Sõru												
max	87	73	70	51	29	43	39	51	68	73	57	84
min	-57	-85	-69	-67	-44	-43	-25	-39	-34	-49	-53	-97
Heltermaa												
max	93	80	85	53	26	47	48	60	71	71	65	84
min	-54	-76	-71	-68	-50	-39	-22	-28	-24	-46	-61	-99

[1] Maximum and minimum sea levels measured in relation to the Kronstadt benchmark.

Table 12.6. Maximum and Minimum Sea Levels Obtained by Long-Term Observations in Pärnu-Ikla Coastal Area (cm)[1]

	Month											
	Jan	Feb	Mar	Apr	May	Jun	Jul	Aug	Sep	Oct	Nov	Dec
Pärnu												
max	159	155	175	179	135	94	97	159	174	179	183	153
min	-107	-87	-103	-115	-85	-65	-39	-51	-61	-95	-100	-120
Kihnu												
max	132	150	135	68	92	59	67	94	99	118	95	127
min	-102	-85	-86	-86	-58	-48	-27	-47	-38	-62	-82	-133

[1] Maximum and minimum sea levels measured in relation to the Kronstadt benchmark.

basins and the isolation of the sea, which has a very narrow connection through the Danish Straits with the Atlantic Ocean, giving the effect of a large lake.

Finally, the data of isostatic land uplift measurements were taken into consideration in land loss estimates.

Preliminary Results and Discussions

Preliminary results of the assessment of Estonia's vulnerability to 1.0 m sea level rise in 2075 were obtained for two areas. The assessments of the other two test areas are still being developed. A 0.3 m scenario would not have a noticeable effect because of the isostatic land uplift.

According to the data obtained from five meteorological and sea level observation stations in Hiiumaa, the highest sea levels have been measured at the stations most exposed to waves and storm surges, reaching 1.53 m above the Kronstadt 0 (local benchmark for the eastern part of the Baltic Sea) in Tahkuna (Table 12.5). At the same time sea level fluctuations are more considerable on the eastern and southern coasts of the island, probably due to the shoals.

Although a slow uplift of the earth's crust is occurring in Hiiumaa today (Vallner et al., 1988), there has been erosion of sandy beaches in recent decades, and in some places the sea is advancing again (Orviku, 1987).

Since 1954, Hiiumaa has been subject to at least eight (in 1954, 1967, 1969, 1971, 1975, 1983, 1986, 1990) extremely strong storms that, according to statistics, should occur only once in 100 years. In autumn and winter, the westerly and southwesterly storm winds raise the sea level up to 2.0 m above its summer level. As there is little evidence of a rising sea level in Estonia over this period, beach erosion appears to be largely due to the recent increased storminess in the eastern Baltic Sea.

Measurements and calculations indicate that the shore types most vulnerable to sea level rise are sandy shoal and silt shores. Sandy beaches are concentrated on the western coast of Hiiumaa, while silt shores prevail in the northwest and southeast. Under a 1.0 m sea level rise scenario, the most extensive coastline recessions (over 500 m) would occur on the silt shores near Kõrgessaare and Käina.

A 1.0 m sea level rise would inundate 38 km² of the coastal area of Hiiumaa (Table 12.7.). The land loss includes 21.5 km² of coastal meadows, 8.3 km² of forests, and 5.6 km² of beaches. Also, 155 low-lying islets off the coast of Hiiumaa would disappear, and 43 new islets would be created by separation from current headlands. A number of valuable coastal ecosystems, plant communities, and plant species would be destroyed. Some of them are located within the limits of the Hiiumaa Biosphere Reserve, which is a valuable breeding ground for birds.

Preliminary estimates of losses indicate that 100% of reed beds and 80% of coastal meadows (salt marshes), including rare saline plant communities of *Salicornia europaea, Glaux maritima-Juncus gerardi* and *Carex glareosa,* are in direct danger. In addition, biotopes of several unique plant species such as *Myosotis laxa, Samolus valerandi, Sagina maritima, Herminium monorchis, Halimione pedunculata, Suaeda maritima, Coeloglossum viride,* and *Polygonum oxyspermum* (Inglelög et al., 1993) would also be affected. The very unusual ecosystems of numerous lagoons and calciphilous meadows rich in orchids on the northwestern coast of Hiiumaa would completely disappear.

Because the villages on the island are usually situated at a distance from the coastline, only 18 dwellings would be in direct danger; 59 boathouses and sheds for nets are located within the limits of inundation zone. Additionally, seven harbors, eight lighthouses, and 6 km of roads connecting harbors with villages and Kärdla would need to be protected by seawalls and dikes. Storehouses in

Table 12.7. Possible Inundation under a 1.0 m Sea Level Rise (km²)

	Hiiumaa	Pärnu-Ikla
Forests	8.3	0.3
Grasslands/meadows	21.5	20.5
Beaches	5.6	1.4
Other	2.6	0.0
Total land loss	38.0	22.2

the harbors would need to be rebuilt. The most useful adaptation option to preserve the natural ecosystems of Hiiumaa would be hardening of headlands.

For the Pärnu-Ikla study area, coastal measurements and calculations using the Bruun Rule indicate that the areas most vulnerable to sea level rise are located in the northern part, where silt shores are predominant, but also on the densely populated Isle of Kihnu. Coastal formations that act as protection against inundation are located at a distance from today's coastline. Dikes built before the last world war are no longer effective. During the most recent strong storms, the waves have reached many dwellings 300 m inland. Where ridges of coastal formations are interrupted by brook valleys or drains, low-lying meadows beyond the ridges and dunes are also vulnerable to being inundated, and when the seas are high, dunes are being eroded.

A 1.0 m sea level rise would inundate 22.2 km² of coastal territory, including 0.30 km² of forests, 20.5 km² of grasslands, and 1.4 km² of beaches (Table 12.7). Almost 2.0 km² of the territory of Pärnu, the largest city in this region, and 3.8 km² of the Isle of Kihnu are located in the zone of potential inundation. The potential reduction of the effects of sea level rise by isostatic land uplift here (about 1.0 mm/year) is less than in

Hiiumaa. That is why the potential land loss compared with that in Hiiumaa is relatively much greater (the coastline length is only 149 km compared to a coastline length of 325 km on Hiiumaa). One small and low islet existing today would submerge, and the 5.0 m high Piklanina headlands would become an islet.

Finally, there are no rare or highly valuable natural ecosystems in the Pärnu-Ikla area that would need special protection. The primary problems here are the socioeconomic impacts and particularly, the protection of recreational areas.

SUMMARY CONCLUSIONS

Because of Estonia's long tradition of interest in its natural resources, and its relatively small territory, all of the country's natural components are well investigated and mapped. Existing large-scale topographical, geological, geomorphological, hydrological, soil, vegetation, and landscape maps together with long-term meteorological and hydrological observation data make it possible to evaluate any aspect of the country's natural resources and ecosystems with a high level of confidence. Extensive paleogeographical data also enable researchers to compare current and predicted scenarios with those of the past.

However, because of the peculiarity of Estonian history in this century, there are comparatively long gaps in reliable socioeconomic data, making it difficult to develop workable solutions. The rapid changes taking place in Estonia's sociopolitical life today make it difficult to predict future scenarios of development.

Despite these difficulties, performing a vulnerability and adaptation assessment is an excellent opportunity for Estonia to look at its future, understand the possible events that could affect the environment, and prepare to respond to them. This chapter contains preliminary results of estimates obtained with the programs and models proposed by the Intergovernmental Panel on Climate Change (IPCC). Because of the great variability of natural conditions in Estonia, there are some deviations between the model calculations and actual parameters. The process of developing Estonia's country study requires new data, and modifications to existing models and programs in order to allow consideration of all the significant aspects of the problem.

REFERENCES

Benioff, R., S. Guill, and J. Lee (eds.) (In press). *Vulnerability and Adaptation Assessments: An International Guidebook.* Dordrecht, The Netherlands: Kluwer Academic Publishers.

Berglund, B.E. (ed.). 1986. *Handbook of Holocene Palaeoecology and Palaeohydrology.* New York: John Wiley & Sons, 869 pp.

Birks, H.J.B. 1981. The Use of Pollen Analysis in the Reconstruction of Past Climate: A Review. In: Wigley, T. et al. (eds.): *Climate and History.* Cambridge Univ. Press, Cambridge, Great Britain, 111-138.

Botkin, D.B., Janak, J.F., and J.R. Wallis. 1972. Some Ecological Consequences of a Computer Model of Forest Growth. *Journal of Ecology,* 60: 849-873.

Chambers, F.M. (ed.). 1993. *Climate Change and Human Impact on the Landscapes.* London, Great Britain: Chapman & Hall, 303 pp.

Delcourt, H.R. and P.A. Delcourt. 1988. Quaternary Landscape Ecology: Relevant Scales in Space and Time. *Landscape Ecology,* 2: 23-44.

Guiot, J., S.P. Harrison, and I.C. Prentice. 1993. Reconstruction of Holocene Precipitation Patterns in Europe Using Pollen and Lake-Level Data. *Quaternary Research,* 40: 139-149.

Holdridge, L.R. 1967. *Life Zone Ecology.* Tropical Science f. San Jose, Costa Rica.

Huntley, B. and I.C. Prentice. 1988. July Temperatures in Europe from Pollen Data 6,000 Years Before Present. *Science,* 241: 687-690.

Ingelög, T., R. Andersson, and M. Tjernberg (eds.). 1993. *Red Data Book of the Baltic Region.* Part 1. Lists of Threatened Vascular Plants and Vertebrates: Uppsala, Sweden; Riga, Latvia. 95 pp.

Jelovicheva, J., I. Bogdel, V. Zernitskaya, and V. Klimanov. 1988. Climate Reconstruction by Pollen Data. In: *Palaeoclimate During the Holocene in the Territory of USSR.* Moscow, USSR: Institute of Geography, 86-103.

Kalda, A. 1995. Broad-Leaved Deciduous Forests in Estonia. In: *Consortium Masingii,* Tartu, Estonia: Tartu University, 89-96.

Karoles, K., A. Leemet, and O. Lugus. 1994. *Forest and Forest Products Country Profile. Estonia.* United Nations, New York, Geneva, 62 pp.

Kask, M. and L. Laasimer. 1975. Botanical Characterization of the Excursion Route in the Estonian SSR. In: *Some Aspects of Botanical Research in the Estonian SSR.* Tartu, Estonia: Estonian Academy of Sciences, 7-26.

Kellomäki, S. and M. Kolström. 1992. Simulation of Tree Species Composition and Organic Matter Accumulation in Finnish Boreal Forests under Changing Climatic Conditions. *Vegetatio,* 102: 47-68.

Kellomäki, S., H. Hänninen, and T. Kolström. 1988. Model Computations on the Impacts of the Climatic Change on the Productivity and Silvicultural Management of the Forest Ecosystem. *Silva Fennica,* 22: 293-305.

Kessel, H. and J.M. Punning. 1984. The Development of the Baltic Sea in the Holocene. In: *Estonia: Nature, Man, Economy.* Tallinn, Estonia: Estonian Geographical Society, 36-47.

Klimanov, V. 1987. About Climatic Changes in the Northern Part of the East-European Plain. *Palaeohydrology of the Temperate Zone III, Mires and Lakes,* Acad. Sci. Estonian SSR, Inst. Geol., 23-37.

Klimanov, V., T. Koff, and J.M. Punning. 1986. Climatostratigraphic Study of Subatlantic Sediments in the Nigula Bog (SW Estonia) and Reconstruction of Palaeoclimatic Conditions by Pollen Data. In: *Isotope-Eochemical Studies in Baltic Countries and Byelorussia*. Tallinn, Estonia, 117-129 (in Russian).

Koff, T. 1992. Palynological Studies of Estonian Holocene. *Publications of Karelian Institute* (University of Joensuu, Joensuu, Finland), 102: 75-83.

Laas, E. 1967. *Dendroloogia*. Tallinn, Estonia:Valgus, 671 pp. (in Estonian).

Laasimer, L. 1965. *Vegetation of the Estonian SSR.* Tallinn, Estonia: Valgus, 397 pp. (in Estonian).

Lõhmus, E. 1984. *Estonian Types of Forest Habitat.* Tallinn, Estonia. 88 pp. (in Estonian).

Orviku, K. 1987. Estonian Seashore and Their Dynamics. *Eesti Loodus*, no. 11: 712-719. (in Estonian).

Orviku, K. 1992. Characterization and Evolution of Estonian Seashores. Tartu, Estonia: Summary of Doctoral Thesis at Tartu University, 20 pp.

Orviku, K. and U. Sepp. 1972. Stages of Geological Development and Landscape Types of the Islets of the West-Estonian Archipelago. *Geographical Studies* (Tallinn, Estonia), 15-25.

Paal, J.L., T.A. Oja, and S.F. Kolodyazhnyij. 1989. *Taxonomic and Temporal Continuum of Plant Cover*. Tallinn, Estonia:Valgus. 182 pp.

Prentice, I.C. 1983. Postglacial Climatic Change: Vegetation Dynamics and the Pollen Record. *Progress in Physical Geography*, 72: 273-286.

Punning, J.M. 1987. Holocene Eustatic Oscillations of the Baltic Sea Level. *Journal of Coastal Research*, 3/4: 505-513.

Punning, J.M. 1993. Sea Level Changes and Paleogeographical History of the Baltic Sea. In: *Sea Level Changes and Their Consequences for Hydrology and Water Management*. Noordwijkerhout, Netherlands, 61-71.

Punning, J.M. (ed.). 1994. *The Influence of Natural and Anthropogenic Factors on the Development of Landscapes.* Tallinn, Estonia: Institute of Ecology, Publ. 2, 227 pp.

Punning, J.M. and A. Raukas. 1985. *Palaeogeography of Northern Europe During Late Pleistocene. Advances of Sciences and Technology.* Moscow, 214 pp. (in Russian).

Punning, J.M., T. Martma, and R. Vaikmäe. 1984. Light Isotope Variations in Carbonate Sediments and Their Palaeogeographical Value. *ZFI Mitteilungen* (Leipzig), 84: 329-336.

Roostalu, H., A. Piho, and P. Saarnak. 1980. Complex Zoning of Soil-Climatic Resources. *Transactions of Estonian Agricultural Academy* (Tartu, Estonia), 122: 101-116.

Shugart, H.H. 1984. *A Theory of Forest Dynamics*. New York: Springer-Verlag.

Solomon, A.M. and W.P. Cramer. 1993. Biospheric Implications of Global Environmental Change. In: *Vegetation Dynamics and Global Change*. New York, USA, Chapman & Hall, 25-53.

Solomon, A.M. and R. Leemans. 1990. Climatic Change and Landscape Ecological Response: Issues and Analysis. In: *Landscape-Ecological Impact of Climate* Change. Amsterdam, Netherlands: IOS Press, 293-316.

Statistical Office. 1995a. *Forestry 1994*. Statistical Office of Estonia, Tallinn. 20 pp.

Statistical Office. 1995b. *Statistical Yearbook 1993*. Statistical Office of Estonia, Tallinn. 79 pp.

Valk, U. and J. Eilart (eds.). 1974. *Estonian Forests*. Tallinn, Estonia: Valgus, 307 pp. (in Estonian).

Vallner, L., H. Sildvee, and A. Torim. 1988. Recent Crustal Movements in Estonia. *Journal of Geodynamics*, 9: 215-223.

Chapter 13

Vulnerability and Adaptation Assessments for Poland

Ryszard B. Zeidler

Department of Coastal Engineering and Dynamics
Polish Academy of Sciences Institute of Hydroengineering, Gdańsk

SUMMARY. *The vulnerability and adaptation assessment for Poland has been completed for the coastal resources sector. Sea level rise scenarios of 10 cm and 30 cm by 2030 and 30 cm and 100 cm by 2100 are assumed (along with some wind effects) for the entire Polish coast. Three adaptation strategies, retreat, limited protection, and full protection, are evaluated and compared quantitatively in physical and socioeconomic terms. More than 2,200 km² and 230,000 people are found to be vulnerable (inundated or at risk of flooding) to a 100 cm sea level rise by 2095, and the total cost of land lost due to a 100 cm sea level rise is assessed at nearly US$30 billion (plus an additional US$18 billion for land at risk of flooding), while the cost of full protection is estimated to be US$6 billion. These estimates for both retreat and protection costs are roughly 50% higher if socioeconomic conditions for 2025 are taken as a reference date (coupled with delays in decision making and inflation of prices). Poland is also currently developing a water resources assessment for the catchment areas of the Warta and Prosna rivers. The preliminary results of this assessment are inconclusive.*

INTRODUCTION

The vulnerability and adaptation assessment for the coastal resources of Poland has been completed, and the results are reported here. The followup research stemming from this study will primarily focus on the climate-change boundary conditions other than sea level rise.

A recent vulnerability and adaptation assessment program for the entire Polish coast (Zeidler, 1992, referred to hereafter as the 1992 assessment) has provided a background for this vulnerability and adaptation assessment. In the past three years there have been a number of physical and socioeconomic developments in Poland that have forced a revision of earlier findings.

Among the physical boundary conditions, the most distinctive departures from earlier assessments stem from (1) recent controversies over the sea level change, (2) a reassessment of Polish sea level change trends and a readjustment of the time horizon to beyond 2030, (3) inclusion of the

Note: This chapter does not contain a detailed description of the assessment methodologies that were recommended by the U.S. Country Studies Program. Refer to Chapter 2 for an overview of the methodologies.

J. B. Smith et al. (eds.), Vulnerability and Adaptation to Climate Change, 277–294.

most recent data for annual sea level maxima (storm surges), and flooding probabilities for the Polish coast, and (4) inclusion of some additional climate change factors such as changes in wind conditions and subsequent shoreline evolution patterns.

The socioeconomic factors analyzed in this assessment also differ from the 1992 assessment in many respects, mostly because of the price variation for both land (in the retreat strategy) and protection measures (in the protection strategy). In addition, the macroeconomic development scenario derived recently for Poland provided a valuable framework that could be used throughout this analysis.

The output required by other Polish studies brought about significant modifications to and expansion of the 1992 assessment. Among other things, in addition to the two basic strategies, retreat and full protection, two other options have been considered here: limited protection, and current shore protection and coastal zone management activities. Integrating the vulnerability and adaptation work with countrywide mitigation strategies (such as reduction of the emission of greenhouse gases) has necessitated the derivation of a number of socioeconomic quantities and estimates such as investment patterns and costs, employment patterns, and production of some construction commodities.

The time span of other Polish studies ends in 2030, while the horizon assumed in most climate change studies is 100 years. To meet the requirements of both horizons, estimates for 2030 and 2100 are used in this assessment (2025 is also used as a starting date for the 30-year development scenario in the 100-year assessment described below).

The Baltic Sea is a shallow, almost landlocked sea; its salinity along the Polish coastline is low, barely 7 to 8 ppt along the Polish coastline (Figure 13.1). The coast forms two major gulfs, the Pomeranian Gulf and the Gulf of Gdańsk, and two water bodies, the Szczecin Bay and the Vistula Lagoon, connected with the sea via narrow straits. Two major estuarine systems on the Polish coast are the Odra River in the west and the Vistula River in the east, respectively about 700 km² and 850 km². A substantial part of the Vistula Delta, the Żuławy, is a depression reaching almost 2.0 m below mean sea level. The open sea coast is almost exactly 500 km long (of which 72 km are on both sides of the Hel Peninsula), while on the inland side the banks of Szczecin Bay measure about 240 km.

The Polish coast consists of two basic formations — dunes and barrier beaches, and cliffs. The former stretch along most of the coast; cliffs make up less than 20% of the total length. Coastal lakes, or ancient lagoons, typically found in the central and eastern parts of the Polish coast, are separated from the sea by narrow, sandy barriers.

The Hel Peninsula is a unique spit system separating the Gulf of Gdańsk from the open sea. Because of its intensive erosion and its importance, it has become a subject of concern to both Polish and international authorities.

The vulnerability and adaptation assessment for Poland's coastal resources presented in this chapter addresses:

- sea level rise and socioeconomic scenarios used in the assessment
- methods used

- estimates of land loss and land at risk with and without protection

- economic assessments of loss.

Poland is also currently developing a water resources assessment for the catchment areas of the Warta and Prosna rivers. The preliminary results of this assessment are inconclusive.

SCENARIOS

Sea Level Rise Scenarios

The Baltic Sea is an unusual body of water with a surplus of precipitation and river discharge over the outflow to the North Sea; this causes a steady sloping of the water surface toward the west-southwest along the

Figure 13.1. Map of Poland

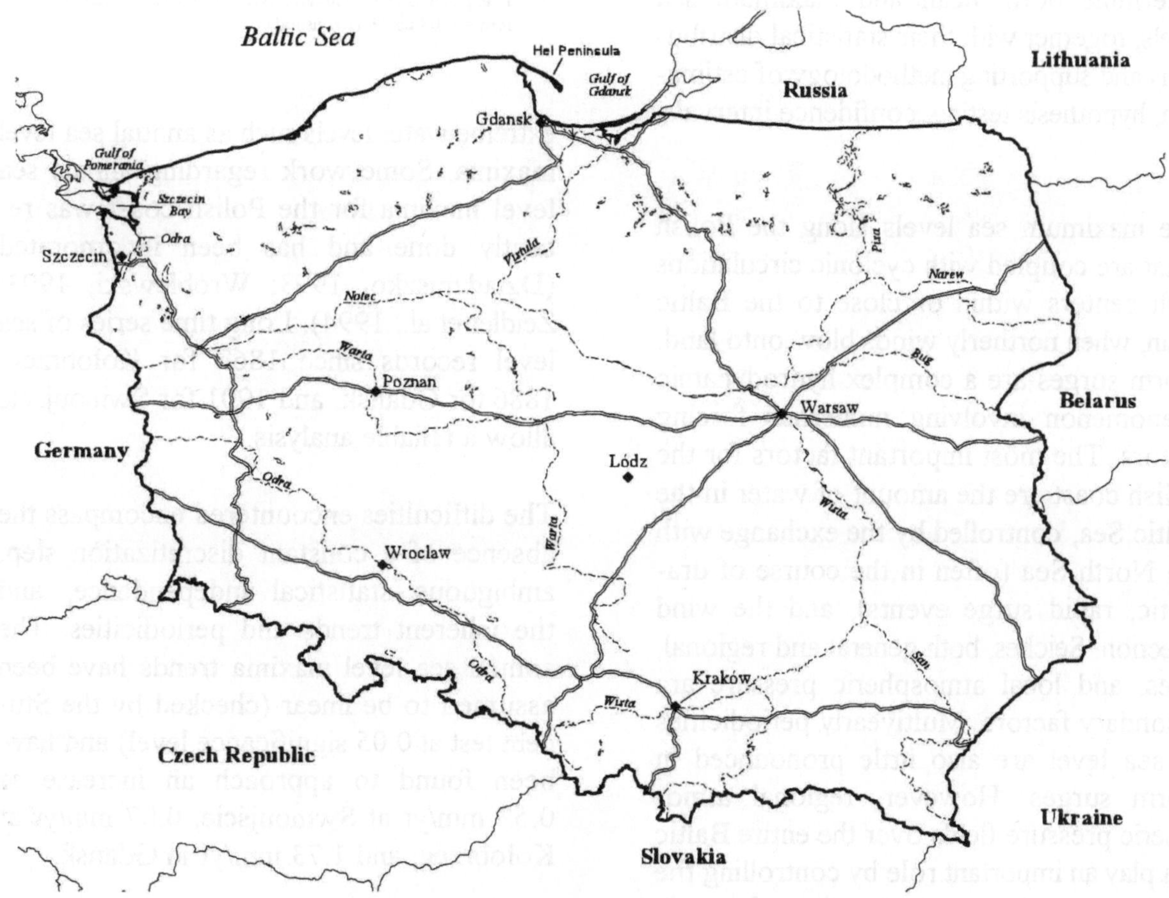

Polish coastline estimated at 9 cm from Świnoujście (491) to Gdańsk (500).

Recent examinations of trends and statistical distributions in sea level datasets, revised and updated for the Polish coast, have partly confirmed some earlier conclusions drawn for mean sea level (and have revealed new findings for annual maximum sea levels). The means are illustrated in Table 13.1.

In multiyear forecasts and vulnerability and adaptation assessments, it is important to determine both mean and maximum sea levels, together with their statistical distributions and supporting methodology of estimation, hypothesis testing, confidence intervals, etc.

The maximum sea levels along the Polish coast are coupled with cyclonic circulations with centers within or close to the Baltic basin, when northerly winds blow onto land. Storm surges are a complex hydrodynamic phenomenon involving numerous forcing factors. The most important factors for the Polish coast are the amount of water in the Baltic Sea, controlled by the exchange with the North Sea (often in the course of dramatic, rapid surge events), and the wind direction. Seiches, both general and regional, tides, and local atmospheric pressure are secondary factors. Multiyearly periodicities of sea level are also little pronounced in storm surges. However, regional atmospheric pressure fields over the entire Baltic Sea play an important role by controlling the wind "setup" (i.e., superelevation of water), which causes a superelevation of water.

Accurate specifications of flooding scenarios require the probability distributions of

Table 13.1. Mean Sea Level and Mean Sea Level Rise Rate for 1951-1985 at 10 Stations[1]

Station	MSL (cm)	MSL Rate (mm/yr)
Gdańsk-Nowy Port	504.0	2.9
Gdynia	502.9	2.2
Hel	501.4	1.7
Kołobrzeg	498.1	0.8
Łeba	500.1	1.5
Świnoujście	497.1	1.4
Tolkmicko	502.2	1.5
Trzebież	504.8	2.0
Ustka	499.9	2.0
Władyslawowo	499.5	1.7

[1] Tide gauge zero = -500 cm NN_{55} (Amsterdam); Tidal gauge zero = -508 cm H_r (Kronshtat).

extreme water levels such as annual sea level maxima. Some work regarding annual sea level maxima for the Polish coast was recently done and has been incorporated (Dziadziuszko, 1993; Wróblewski, 1993; Zeidler et al., 1994). Long time series of sea level records since 1867 for Kołobrzeg, 1886 for Gdańsk, and 1901 for Świnoujście allow a reliable analysis.

The difficulties encountered encompass the absence of a constant discretization step, ambiguous statistical independence, and the inherent trends and periodicities. The annual sea level maxima trends have been assumed to be linear (checked by the Student test at 0.05 significance level) and have been found to approach an increase of 0.53 mm/yr at Świnoujście, 0.07 mm/yr at Kołobrzeg, and 1.73 mm/yr in Gdańsk.

Annual sea level maxima exceedence curves (detrended and fitted to the Gumbel distribution) for three locations have been employed in this analysis for Poland.

Aside from sea level change and storminess (annual sea level maxima), the effect of wind change was also analyzed. Climatic changes in temperature, precipitation, evapotranspiration, and their derivatives are also being analyzed.

For analysis of wind change in the Baltic Sea, some recent Polish findings can be used. Zeidler (1995) used wind change trends and computed sediment transport along the Polish coast for 50 years. Despite a sharp change in wind direction (but not speed), the changes in sediment transport and subsequent shoreline evolution due to the wind change alone are not dramatic. Yet they might become quite conspicuous if the wind change is added to the accompanying storm surge, which seems to be correlated with wind direction.

Physical Boundary Conditions and Impact Zones

The sea level rise scenarios assumed for 2030 (denoted by SLR) and 2100 (denoted by ASLR), and shown in Table 13.2, have been based on the IPCC common methodology assumptions. The lower bound estimates (SLR1 and ASLR1) are close to the present trends measured on the Polish coast, while the upper bound estimates (SLR2 and ASLR2) are almost twice as high as the predictions based on present trends (about 60 cm per century; Titus and Narayanan, 1992).

The +2.5 m altitude contour was chosen as the inland boundary of the study area for both the 1992 assessment and this assessment. This contour represents the sum of a sea level rise of 1.0 m and a 100-year storm surge for the 1992 assessment as well as the sum of a 30 cm sea level rise and an approximation of a 10,000-year storm surge for this assessment. Hence the uppermost landward boundary is identical for both assessments for the sake of simplicity and to make the present analysis on the safe side of an upper bound of all possible hazards and risks that might happen by 2030. Yet the internal boundaries of impact zones are different in the two studies. Three impact zones have been selected, as described in Table 13.3, and their vulnerability has been assessed as at loss or risk (Table 13.4).

Table 13.2. Sea Level Rise Assumptions

Sea Level Rise Scenario	2030 (cm)	2100 (cm)
SLR1	10	—
SLR2	30	—
ASLR1	—	30
ASLR2	—	100

Table 13.3. Contours of Impact Zones for Poland's Climate Change Coastal Study Area

Impact Zone	(ASLR1/ASLR2)	(SLR1/SLR2)
I	0.0-0.3 m	0.0-0.1m
II	0.3-1.0 m	0.1-0.3 m
III	1.0-2.5 m	0.3-2.5 m

Table 13.4. Impact Zone Inundation and Risk by Scenario for Poland's Climate Change Coastal Study Area

Scenario	Area Inundated	Area at Risk
ASLR1, SLR1	Zone I	Zones II and III
ASLR2, SLR2	Zones I and II	Zone III

The data collected from maps (mostly 1:25,000 scale), reports, computations, studies, and other sources (e.g., land use, socioeconomic) were entered according to impact zone in working tables, and incorporated with information on the probability distribution of extreme water levels. Under ASLR1, SLR1, the area that is estimated to be inundated or permanently flooded (i.e., lost) is identified as Impact Zone I, while the area at risk is defined as Impact Zones II and III. Under ASLR2, SLR2, the area that is estimated to be lost is defined as Impact Zones I and II, while the area at risk is defined as Impact Zone III.

Baseline Socioeconomic Scenarios

Poland versus the Study Area

Vulnerability and adaptation assessments require quantitative assessments of costs and benefits coupled with climate change vulnerability and adaptation strategies. The IPCC common methodology imposes the 30-year scenario of likely socioeconomic developments, to generate another reference date (2025 instead of 1995) and boundary conditions as a starting point for vulnerabil-ity and adaptation assessments. Both 1995 and 2025 should be looked at in the broader context of Poland, not just the coastal zone.

Major socioeconomic characteristics of the study area are summarized in Table 13.5 and compared to nationwide figures for Poland. The 30-year scenario of socioeconomic growth has been analyzed to serve as reference for assessment of vulnerability in the future, within a newly transformed coastal zone. In addition, activities in the coastal zone, as well as their neglect, add to the emissions or reduction of greenhouse gases.

Accordingly, a review of possible development patterns has been provided for socioeconomic developments over all the Polish coast, with emphasis on recreation and tourism; ports, harbors, and shipyards; mining and industry; urban and rural development; transportation; and agriculture, fisheries, and forestry.

Prices and Land Cost Structure

It has been quite an exercise to determine the current prices of different land-use categories in Poland. An almost balanced market does exist for agricultural land, forests, and other natural resource lands. Although these prices vary across the country, they have been established with reasonable accuracy.

More variation and less information were encountered for urbanized lots, and particularly various sectors of industry. Larger lots of established housing and industry are not often offered for sale, and the prices may be distorted due to the transient nature of the Polish economy, the debt burden of many companies for sale, and uncertainties about the shape and future of nationwide privatization. One is forced to rely primarily on estimates available from internal revenue offices, insurance companies, advertisements, and inquiries to property dealers offering houses, workshops, and factories. These prices must then be extrapolated to categories of interest (i.e., larger urban, industry, rural). Substantial assistance has been provided by design offices for roads, urban planning, railways, and ports.

The capital value of the study area was found by multiplying the current market prices by respective categories of land use. The prices in 35 years have been adopted from the Netherlands market upon the assumption that Poland's economy in the year 2025 will approach the Dutch economy today. Table 13.6 summarizes the findings and assumptions for land values in the various categories.

The Polish Coast in 2025

The year 2025 becomes a starting point for an alternative assessment (in parallel with

Table 13.5. General Socioeconomic Data for 1995 and 2025

		1995		2025	
Type of Data	Units	Country	Study Area	Country	Study Area
GDP	10^9 $US[1]	80[1]	0.7[3]	315[1]	2.8[4]
Agriculture and Fishery	%	7[2]	10[3]	4[2]	7[5]
Forestry	%	1[2]	1[3]	1	1[5]
Building	%	10[2]	10[3]	7[2]	10[5]
Transportation and Communication	%	3[2]	5[3]	2[2]	5[5]
Industry	%	38[2]	35[3]	26[2]	28[5]
Services, Trade, Municipal, and All Other	%	41[2]	39[3]	60[2]	49[5]
Population[6]	thou	38,600	—	41,500	—
Area 1 (Zone I/II/III)	thou	—	5.1/30.7/31.9	—	8/48/51
Area 2 (Zone I/II/III)	thou	—	0.3/5.8/3.8	—	0.4/9/6
Area 3 (Zone I/II/III)	thou	—	0.1/10.1/4.0	—	0.2/16/7
Area 4 (Zone I/II/III)	thou	—	35.4/58.6/49.1	—	56/93/77
Surface Area	10^3 km^2	312.70	2.20	312.70	2.20
Agriculture	10^3 km^2	150	1.93	—	1.20
Forestry	10^3 km^2	86.00	0.20	—	0.30
Fishery	10^3 km^2	2.00	0.00	—	0.01
Industry	10^3 km^2	5.00	0.01	—	0.01
Recreation	10^3 km^2	(28.00)	0.01	—	0.30
Urban	10^3 km^2	(20.00)	0.02	—	0.10
Rural	10^3 km^2	18.00	0.10	—	0.20
Nature Reserve	10^3 km^2	3.00	0.01	4.00	0.05
Capital Value[7]	10^9 $US	240	3	1,000	12
Growth Rate					
GDP	%/yr	4.00	4.00	—	—
Population	%/yr	0.25	0.50	—	—
Capital Value	%/yr	4.00	4.00	—	—

[1] US$ 1992.
[2] GDP for Poland in 1992 at the exchange rate of 14,000nl/$US assumed for 1995 while GDP for 2025 extrapolated from 1995 by SE1.
[3] Shares of sectors derived from SE1.
[4] Shares of sectors in GDP derived from statistical yearbooks for the Gdańsk Province taken as representative of the study area.
[5] Study area's growth at pace with the country.
[6] Redistribution of share due to restructuring of land use patterns.
[7] As predicted for 1995.
[8] Gross as of 1993; CV for 2025 predicted to grow at the rate of GDP.

Sources: Statistical yearbooks and macroeconomic forecast by U.S. CSP Poland.

Table 13.6. Land-Use Values[1]

	1995 ($US/m²)					2025 ($US/m²)				
	Area 1	Area 2	Area 3	Area 4	Mean	Area 1	Area 2	Area 3	Area 4	Mean
Beach	10	10	10	10	10	20	20	20	20	20
Agric. — High	3	3	3	3	3	6	6	6	6	6
Agric. — Low	1	1	1	1	1	2	2	2	2	2
Wasteland	0.5	0.5	0.5	0.5	0.5	0.7	0.7	0.7	0.7	0.7
Forest	2	2	2	2	2	3	3	3	3	3
Recreation	20	20	20	20	20	33	33	33	33	33
Nature Resource	20	20	20	20	20	20	20	20	20	20
Rural	120	120	120	120	120	250	250	250	250	250
Urban	700	200	500	900	700	1,300	600	800	1,500	1,300
Industry	600	200	300	600	600	600	200	300	600	600
Harbors	1,500	1,500	1,500	1,500	1,500	1,500	1,500	1,500	1,500	1,500
Roads	900	500	500	900	900	1,200	800	800	1,500	1,200

1 All values in US$ 1992.
2 For nonmarket value. 2.35 zł ≈ $1.00 as of June 1995. Old Polish zloty, in circulation by 31 December 1996, is 0.0001 zł (PZN, from 1 January 1995).

1995), in which land-use structure, prices, etc., are used for the comparison of adaptation strategies. (The year 2025 should not be confused with 2030, which marks the end of the 35-year time span under consideration with SLR1 and SLR2.)

In the 30 years to come, the present rate of population growth for the country is expected to slow. We have taken an average trend of the last 5 and 10 years for the nation. For the study area, the anticipated growth is faster, as the worldwide tendency shows a growing population of coastal communities. Moreover, some centers such as Elbląg will also grow faster than average.

As to the land-use patterns over 30 years, a general transition will occur from agriculture, particularly low intensity, to recreation, services, and infrastructure. Moreover, for the study area, severe cuts in agriculture were assumed since the area is a rather narrow strip of coast of specific destination and many uses.

On the central coast (Areas 2 and 3), the present functional links between Koszalin and Kołobrzeg, and Ustka and Słupsk, are estimated to give birth to two additional urbanized units of diversified, marine-oriented functions. On the east coast (Area 4), Elbląg will play an increasingly important role, not only for its closeness to the developing Baltic states of Lithuania, Latvia, and Russia, but also because it is well industrialized and has good access to the sea (via the Vistula Lagoon). The following maritime agglomerations are likely to materialize from east to west: Elbląg, Gdańsk-Gdynia, Ustka-Słupsk, Koszalin-Kołobrzeg, and Szczecin-Świnoujście. Several major recreation strips are expected to develop rapidly in the coming years.

Oil and gas have been prospected in Poland's coastal zone; it is estimated that the amounts extracted off Rozewie and a few other sites will secure about 5% of Poland's demand for liquid fuel. The inherent infrastructure such as pipelines, transportation, standby vessels, and storage areas uses some of the land in the study area.

METHODS USED

Selection of Study Sites

Because of its diversified features, Poland's coast, referred to as the study area henceforth, has been split up into the four segments described in Table 13.7 and shown in Figure 13.2. The four major coastal areas distinguished are the Odra Estuary (Area 1), the western dunes and barrier beaches (Area 2), the central-east dunes and barrier beaches with the Hel Peninsula (Area 3), and the Vistula Delta (Area 4). Area 1 includes the agglomeration of Szczecin (and Świnoujście), while Area 4 encompasses the contiguous urban areas of Gdańsk, Gdynia, and Sopot, together with Elbląg. The areas were chosen in such a way as to reflect the major geomorphic differences along the Polish coast.

Table 13.7. Division of Poland's Climate Change Coastal Study Area

Area	Description	Length (km)	Surface Area (km²)
1	Western state border to River Rega	320[1]	595
2	River Rega to Łeba	170	244
3	Łeba to Gdynia North	160	173
4	Gdynia North to Eastern state border	190	1,217

[1] Including 240 km of Szczecin Bay.

Figure 13.2. Poland's Coast Split Up into Four Areas

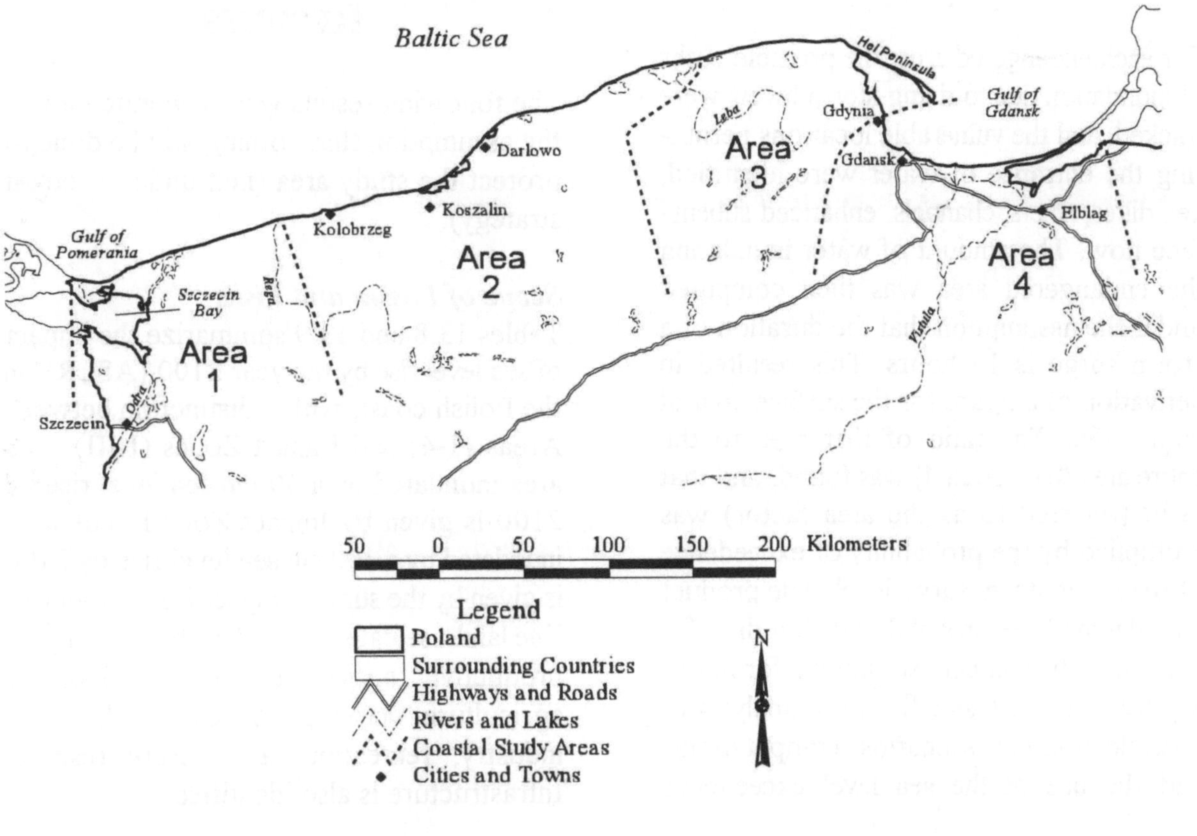

Methodology

The 1992 assessment was based on the IPCC common methodology derived by IPCC. The present assessment includes new information such as a reassessment of Polish sea level change trends; the most recent data for annual sea level maxima, storm surges, and flooding probabilities for the Polish coast; some other climate change factors such as the effects of wind; and a revised and updated set of prices and socioeconomic estimates.

One major methodological addition to the IPCC common methodology consists of the provision of feedback and input data for other climate change studies in Poland, in particular the provision of some coastal data permitting the assessment of the coastal impact on the emission and reduction of greenhouse gases.

For each endangered area, the possible paths of inundation due to rising storm levels were tracked, and the vulnerable locations permitting the entrance of water were identified, i.e., dikes, rivers, channels, enhanced subsurface flow. The amount of water inundating the endangered area was then computed under the assumption that the duration of a storm surge is 16 hours. This resulted in derivation of a figure for the surface area of inundation. The ratio of that area to the entire area (e.g., Area 4) was found, and that ratio (referred to as the area factor) was multiplied by the probability of exceedence of the input storm surge level. The product was deemed the probability of flooding for that area. The numbers so derived for flooding frequencies result from the analysis of possible flooding scenarios, computations, and the use of the sea level exceedence curve. Of course, in this analysis of areas at risk, distinction is made between the area at risk under the sea level rise scenario and the area at risk under the accelerated sea level rise scenario.

Other climate change effects that were analyzed and assessed include changes in wind direction and speed, which in turn cause changes in sedimentation and shore evolution patterns. The latter have been computed by a variety of computer-aided beach and dune erosion models (Zeidler, 1990), among which the Bruun Rule provided the simplest estimates.

In addition to the biophysical assessment, the socioeconomic scenario of the Polish coast in 2025 was used when assessing potential economic impacts.

RESULTS

The following results were generated under the assumption that nothing will be done to protect the study area (i.e., under a retreat strategy).

Scope of Losses and Risks

Tables 13.8 and 13.9 summarize the impact of sea level rise by the year 2100 (ASLR) on the Polish coast, with a distinction between Areas (1-4) and Impact Zones (I-III). The area inundated by a 30 cm sea level rise by 2100 is given by Impact Zone I. The area inundated by a 100 cm sea level rise by 2100 is given by the sum of Impact Zones I and II. The land-use categories identified are highly productive agriculture, low-productivity agriculture, wasteland, forest, rural, urban, industry, recreation, and nature reserve. Infrastructure is also identified.

Table 13.8. Distribution of Land-Use Categories[1] in Impact Zones I, II, and III for 2100 (area in ha, length in km)

Area	AH	AL	AO	F	W	U	I[2]	R	NR	RD	LR	TL	RL	NT	B
Impact Zone I — 0.0-0.3 m															
1	92	11,955	2,635	1,415	30	40	—	—	—	30	25	41	—	—	—
2	—	30	400	—	—	—	—	—	430	—	—	—	—	—	—
3	—	130	40	—	—	—	—	—	—	—	—	—	—	—	—
4	64,870	130	1,685	—	352	130	150	—	—	291	108	302	29	6	—
Total	64,960	12,245	1,760	1,415	382	170	150	—	430	321	133	343	29	6	—
Total Impact Zone I surface area = 84,512 ha															
Impact Zones I and II — 0.0-1.0 m															
1	297	34,338	5,435	7,290	550	640	818	35		70	89	74	15	1	17
2	—	11,100	995	2,745	110	90	—	—	465	27	11	30	1	1	4
3	—	11,531	118	24	17	5	8	12	—	6	8	7	2	—	—
4	86,910	1,930	2,265	1,260	2,590	705	430	95	275	428	179	415	61	65	26
Total	87,207	58,899	8,813	11,319	3,267	1,440	943	142	740	531	287	525	79	66	47
Total Impact Zones I and II surface area = 172,770 ha															
Impact Zone III — 1.0-2.5 m															
1	—	7,935	145	1,690	—	—	300	—	—	36	9	35	6	2	4
2	—	5,380	—	2,315	370	—	—	—	—	36	7	5	3	—	1
3	—	3,210	—	330	155	85	—	730	—	28	10	7	11	—	—
4	21,520	200	—	795	1,215	815	370	—	—	97	12	60	8	9	—
Total	21,520	16,725	145	5,130	1,740	900	670	730	—	197	38	107	28	11	5
Total Impact Zone III surface area = 47,560 ha															

[1] (AH) Agriculture — High, (AL) Agriculture — Low, (AO) Wasteland, (F) Forest, (W) Rural, (U) Urban, (I) Industry, (R) Recreation, (NR) Nature Reserve, (RD) Road, (LR) Local Road, (TL) Power Transmission Line, (RL) Railway, (NT) Narrow-Track Rail, (B) Bridges.

[2] The land-use category Industry (I) includes Ports.

Vulnerability of Specific Systems

Area 1. The Odra estuary is a hydrographic system consisting of many natural and man-made channels and waterways. The major impact of accelerated sea level rise scenarios (30 cm to 100 cm by 2100; ASLR) in this estuary will be the inundation of numerous lowlands and aquatic bodies, extending under the ASLR2 scenario up to River Rurzyca (695 km). The crest of the polder dikes in this area is about 1.5 m to 2.0 m above mean sea level. (A polder is a tract of low land reclaimed from the sea.) At higher stages, most polder dikes will be overtopped. The inundated polders will include those of Miedzyodrze (between East

Table 13.9. Population of the Study Area

Area	Zone	Urban	Rural	Total
1	I[1]	1,630	3,450	5,080
	II[2]	21,080	9,630	30,710
	III[3]	28,960	2,940	31,900
2	I	0	270	270
	II	1,210	4,580	5,790
	III	870	2,880	3,750
3	I	0	110	110
	II	740	9,310	10,050
	III	1,550	2,490	4,040
4	I	9,590	25,810	35,400
	II	40,980	17,650	58,630
	III	36,080	13,030	49,110
Total		142,690	92,150	24,840

[1] 0.0 m to 0.3 m.

[2] 0.3 m to 1.0 m.

[3] 1.0 m to 2.5 m.

and West Odra), and those around the Szczecin Bay, Świna, and Dziwna.

Equally endangered by flooding are the urban areas of Szczecin (primarily Dąbie district), Świnoujście, Trzebież, and other towns. Indirect effects will include damage to sewage systems and other infrastructure. Parts of the ports of Szczecin, Świnoujście, and Police are also vulnerable. Increased flow velocities will pose a hazard to the stability of the Świnoujście wharves.

Impact Zones I and II within Area 1 measure in total about 162 km² and 330 km², respectively. These areas are vulnerable to permanent flooding under the ASLR1 (Zone I) and ASLR2 (Zones I and II) scenarios.

The Polder Systems in Poland. There are two general groups of Polish coastal polders — without natural drainage and with temporary natural drainage. The first group encompasses depressions and cryptodepressions such as the Vistula delta.

All Polish coastal polders can be arranged in eight complexes of 243 polders measuring 183,557 ha. They all may be affected by accelerated sea level rise scenarios (30 cm to 100 cm by 2100; ASLR), mostly ASLR2, and much less by sea level rise scenarios (10 cm to 30 cm by 2030; SLR).

Coastal Barrier Lakes. A very particular feature of the Polish coastline is the presence of a chain of coastal barrier lakes. Narrow coastal barriers between the sea and low-lying lakes, lagoons, or embayments might erode, making inland areas widely accessible to seawater. This effect is estimated primarily under ASLR2 but not under SLR2 or SLR1.

The higher levels attained by storm surges in ASLR scenarios, and perhaps partly at SLR2, could destroy foredunes and erode barrier islands and spits. The lowered barriers would then be susceptible to storm washover, which will lead to inlet breaches and sedimentation on the backside of barriers. The lakes will grow more saline.

Słowiński National Park and Łeba Dunes. Situated to the west of the old fishing village of Łeba are the lakes and sand dunes of Słowiński National Park. Słowiński National Park is one of the country's strangest but most memorable natural landscapes, and special enough to be included on UNESCO's list of the world's biosphere reserves. The park's natural components encompass waters (9,835 ha), swamps (1,410 ha), sands (953 ha), and some other areas of ecological importance (1,571 ha). Since it is low-lying, the park is vulnerable to both ASLR and SLR2.

Hel Peninsula. The Hel Peninsula is a unique spit formation separating the open sea from the Gulf of Gdańsk. Its shoreline is 72 km long. The peninsula is narrow (less than 300 m) and very low in places, so it is already vulnerable at higher water levels. Accordingly, increased erosion is expected under all four scenarios of ASLR and SLR.

Gdynia and Gdańsk Ports. Gdańsk is an important Baltic port near the mouth of the Vistula River. It is a universal port for medium-sized vessels in the old parts of the port and for the largest Baltic vessels at the deep docks of the Northern Harbor. The Port of Gdynia is a universal cargo port built in the period between wars when Poland had no other access to the Gulf of Gdańsk. The internal segments of the port were dug out of land. The altitude above sea level of wharves, jetties, breakwaters, and other facilities varies from port to port, but many locations are vulnerable to ASLR.

Gdańsk City Area. Gdańsk was already known in the 10th century for its castle and it was a settlement of the princes of Pomerania. It became a city in 1235, and was in the possession of the Teutonic knights from 1308 to 1454, when it was returned to Poland.

Gdańsk is now a lively industrial, scientific, and cultural center of northern Poland. It is the seat of a dozen institutions of higher learning, four museums, several theaters, and the Baltic Symphony Hall. The Gdańsk shipyard is one of the oldest and best established enterprises in the city. Some port areas, urban districts, industry, trade, and transportation routes of the Gdańsk area are vulnerable to ASLR impacts. The industry (ports included) occupies 760 ha in Impact Zones I-III, while the endangered urban areas cover 1,130 ha.

Żuławy Polder Area. The Vistula Delta is referred to as Żuławy. It covers an area of 2,320 km^2 and contains the major cities of Gdańsk, Elbląg, and Malbork. The gross agricultural area of the Żuławy polder district, which is quite similar to some Dutch areas, covers 180,000 ha (0.6% of Poland's land area). Most of its depressions and low land are estimated to be damaged under both ASLR and SLR. The dikes already built will be too low for the new hydrological conditions, especially ASLR2. They have been designed to last only a few days under water. The dikes will be subjected to excessive saturation caused by higher water levels, increased groundwater flow due to higher water heads, and ice jamming at the river mouth.

Adaptation Alternatives

Retreat Strategy

Under a retreat strategy, nothing will be done to protect the study area and its natural systems, cities, ports, and industrial areas.

Full Protection Strategy

The "full protection" adaptation strategy is defined as the implementation of all feasible protection measures to minimize losses of any coastal land or values.

Zeidler (1992) described various protection techniques together with an overview of specific subareas and the solutions proposed to secure their protection. Identification of specific systems on the inland side was followed by an examination of the protection strategy along the entire open sea coast, scanned kilometer by kilometer, from east to west.

The total length of new measures along the coastal (open sea) segments for SLR and ASLR would obviously be the highest at ASLR2: 16.3 km of dikes, 21.7 km of seawalls, and 1 km of offshore breakwaters (including adapted measures). The total length of adapted measures along the open sea segments for SLR2 would be 0 km for dikes and 6.7 km for seawalls. The maximum total volume of artificial nourishment is proposed as slightly above 1.2 million m^3 per year per 1 km along shore in the ASLR2 scenario.

Specific protection measures would be needed in each of the four areas. Inter alia, full protection of the Odra estuary means preservation of the polders on peripheries of the estuary. Also, 107 km and 280 km of new dikes would have to be constructed in Area 4 under ASLR1 and ASLR2, respectively; the complete lengths of adapted dikes would then be 243 km and 324 km.

In Areas 2 and 3, new polders would require new facilities such as pump stations, drainage and other infrastructure, and older polders would need to be redesigned. The presence of polders is a typical feature arising on the inland segments of the study area. Under the sea level rise scenarios, the polder system would have to be redesigned by raising the level of the dikes, adding new pump stations, increasing the rate of pumping, and applying other measures to preserve, or even improve, the present condition of polders.

Protection of the Hel Peninsula would largely rely on artificial nourishment.

If one takes Area 4 as a broader illustration, it seems imperative to keep the present agricultural use of the Vistula Delta (much as the lower Odra Valley in Area 1), especially in the polder areas. Hence full protection of the Żuławy polders at ASLR2 would require a bulk adaptation of the existing dikes and the construction of some new dikes and storm and flood prevention facilities. In addition, full prevention measures should be taken to avoid any loss of the valuable and historical development in the city of Gdańsk.

Limited Protection Strategy

The limited protection strategy consists of protecting the most valuable coastal segments endangered by SLR where the losses would be most conspicuous. The segments are identified primarily in Areas 1 and 4 (the Odra and Vistula systems), and only some in Areas 2 and 3 (Hel Peninsula among them), where the cost of protection is high compared to the potential loss (in the retreat strategy).

ECONOMIC ANALYSES

The actual adaptation strategies to be followed will depend on a careful assignment of funds to the areas at major risk. An in-depth analysis of very local areas and priorities will be required to define the most realistic response strategy; however, this is beyond the scope of this study. Tables 13.10, 13.11, 13.12, and 13.13 summarize the vulnerability and adaptation assessment results for the three strategies. In these tables the business as usual column represents present conditions without implementation of adaptation policies.

Upon comparison of the strategies, the following conclusions can be drawn.

- In all cases but Areas 2 and 3 for SLR1, SLR2, and ASLR1, the protection cost is substantially lower than the value potentially lost. The costs to protect Area 3 in the cases of ASLR1, SLR1, and SLR2 are much higher than the capital value of this area, and this could affect any motivation to spend scarce funds on protective measures under these ASLR and SLR scenarios.
- The cost to fully protect Areas 1 to 4 under ASLR2 appears justified in terms of the low cost relative to the capital value saved. For Area 3, this justification is slightly lower.

Table 13.10. Poland's Coastal Zone Vulnerability: ASLR1

Impact Category	Unit	Business as Usual	Retreat 1995	Retreat 2025	Full Protection 1995	Full Protection 2025
Capital Value Lost	10^9 $US	—	6.0	9.8	—	—
People Relocated	thou	—	8.5	10.3	<0.1	<0.1
People at Risk	thou	5.8	31.7	41.2	<1.0	<1.0
Capital Value at Risk	10^9 $US	10.0	40.0	62.0	<5.0	<5.0
Damage (salinity)	10^9 $US	—	<0.01	<0.01	<0.01	<0.01
Ecological Areal Lost	km²	—	1.2	3.1	—	—
Special Area Lost	km²	—	2.7	7.3	1	1
FP Implement Cost	10^9 $US	—	—	—	2.30	3.80
Annual Maintenance	10^9 $US	—	—	—	0.01	0.02

Table 13.11. Poland's Coastal Zone Vulnerability: ASLR2

Impact Category	Unit	Business as Usual	Retreat 1995	Retreat 2025	Full Protection 1995	Full Protection 2025
Capital Value Lost	10^9 $US	—	28	46	—	—
People Relocated	thou	—	146.1	176.3	0.1	1.0
People at Risk	thou	5.8	50.3	108.5	<5.8	<10.0
Capital Value at Risk	10^9 $US	10.0	18.0	26.0	<5.0	<5.0
Damage (salinity)	10^9 $US	—	<0.01	<0.01	<0.01	<0.01
Ecological Areal Lost	km²	—	3.4	10.7	1.0	2.0
Special Area Lost	km²	—	7.9	25.0	3.0	5.0
FP Implement Cost	10^9 $US	—	—	—	6.10	9.80
Annual Maintenance	10^9 $US	—	—	—	0.03	0.50

Table 13.12. Poland's Coastal Zone Vulnerability: SLR1 and SLR2 for Retreat R, Limited Protection LP, and Full Protection

Impact Category	Unit	Business as Usual 1995	Retreat SLR1	Retreat SLR2	Limited Protection SLR1	Limited Protection SLR2	Full Protection SLR1	Full Protection SLR2
Capital Value Lost	10^9 $US	—	2.3	6.0	1.0	2.0	—	—
People Relocated	thou	—	18.4	40.9	0.1	1.0	<0.1	<0.1
People at Risk	thou	5.8	6.5	17.3	<5.0	<5.0	<1.0	<1.0
Capital Value at Risk	10^9 $US	17.5	18.0	40.0	<2.0	<5.0	<1.0	<5.0
Damage (salinity)	10^9 $US	—	<0.01	<0.01	<0.01	<0.01	—	—
Ecological Area Lost	km²	—	0.4	1.2	0.3	1.0	—	—
Special Area Lost	km²	—	7.9	25.0	1.0	1.0	1.0	1.0
LP/FP Implement Cost	10^9 $US	—	—	—	0.6	1.2	0.9	2.3
Annual Maintenance	10^9 $US	—	—	—	0.003	0.006	0.005	0.01

Table 13.13. Gross Evaluation of CC Adaptation Strategies for Areas 1-4 (values in 1992 10^9 $US)

	Area 1	Area 2	Area 3	Area 4
Total Capital Value	6.90	0.28	0.48	12.60
Cap. Val. Lost — SLR1	1.00	0.21	0.02	1.10
Protection Cost — SLR1	0.22	0.18	0.03	0.46
Protection Cost as Percent of Cap. Val. at Loss	22%	86%	150%	42%
Total Capital Value	15.60	0.55	1.10	28.60
Cap. Val. Lost — ASLR1, SLR2	2.70	0.54	0.06	2.70
Protection Cost — ASLR1, SLR2	0.57	0.46	0.09	1.18
Protection Cost as Percent of Cap. Val. at Loss	21%	85%	150%	43%
Total Capital Value	15.60	0.55	1.10	28.60
Cap. Val. Lost — ASLR2	7.40	5.70	0.84	14.10
Protection Cost — ASLR2	2.07	1.40	0.30	2.31
Protection Cost as Percent of Cap. Val. at Loss	28%	24%	35%	16%

- Area 2 contains high-priority nature reserve areas (e.g., Słowinsky Park) and these are only protected at a high cost relative to their capital value. It is unlikely that these areas will receive priority in a climate of very scarce funds where strategic economic areas facing equal risks are likely to capture the available funds. Moreover, the transformation of a nature reserve into an artificially protected area can be controversial.

- Limited protection for both SLR1 and SLR2 yields protection costs comparable with the value at loss, so the motivation to apply limited protection could stem primarily from the high value at risk, not the value lost alone.

Similar trends result from the analysis for ASLR1 and ASLR2 in the socioeconomic scenario assuming 2025 as a reference (the 30-year scenario).

The above demonstrates the likely outcome of a very coarse cost-benefit analysis. Specific decisions for particular sites must be made on the basis of detailed local analyses.

Discussion and Conclusions

For general assessment and comparison of various strategies and scenarios, four vulnerability classes were specified in accordance with the IPCC common methodology, and some amendments were made. The classes chosen for values at loss, values at risk, ecological and other impacts, are labeled "low," "medium," "high," and "critical."

For the retreat strategy, losses are expected to be critical in ASLR2 if expressed as a percentage of GDP. Losses are low for SLR1 but become medium to high for ASLR1 and SLR2 with present development. Of note is that there is already a medium risk classification for the numbers of people flooded annually. This rises to high in ASLR2 with 30-year development. Salinity and ecological areas show low vulnerability.

For the full protection strategy, losses are reduced to the low classification. The cost of protection measures as of 1995 varies from low for SLR1 through medium for ASLR1 and SLR2, but would be high for ASLR2 based on present development, and critical based on the 30-year development. This clearly shows the need to act sooner rather than later.

Preference should be given to protection strategies, yet some limitations arise for Areas 2 and 3 subject to SLR1 and SLR2, because the protection costs are estimated to be higher than the value to be lost in Area 3 or comparable with the value to be lost in Area 2. Still, the need for protection is reinforced by the fact of having high value at risk if unprotected.

There is a real danger that funds will not be allocated to coastal defense measures to protect against sea level rise until serious losses occur and until a higher degree of confidence in a predicted sea level rise of greater than 0.3 m and closer to 1.0 m can be obtained.

Inter alia, some constraints stem from the medium to high cost of protection compared to Poland's GDP. Assuming a 30 cm sea level rise (SLR2 or ASLR1), the cost of protecting Poland's coast is estimated at 3% of Poland's GDP (1992). To reduce the burden on the GDP, it seems reasonable to adopt limited protection for selected areas, such as Areas 1 and 4.

Without additional protective measures and a real awareness among coastal managers of the dangers and possible losses associated with sea level rise, a serious loss of the viability of important coastal areas, both economically and ecologically, will occur. This will hamper growth of an emerging Polish economy and its coastal regions.

REFERENCES

Dziadziuszko, Z. 1993. Yearly Mean and Maximum Sea Level in Polish Ports (in Polish). Typescript IMGW Gdynia.

Titus, J. and V. Narayanan. 1992. The Probability Distribution of Future Sea Level Rise. The Rising Challenge of the Sea. *Proceed. IPCC Conference*. Margarita Island, Venezuela.

Wróblewski, A. 1993. Analysis and Long-Term Sea Level Forecast at the Polish Baltic Sea Coast. Part I. Annual Sea Level Maxima. *Oceanologia* 33.

Zeidler, R.B. 1990. Shore Evolution Due to Waves and Sea Level Changes. *Hydrotech. Transactions* 53:41-62.

Zeidler, R.B. 1992. *Assessment of the Vulnerability of Poland's Coastal Areas to Sea Level Rise.* H*T*S Gdańsk. ISBN 83-85708-01-4.

Zeidler, R.B. 1994. Sea Level Rise and Coast Evolution in Poland. *Proceed. 24th Intern. Conf. Coastal Eng.* pp. 3462-3475. ASCE, New York.

Zeidler, R.B., A. Wróblewski, M. Miętus, Z. Dziadziuszko, and J. Cyberski. 1994. Wind, Wave, and Storm Surge Regime at the Polish Baltic Coast. *J. Coastal Research* 22:33-5.

ACKNOWLEDGMENTS

This study has been carried out under the Country Studies Program, Poland. Some components and findings of the Polish Scientific Research Committee KBN-funded grant 6P 202-004-04 have been included, and the financial support by KBN is hereby gratefully acknowledged.

Chapter 14

Vulnerability and Adaptation Assessments for Slovakia

Milan Lapin,[1] Dušan Závodský,[1] Olga Majercáková[2]
Jozef Mindáš,[3] František Špánik[3]

[1] **Slovak Hydrometeorological Institute**
[2] **Forest Research Institute**
[3] **Agricultural Technical University**

SUMMARY. This contribution is partly based on Slovakia's National Climate Program reports and on the expertise of those listed in the References section. A preliminary assessment of potential climate change impacts on the agriculture, forest, and water resources and the consequent changes in selected socioeconomic sectors is presented. More complete assessments are expected to be available in 1996. Simple model calculations for the agriculture sector indicate that a 1 ° to 2 °C warming could potentially increase the solar radiation, and vegetative periods. However, the evapotranspiration deficit is expected to increase and soil moisture is expected to decrease. The assessment for the forestry sector will examine changes in species composition, bioproduction, soils, and injurious agents. It is expected the European fir and Norway spruce will be the species in Slovakia most affected by climate change. Preliminary estimates from scenarios for 1990-2030 indicate that water resources may potentially be decreased by approximately 20%.

INTRODUCTION

The Slovak Republic lies at the heart of Europe (Figure 14.1). It occupies the territory between the Danube River to the south and the Tatra Mountains to the north. The area of the country is 49,036 km^2, including 24,471 km^2 (50%) of agricultural land, 14,860 km^2 (30%) of arable land, 19,911 km^2 (41%) of forests, 940 km^2 (2%) of water surface, and 1,275 km^2 (3%) of built-up areas.

Slovakia is a mountainous country, and part of the Carpathian mountain system. The Danube and East Slovakian lowlands are the northern parts of the Panonian plains. Sixty percent of Slovakia's surface is over 300 m above sea level, 15% is above 800 m, and 1% is above 1,500 m. The lowest point

Note: This chapter does not contain a detailed description of the assessment methodologies that were recommended by the U.S. Country Studies Program. Refer to Chapter 2 for an overview of these methodologies.

J. B. Smith et al. (eds.), Vulnerability and Adaptation to Climate Change, 295–312.

Figure 14.1. Map of Slovakia

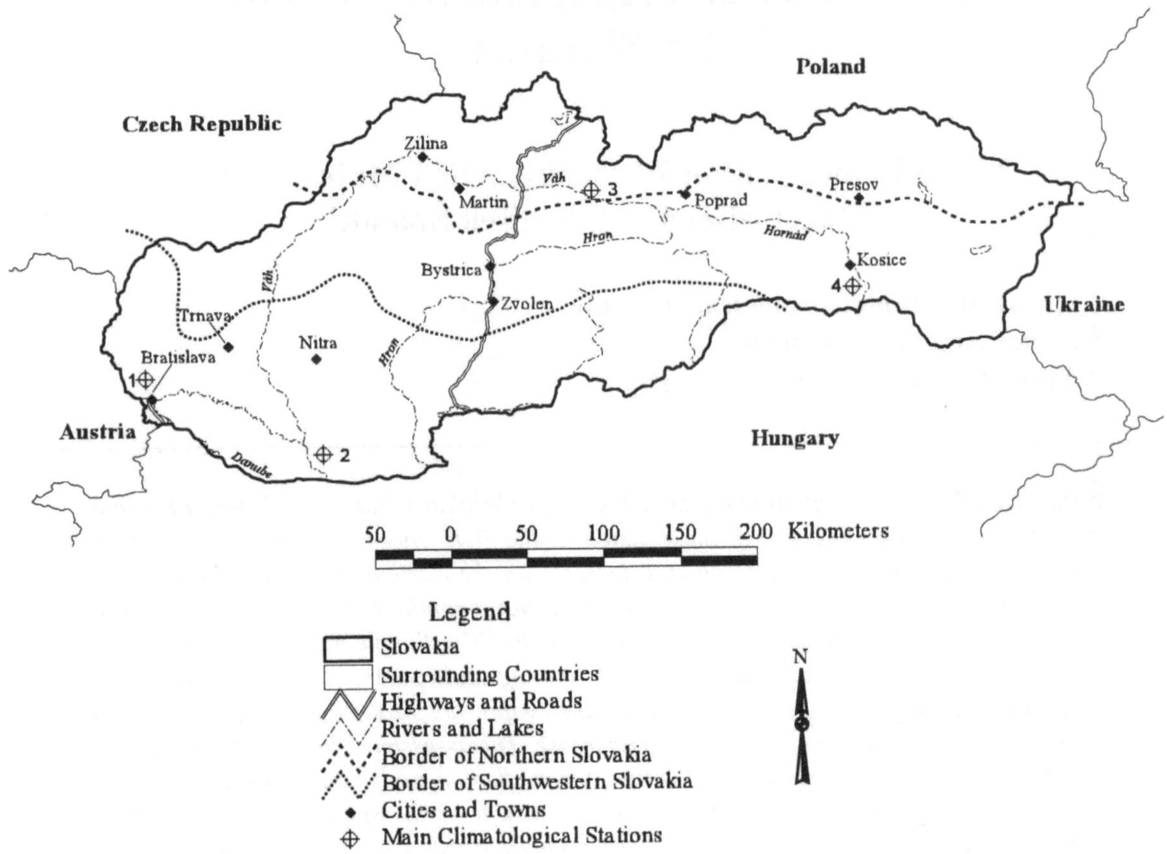

in Slovakia is 94 m and the highest (Gerlach Peak) is 2,654 m.

Most of the country belongs to the Danube River drainage basin; only a small part in the north drains into the Baltic Sea. The Danube River is part of the boundary with Austria and Hungary. The major rivers in the country are the Váh and the Hron. The capital of Slovakia, Bratislava, is the largest city, the center of political and cultural life, an important industrial center, and a Danube river port.

According to the global climatological classification, Slovakia is in a mild climate zone. A regular rotation of four seasons and variable weather throughout the year are typical for this country. The climate in Slovakia is influenced predominantly by its position in Central Europe, by the topography of the western Carpathian Mountains and the Alps, and by the prevailing western zonal atmospheric circulation. The southwestern and southern parts of the country belong to the moderate climate region with mild winters, and the remaining part belongs to the moderate climate region with colder winters.

Compared to the Czech Republic and Austria, which lie to the west, the climate in Slovakia has more continental features. Winters are colder by about 3°C and summers are warmer by about 2°C. These differences increase from west to east within the country. Average January temperatures range from -1°C in the Danube lowlands to -12°C in the Tatra Mountains. Average temperatures in July exceed 20°C in the Slovak lowlands, while at an elevation of 1,000 m they reach about 14°C. Southern Slovakia receives about 2,000 hours of sunshine each year, while the northwest area of the country receives only 1,600 hours. Average annual precipitation for Slovakia is 743 mm, of which 65% is evaporated and 35% represents runoff. The smallest precipitation mean (550 mm annually) is observed in the Danube lowlands, while in the highest elevations of the Carpathians it usually exceeds 1,500 mm annually. Snow cover is not stable, and winters in the lower altitudes are usually without permanent snow cover.

The annual distribution of mean monthly precipitation totals is equable throughout the year with the maximum in June or July and the minimum from January to March; there are no regular wet or dry periods in Slovakia. The period from July to October is relatively dry in the southern part of Slovakia. However, southern and southwestern Mediterranean cyclones cause heavy precipitation in the southern and eastern parts of Slovakia, mainly from October to December and in May and June in some years. Western and northwestern atmospheric currents bring precipitation to the mountains of western and northern Slovakia. Any change in atmospheric circulation may have a dramatic effect on this simplified scheme of climate conditions in Slovakia.

The preliminary vulnerability and adaptation assessments presented here are based on current knowledge, existing climatic and hydrologic trends, and expected changes in the next decades as furnished by results of the Slovak National Climate Program and other research projects in Slovakia. These preliminary results can be used as a basis for the first framework strategies. More complete assessments using the general circulation models (GCMs) are currently in progress. This chapter describes the scenarios being used. Because the greatest progress in conducting impacts assessments has been made in the water resources, forestry, and agriculture sectors, only these three sectors were included in this preliminary report. In addition, some of the recent work on adaptation alternatives is presented for each of the three sectors.

Since there is uncertainty about climate change impacts on some socioeconomic sectors, there was not enough cause for concern to prepare adaptive strategies before 1994 (Moldan et al., 1993; Lapin et al., 1994). However, the negative impacts of the warm and dry 1988-1993 period on water management, forests, and crop production changed this situation remarkably, and several reports on preliminary strategies have been issued.

SCENARIOS

Global warming will probably alter atmospheric circulation and thus affect precipitation, cloudiness, snowcover, relative humidity, and other climatic elements all over the world (Bultot and Gellens, 1989; Lapin et al., 1994). These changes can seriously influence the natural environment and socioeconomic activities in many countries. Some analytical results and scientific assessments of the impact of an increase in temperature on precipitation, relative humidity, daily temperature range, and snow cover in Slovakia were presented in the Slovak National

Climate Program reports (Brázdil et al., 1994; Lapin, 1992; Lapin et al., 1994; Huzulák and Matejka, 1994; Mindáš, 1995; Špánik, 1995; Majercáková, 1995). The following outlines some of these correlations and describes the use of GCMs to estimate potential climate change for Slovakia.

Historical Changes in Temperature and Other Climate Variables

Temperature and precipitation in this region are highly variable, especially in the winter and spring. The absolute daily maximum and minimum temperatures are 39°C and -41°C, respectively. Annual precipitation in lowlands is 450 to 650 mm at 75% certainty of higher totals (lower quartile). Snowcover in winter is stable for more than one month at altitudes above 300 m.

Climate warming may change the climate in Slovakia toward higher variability, and this could change the limits for both natural ecosystems and socioeconomic activities.

Table 14.1 presents selected annual climatic variables from 1901 to 1994 for the Hurbanovo Observatory in the southern part of Slovakia (Climatological Station 2 on Figure 14.1).

Figure 14.2 shows the trends and values of individual years for annual temperature means and precipitation totals in the vegetation period at the Hurbanovo station (including polynomial trendlines of 4th order). By comparing this data to information on relative humidity and April to September temperature means, it can be seen that evaporation and precipitation trends are caused mainly by climatic characteristic changes from April to September, and that the decrease in relative humidity contributes considerably to the rise in potential evaporation. Exceptional means and totals (extremes) occurred more frequently in the 1981-1994 period than before 1981.

Table 14.1. Selected Average Annual Climate Values at the Hurbanovo Observatory

Period	Temperature (°C)	Precipitation (mm)	Potential Evaporation (mm)	Actual Evaporation (mm)	Soil Moisture[1] (mm)
1901-1930	9.5	586	735	480	95
1931-1960	9.9	572	773	469	88
1961-1990	10.0	523	806	451	81
1981-1990	10.1	496	835	437	75
Summary					
1901-1990	9.8	560	771	466	88
1988-1994	10.7	508	859	419	78

[1] Mean calculated usable soil moisture in upper 1-m soil layer.

Source: Lapin et al., 1994, pp. 35-77.

Figure 14.2. Annual Means of Temperature and April-September Precipitation Totals at Hurbanovo in 1871-1994

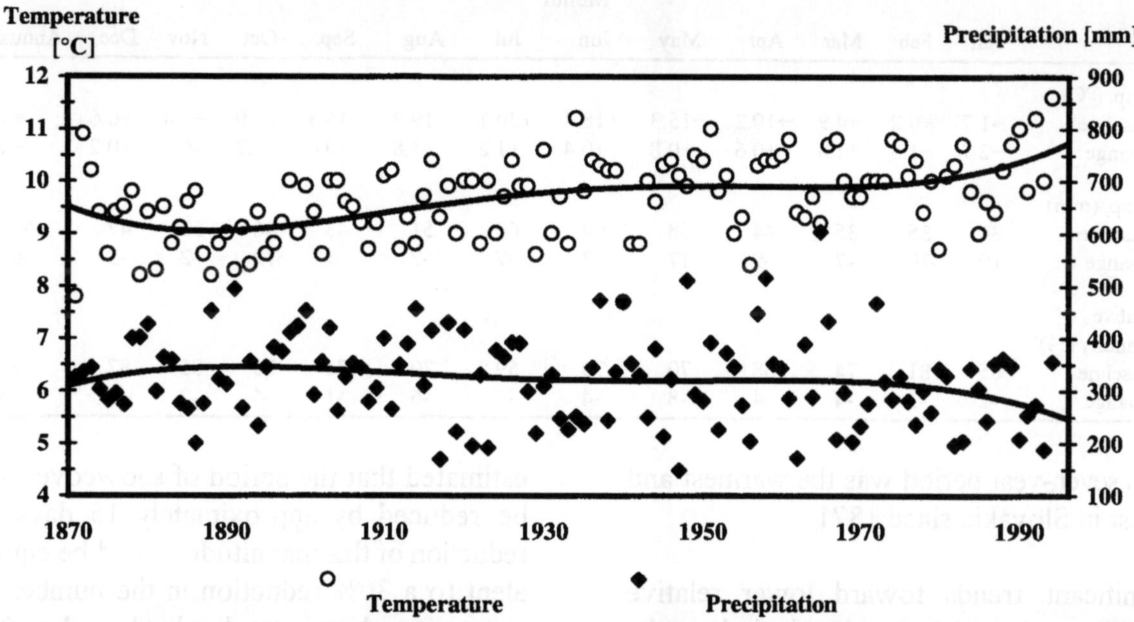

An analysis of polynomial trends and values by individual years for water balance elements: precipitation, potential evaporation, actual evaporation, soil moisture, and balance runoff at Hurbanovo. The values for potential evaporation, actual evaporation, and soil moisture were calculated according to the Zubenok-Budyko complex method (Budyko and Zubenok, 1961). In the mountainous part of Slovakia, there was a great surplus of precipitation compared to potential evaporation in the first three decades of this century. However, since 1980 there has been a precipitation deficiency, primarily in the lowlands, due to a decrease in precipitation and an increase of potential evaporation in all of Slovakia.

These analyses of climate are considered preliminary because of the simple methods used. In spite of this, trends toward higher aridity can be seen clearly, and predominantly in the southern part of Slovakia, which has the most important agricultural regions. In the mostly mountainous northern regions, the increase in actual evaporation and the decrease in precipitation have caused a decrease in runoff: the mean discharges of the major Slovak rivers have decreased by 10% to 30% (and by more than 40% in the smaller rivers in the south) since 1931. The values of these trends toward increased mean annual temperatures, decreased annual precipitation, and decreased relative humidity are most significant in the 1971-1994 period (Lapin et al., 1994).

Detailed analysis of the effect of increased temperature on other climate elements in Slovakia has identified a number of correlations. The analysis indicates that a warming of 1° to 2°C could cause a similar change in precipitation and relative humidity as that observed in the last several decades. Table 14.2 shows the deviations of some climatic means in the 1988-1994 period from the normal means of the 1901-1980 period.

Table 14.2. Deviations of Selected Climatic Means of the 1988-1994 Period from Normals of the 1901-1980 Period (baseline) at Hurbanovo

	Month												
	Jan	Feb	Mar	Apr	May	Jun	Jul	Aug	Sep	Oct	Nov	Dec	Annual
Temp. (°C)													
Baseline	-1.7	+0.2	+4.9	+10.2	+15.3	+18.4	+20.1	+19.3	+15.3	+9.9	+4.4	+0.6	+9.7
Change	+2.7	+1.5	+1.8	+0.6	+0.8	+0.4	+1.2	+1.8	+0.6	+0.2	-0.5	+0.2	+0.9
Precip. (mm)													
Baseline	36	35	35	44	58	62	60	51	43	46	53	47	570
Change	-19	-10	-7	-9	-17	3	-7	-2	-1	12	-2	-3	-62
Relative Humidity (%)													
Baseline	85	81	74	68	70	70	69	70	74	80	85	87	76
Change	-2	-3	-4	-4	-8	-4	-7	-6	-1	-4	-2	-3	-4

This seven-year period was the warmest and driest in Slovakia since 1871.

Significant trends toward lower relative humidity and precipitation (particularly in the spring and summer months) were recognized in Slovakia. Significantly close correlation was found between mean relative humidity and precipitation totals in the warm half of the year (April to September) in all of Slovakia. According to this analysis, if precipitation totals decrease by 100 mm on average, relative humidity is estimated to decrease by 2% in the lowlands and by 1% in other regions. Acceptable correlations were also found between relative humidity and mean temperature in the warm half of the year (April to September). Across all of Slovakia, a temperature increase of 1°C is estimated to decrease relative humidity by about 2%.

In the cold half of the year, no significant correlations were found between relative humidity and precipitation or between relative humidity and temperature. However, snow characteristics were found to be closely correlated with mean temperature. If the temperature increases by 1°C, it is estimated that the period of snowcover will be reduced by approximately 15 days. A reduction of this magnitude would be equivalent to a 30% reduction in the number of snowcover days in the lowlands and a 10% reduction in the number of snowcover days in the mountains. In addition, a 1°C rise of temperature in the cold half of the year is estimated to decrease the total sum of daily snowcover depth by 100 to 300 cm in lowlands (approximately 50% of the long-term mean) and by 400 to 1,400 cm in the middle altitudes (approximately 35% to 15% of the long-term mean) (Lapin et al., 1994).

Correlations of snowcover and snowdepth are more significant with temperature than with precipitation up to 900 m altitude. Above this elevation the change in precipitation is more important than the change in temperature. For example, in higher altitudes (the Slovak mountains) the increase in precipitation in the winter can partly compensate for temperature rise impacts, which would decrease snowcover (Lapin et al., 1994). These correlations can be considered a rough prediction or scenario for climate change in the near future.

Some preliminary results of an analysis of atmospheric circulation change in the region were also analyzed. The daily circulation classification prepared by T. Niedzwiedz in Cracow, Poland (Niedzwiedz, 1993; Niedzwiedz and Ustrnul, 1994) can be used also in Slovakia. There has been a noticeable increase of the westerly zonal and decrease of the southerly meridional direction of atmospheric circulation after 1960. This change was correlated with significant changes in precipitation and temperature in the large area between the Carpathians and the Alps. The influence of mountain topography along with a change in atmospheric circulation can cause a climate change in the surrounding central European area that may modify the large-scale impacts of climate change caused by the greenhouse effect. The correlation of precipitation and relative humidity changes with changes in atmospheric circulation seems to be very close in Slovakia.

Climate Change Scenarios

Climate change scenarios that project an annual increase in average temperature of 1° to 2°C for the year 2025 (in comparison with the means of the 1951-1980 period) were used in this analysis (Scientific Assessment of Climate Change, 1990; Brázdil, 1991; Kalvová and Vanícek, 1991). Preliminary regional temperature change scenarios based on GCM results (annual means) and historical analogues (annual fluctuations) were prepared for the Czech and Slovak republics in 1991 (Table 14.3).
Because the applicability of the GCM scenarios to central Europe is uncertain, the preliminary regional precipitation and relative humidity scenarios (Table 14.3) are based on estimated temperature change and historical analogues, and only partly on GCM scenarios (precipitation totals) (Bultot and Gellens, 1989; Lapin, 1992 and 1994).

Three temperature and precipitation change scenarios were prepared by Nieplová et al. (1994) based on GCM outputs (Tables 14.4 and 14.5). The selection of GCM scenarios for Slovakia was made by comparing the $1\times CO_2$ outputs for the five GCMs evaluated: the Canadian Climate Centre Model (CCCM), the Goddard Institute for Space Studies (GISS), the United Kingdom Meteorological Office (UKMO), and the Geophysical Fluid Dynamics Laboratory (GFDL) equilibrium (GFD3) and 1% transient (GF01) models with 1951-1980 means and annual fluctuations (Benioff et al., in press). None of the $1\times CO_2$ outputs corresponded well with the current temperature and precipitation means and annual fluctuations in Slovakia, but the deviations of the GISS, CCCM, and GFD3 models are the smallest among the models evaluated.

The poor correlation is one of the reasons Slovakia has developed a plan to prepare several updated alternative regional scenarios of temperature, precipitation, relative humidity, and snowcover based on historical climate change analogues. These alternative scenarios will consider relatively warmer periods since 1871, analysis of atmospheric circulation change, correlation and trend analyses, and the GCM scenarios of mean annual temperature rise. In addition, incremental scenarios will be constructed for temperature, precipitation, and relative humidity.

Table 14.3. Deviations of Preliminary Mean Monthly Analogues Climate Change Scenarios for 2025 from Normals Calculated for the 1951-1980 Period

Month	Jan	Feb	Mar	Apr	May	Jun	Jul	Aug	Sep	Oct	Nov	Dec
Change from Baseline Mean Monthly Temperature (deviations in °C) in All Slovakia												
Temp. 1	1.5	1.4	1.2	1.0	0.8	0.6	0.5	0.6	0.8	1.0	1.2	1.4
Temp. 2	3.0	2.9	2.5	2.0	1.5	1.1	1.0	1.1	1.5	2.0	2.5	2.9
Change from Baseline Mean Monthly Relative Humidity (deviations in %) in Mountainous Areas of Slovakia												
Relative Humidity 1	-1	-1	-1	-1	-1	-1	-1	-1	-1	-1	-1	-1
Relative Humidity 2	-2	-3	-3	-3	-3	-2	-2	-2	-3	-3	-3	-2
Change from Baseline Mean Monthly Relative Humidity (deviations in %) in Lowlands of Southwestern Slovakia												
Relative Humidity 1	-2	-2	-3	-3	-2	-1	-1	-1	-2	-2	-2	-2
Relative Humidity 2	-3	-4	-5	-6	-5	-3	-2	-2	-3	-4	-3	-3
Change from Baseline Mean Monthly Precipitation Totals (deviations in %) in Southern Half of Slovakia												
Precipitation 1	15	10	-15	-20	-10	10	5	0	-10	-10	0	5
Precipitation 2	0	0	-15	-20	-15	0	-15	-10	-10	-20	0	0
Change from Baseline Mean Monthly Precipitation Totals (deviations in %) in Northern Half of Slovakia												
Precipitation 1	15	10	-10	-15	-5	15	10	5	-5	-5	5	10
Precipitation 2	10	10	-5	-10	-5	10	-5	0	0	-10	10	10
Precipitation 3	20	20	10	5	5	10	5	5	5	5	10	20

Source: Kalvová and Vanícek, 1991.

Baseline Socioeconomic Scenarios

The Slovak vulnerability and adaptation assessment has not yet incorporated socioeconomic information. However, the information presented below provides a sense of how future changes in socioeconomic characteristics may affect the sensitivity of the vulnerability and adaptation assessments. The population in Slovakia has grown from 3 million inhabitants in 1920 to 5.3 million in 1992. The highest natural increase in population, over 1.7%, occurred in 1950. Since then the natural population increment has systematically decreased. Since the beginning of the economic transition process in Slovakia in 1989, the real average population of Slovak Republic has stagnated. The assumed official net population increase of about 0.3% to 0.4% up to 2010 might be overestimated. The average life expectancy at birth for men (66.5 years) is 6 to 7 years less than in developed countries and for women (73.5 years) 5 to 7 years less. Mortality, which increased by 2.2 per

Table 14.4. Mean Monthly Precipitation Totals (mm) for Slovakia for 1951-1980 and Adjustment Statistics from 2×CO_2 GCM Scenarios for North and South Slovakia

		Month											
		Jan	Feb	Mar	Apr	May	Jun	Jul	Aug	Sep	Oct	Nov	Dec
Baseline		44	44	42	53	71	98	93	79	54	53	61	56
CCCM	North	1.32	1.17	1.01	1.10	0.94	0.99	0.88	0.87	0.87	1.18	1.27	1.35
	South	1.29	1.02	1.15	1.04	0.85	0.83	0.79	1.03	0.91	1.11	1.16	1.22
GISS	North	1.16	1.31	1.27	1.34	1.18	1.07	1.08	1.06	0.84	1.25	1.31	1.04
	South	1.16	1.24	1.26	1.28	1.18	1.07	1.20	1.19	0.80	1.28	1.29	1.06
GFD3	North	0.95	1.44	1.08	1.13	1.07	0.99	0.84	1.19	1.36	1.45	1.44	1.01
	South	0.90	1.50	1.06	1.08	1.03	0.99	0.81	1.39	1.28	1.33	1.43	0.96

Table 14.5. Mean Monthly Temperature (°C) for Slovakia for 1951-1980 and Adjustment Statistics from 2×CO_2 GCM Scenarios

	Month											
	Jan	Feb	Mar	Apr	May	Jun	Jul	Aug	Sep	Oct	Nov	Dec
Baseline	-3.8	-1.8	2.2	7.7	12.5	16.1	17.5	16.8	13.0	8.0	3.0	-1.5
CCCM	3.9	4.6	5.0	3.0	2.6	3.0	3.6	3.2	3.5	3.0	3.4	2.2
GISS	5.5	6.0	5.0	3.5	2.9	2.8	2.9	2.7	3.4	3.9	4.5	6.0
GFD3	2.6	4.1	5.1	4.9	4.8	4.5	4.3	4.0	4.6	4.7	4.1	3.8

thousand between 1960 and 1980, has since stagnated.

The gross domestic product (GDP) at constant market prices (1984) in Slovakia amounted to 201 billion Slovak crowns in 1985 and 215.4 billion Slovak crowns in 1990 (approximately US$7.4 billion in 1990). The average annual growth rate for GDP between 1985 and 1990 was 1.39%. The political changes in central Europe during this period influenced the general social and economic situation in Slovakia. Between 1990 and 1992, the economy declined almost 20%, but a preliminary assessment shows about a 5% GDP increase in 1994 (compared to 1993) and the same increase is expected in 1995 (compared to 1994).

Slovakia is an industrial country. Table 14.6 presents the percent contribution of various economic sectors to the GDP in 1991. High demand for energy and raw materials is a key feature of the Slovak economy. However, there is a shortage of domestic sources of high-quality raw materials (excluding non-ore material and magnetite).

The transition of Slovak industry to a free market economy is a long-term process. Since the transition began in 1990, industrial production has dropped dramatically.

Table 14.6. GDP in Slovakia by Sector (1991)

Material Sectors	
agriculture	4.9%
forestry	0.8%
water works and supply	0.9%
industry	51.8%
construction	7.4%
freight transport	5.8%
trade and catering	5.5%
other	3.8%
Total	80.9%
Nonmaterial Sectors Total	19.1%

Source: Statistical Yearbook of the Slovak Republic 1993, 1994; The Slovak First National Communication on Climate Change, 1995.

AGRICULTURE SECTOR

In Slovakia 2.45 million ha of land are being used for agriculture, including 1.49 million ha of arable land. Per capita acreage is 0.46 ha of farmland and 0.28 ha of arable land.

Until 1989, by bureaucratic directive, agriculture was concentrated into large organizational units (cooperatives) with complicated systems of management, which often neglected specific soil, climate, geographic, and regional factors. This led to ecologically irrational and economically inefficient agricultural and food production. The policy of "cheap" food, based on increasingly subsidized consumer prices of foodstuffs, distorted food production markets.

The privatization and transformation process of cooperatives began (before 31 December 1992) by means of a special law. The majority of the transformed cooperatives still have not carried out an internal economic transformation and commercialization of their production. In 1992, there were 8,727 private farmers registered, with an average farmland area of 7.6 ha.

There has been no marked change in crops production; however, some changes have been observed. An increase in cereals acreage has been recorded, and sugar beet and bulky crop production has decreased. All species of animal production has decreased. Compared to 1989, the inventory of slaughter animals in 1992 dropped by 21% because of the declining purchasing power of the population; inventories of slaughter beef cattle fell by 16%, pigs by 22%, slaughter poultry by 23%, and milk production declined by 39%. In 1992 the number of heads was 1.2 million beef cattle, 2.3 million pigs, 0.5 million sheep, and 13.4 million poultry.

Slovak's agrarian policy will, in the near future, focus on the following strategic goals:

- state food security through the availability of healthy, quantitatively and qualitatively balanced food for the population
- economic stability, appropriate agricultural income, and the balanced development of regions
- improvement and protection of farmland through ecological cultivation and the prevention of food chain contamination with harmful substances
- preservation of agriculture in the less competitive, hilly areas to maintain the social functions of agriculture and to maintain rural settlement in the countryside.

Preliminary Results

The Decision Support System for Agro-technology Transfer version 3 (DSSAT 3) is currently being used to develop a specific assessment of agriculture sector vulnerability for Slovakia (Benioff et al., in press).

Slovakia is a hilly country; only 39.7% of its surface is less than 300 m above sea level, 55.3% is considered mountainous, and 8.5% is made up of high mountains. The share of arable land is 30%, that of grasslands and pastures is 17%, and gardens and vineyards constitute 2.4% of the territory. Cereals, maize, potatoes, and sugar beet are the principal agricultural products.

Simple model calculations using preliminary climate change scenarios (Lapin et al., 1994; Špánik, 1995) indicate that an estimated warming of 1° to 2°C and a small change in precipitation (a winter increase of 10% and a decrease during the long vegetative period of 10% to 20%) would be accompanied by an increase in the probability of the temperatures being in an optimal range for agricultural production of 7% to 11% and an increase in solar radiation of 2% to 23%. It is also estimated that there would be an increase in the evapotranspiration deficit by 13% to 26% and a decrease of snow cover occurrence and depth. Prolongation of vegetative periods by 6% to 8% is expected. Usable soil moisture will probably decrease by about 20% in the main agricultural fields in southern Slovakia, which is critical from March to May (Huzulák and Matejka, 1994; Lapin and Šipocz, 1991; Lapin, 1992; Lapin, 1994; Lapin, 1995; Špánik, 1995).

Impacts of climate change in agriculture can be summarized as follows (Špánik, 1995):

- changed average distribution of temperature
- changed phenological conditions of plants
- changed solar radiation certainty of plants for photosynthesis
- changed agroclimatic production potential
- changed average distribution of water (soil moisture)
- changed winter condition of plants
- changed CO_2 concentration and biomass production
- changed physical and chemical characteristics of soil
- changed occurrence of diseases, pests, and weeds.

Preliminary Adaptation Alternatives

Climate changes could lead to plant species selection pressure. Because of the complex interactions involved, the results of such pressure within the adaptation processes cannot currently be predicted. One thing is certain, however: during these adaptation processes, the stability of the ecosystem's structure can be negatively affected. For example, by the appearance of new competing species, new pests, or increased pest abundance; by new pathogens; and by increased fire risk and soil erosion. With rising temperature, the natural range of tropical-disease pathogens spreads northward. Of special importance among the direct effects of a change in atmospheric chemistry are the effects caused by the rise of CO_2 and by changed oxidant content (Huzulák and Matejka, 1994; Mindáš and Škvarenina, 1994; Závodský and Závodská, 1992, 1994; Mindáš, 1995; Špánik, 1995).

Climate change impacts on agriculture are expected to be complex, partly positive, and primarily negative. Mitigation of the negative impacts and utilization of the positive ones should be prepared long in advance, because of the long-term character of the adaptation processes in agricultural systems as well as in agricultural ecosystems. The adaptive measures under consideration are as follows (Špánik, 1995).

Long-term strategies in agricultural planning:

- change crop cultivation technologies
- change agroclimatic regionalization and structure of crops and varieties cultivated
- change breeding objectives
- change plant protection.

Agricultural practice:

- regulate water supply, taking into consideration the expected hydrological cycle changes
- use mulch to preserve water and energy
- improve soil fertility and utilize sustainable agriculture techniques
- educate the public on climate change issues.

Forestry Sector

The Holdridge Life Zone Classification Model is currently being used in Slovakia to assess vulnerability (Benioff et al., in press). Forestry and agriculture in Slovakia employed about 250,000 people in 1992, which is equivalent to 12.1% of the labor force or 4.7% of the population. Forests are one of Slovakia's most important natural resources. The forests cover 19,911 km^2 or 41% of the country's surface area. Broad-leaved trees prevail in the forests of Slovakia (57%). Conifers represent 43% of forest inventory.

A comparison of forested land in 1920 and in 1990 indicates that forested land increased by more than 20% because of afforestation of farmland and acreage adjustments of agricultural crops. Positive changes were also recorded in the categorization of forests. Today, managed forests represent approximately 76% of the total woodland area, with a marked increase in the land area of protected forests (13%). Also, the land area of forests in national parks, and forests protected to maintain water resources, etc., has increased (11%).

The age composition of forest stands in the total forested area is also quite favorable. Forest stands up to the age of 40 years represent 33%, 41-80 year old trees about 43%, 81-120 year old trees 19%, and over 120 years old approximately 5% of the total forest. Slovakia's timber resources went from 193.5 million m^3 in 1950 to 352.2 million m^3 in 1991. The timber-growing stock increased from 140 to 189 m^3/ha.

There are also negative trends. In the last decades, the health of the forests of Slovakia has markedly worsened. After 1985, the forest damage dramatically increased and 85% of all forest damage was identified as the first and fourth stage of forest damage severity (according to the European 5 stage classification of forest damage; stage 4 represents a maximum of forest damage). Unfavorable climate conditions (high temperatures, low precipitation, changes in snowcover) together with emissions have contributed to the present unsatisfactory health of the Slovak forests. Current monitoring reveals a relative stabilization of damage.

Priority of tenancy relations, made legal and institutionalized by Forest Act No. 61/1977, resulted in total state tenancy of all forests in Slovakia until 1991. The return of tenancy to original owners represents a new situation, and one in which the best forestry management procedures and the protection of forest ecosystems are of utmost importance. In realization of these tasks, it will be necessary to take note of the economic transition. A harmonization of public welfare and productive functions of forests is important. The important principles of the State Forestry Policy in Slovakia are inter alia gradual afforestation of farmland area unsuitable for agriculture and overall ecologization of forestry.

Preliminary Results

Based on the effects identified in the preliminary regional climate scenarios as well as concentrations of some gases in the atmosphere (greenhouse gases, photooxidants, etc.), the following topics should be examined (Mindáš, 1995):

- changes in species composition
- changes in bioproduction
- forest soils
- injurious agents
- forests as a source and sink of greenhouse gases.

Based on the analyses done within the National Climate Program (Mindáš and Škvarenina, 1994; Mindáš, 1995) and the literature, the following preliminary conclusions can be made:

- European fir (*Abiesalba* A.) and Norway spruce (*Picea abies* L. Karst.) will probably be the species most affected by climate change.

- Expansion of beech (*Fagus sylvatica* L.) into upper altitudes can be expected.

- The estimated intensity and time scale of climate change will probably exceed the ability of existing forest ecosystems to adapt.

- Soil moisture will be unfavorable in lower altitudes, and this will result in physiological weakness of oaks (*Querems* spp.) and more xerophyte forests (dry forests) or grasslands. Approximately 20% of the forests are located on "carbonate" soils. These forests are highly vulnerable to a decrease in soil moisture.

- Increases in production of all forest stands (10% to 30%) can be expected, depending on the vitality and existence of productionally stable stands.

- Changes in the dynamics of nutrient cycles (especially nitrogen and carbon) are expected, especially in upper altitudes, where the production optimum of the forests will probably be shifted.

- Changes in biotic and abiotic injurious agents are unknown, especially for insects, where a very flexible adaptation to changing environmental conditions can be presumed.

Preliminary Adaptation Alternatives

Considering the serious risks to forests from the estimated climate change in Slovakia, it is necessary to identify and to implement measures to minimize the negative impacts of global change. Within a short period of time, the following measures may be necessary (Mindáš, 1995):

- hasten the change of Norway spruce monocultures to mixed stands of pine, oak, beech, and other broad-leaved species

- preserve the biodiversity of forest communities through forest management

- identify species or genotypes that can tolerate/adapt to climate change

- reevaluate forest management plans in light of climate change

- increase carbon dioxide fixation by afforestation of lands not used for agriculture.

Water Resources Sector

Slovakia has limited natural water resources. This fact underlines the importance of water management in the country. Of the 8,437 km of water courses, 3,156 km are regulated. Based on the 1931-1980 means, the annual water balance of Slovakia is as follows:

- precipitation, 36,898 million m^3
- runoff, 12,784 million m^3
- evaporation, 24,114 million m^3.

The average discharge from runoff is 405 m^3/s. During the last decades, a significant decrease in discharge has been observed in Slovakian rivers. Except for the southwestern Danubian region, Slovakia suffers from a sizeable deficiency in useable stocks of underground water. The usable capacity of the Danube alluvia represents about 23 m^3/s of drinking water. Protection of this highly valuable natural resource from anthropogenic pollution is one of the most important goals of the state environmental policy. Most of the Slovak territory relies on surface waters. The volume of water reservoirs increased from about 300 million m^3 in 1975 to 1.6 billion m^3 in 1992.

Several regions of Slovakia severely lack soil moisture, and mainly during the vegetation period. More than 800,000 ha of arable land need irrigation systems; at present only 328,000 ha can be irrigated.

The general principles and major goals of Slovak water management policy are documented in Master Water Management Plans (Ministry of Forestry and Water Management of the Slovak Republic, 1975). Most activities are executed by four river basin agencies. In March 1994, the Slovak

Government accepted the new water management policy of the Slovak Republic.

Preliminary Results

Scenarios estimating approximately 20% decreases in water resources were preliminarily assessed for 1990-2030. The WATBAL model, as well as statistical models are being used to generate more specific results (Benioff et al., in press). In addition to the preliminary assessment using climate scenarios, historic data indicates that the mean river discharges in Slovakia (measured since 1931) have generally been decreasing in the last several years when compared to long-term means calculated for the 1931-1980 period. The degree of decrease varies throughout the territory of Slovakia. The greatest decrease (by more than 40%) occurred in southern and southeastern Slovakia (Ipel, Slaná, Bodva river basins). The largest decrease in discharge has occurred in November and the following autumn-winter months, and this decrease has been significant during the last 10 to 15 years (Majercáková and Šedík, 1994; Majercáková, 1995).

The trend analysis identified a decreasing monthly mean runoff at about 80% of the gauging stations in Slovakia. In the relatively warmest and driest period (1988-1993), spring runoff decreased by 6% to 7% in some regions. The most significant decrease in past years occurred in the central and southern part of Slovakia (approximately 1.4% to 1.5% per year). In the crystalline range of mountains (predominantly in the northern and northwestern part of Slovakia) the annual decrease did not exceed 0.8%.

The results of shorter periods of observation for groundwater obtained by the Slovak Hydrometeorological Institute for some Slovak regions need to be reevaluated. Limited analysis showed significant decreases between 1971 and 1993 at most stations with groundwater level measurements (Majercáková, 1995).

Some experts who have analyzed the areal water balance change in Slovakia have stated that in some localities, a significant decrease of spring yields or lowered groundwater stages have occurred. This decrease represents 7.9% to 11.3% compared with values assumed in the calculation of usable groundwater resources under the State Water Management Balance Plan (Majercáková, 1995). For more accurate estimations, extensive observations are needed to exclude the possibility that this decrease is temporary, caused by a dry period.

According to simple model calculations performed using preliminary climate change scenarios (Lapin et al., 1994), the present trends are expected to continue. This may negatively influence not only the water supply for the general public, industry, and agriculture, but also the operation of hydropower plants and river transport systems in Slovakia.

Preliminary Adaptation Alternatives

Possible hydrological cycle and water balance changes in Slovakia due to climate change will need to be evaluated using more detailed models under different geographical conditions. Therefore, the first step in adapting to potential climate change will be to collect the information necessary to assess vulnerability and to make decisions about adaptation alternatives.

Further evaluations should be done on the stationarity of hydrological processes (continual decrease of river discharges and spring yields, for example) and on low river discharges (magnitude, frequency, and duration of low river discharges). These changes can seriously influence the routine utilization of measured hydrological data and calculated characteristics for decision making (Majercáková, 1995).

In the water resources sector, it is recommended that generally binding regulations be issued to ensure the collection of information needed for unambiguous identification and quantification of climate change impacts on groundwater resources (Majercáková, 1995). Such regulations include requirements for:

- continual measurement of groundwater consumption
- continual measurement of spring yields, including the utilized portion and the unutilized portion that returns to the environment
- compulsory measurements of groundwater levels in areas with water resources exploitation to evaluate disposable water amounts
- modification of water quality measurement
- more rigorous governmental inspection of water resources exploitation
- establishment of a center for controlling groundwater exploitation, for balancing water amounts and quality, and for consulting.

Either a very detailed water management balance of selected regions or a concise water management balance of all of Slovakia with defined water surplus and deficiency regions needs to be prepared.

SUMMARY CONCLUSIONS

This first, preliminary synthesis was prepared by the experts of the Slovak Republic's Country Study Program; it synthesizes the findings presented in National Climate Program reports and in some of the literature cited. The National Climate Program project comes under the aegis of the Ministry of Environment. These preliminary results may be used as a framework for strategies; however, more detailed results are expected to be available in 1996.

Slovakia is highly vulnerable to changes in water balance parameters (precipitation, evaporation, runoff, soil moisture, and snow cover). This will cause serious negative impacts predominantly in southern Slovakia. The primary sectors affected would be agriculture, forestry, and water resources, but consequently all other sectors would be affected. More precise evaluation of vulnerability will require more detailed and precise climate change scenarios as well as new methods or models for impact assessments.

REFERENCES

Benioff, R., S. Guill, and J. Lee (eds.) (In press). *Vulnerability and Adaptation Assessments: An International Guidebook.* Dordrecht, The Netherlands: Kluwer Academic Publishers.

Brázdil, R. 1991. Climate Scenarios for the Global Warming Model on the Territory of the CFSR (in Czech). *Meteorol. Zprávy*, 44(6):170-181.

Brázdil, R., M. Budíková, P. Faško, and M. Lapin. 1994. Fluctuation of Daily Maximum and Minimum Air Temperatures in the Czech Republic and in the Slovak Republic (in Czech). *Meteorol. Zprávy*, 47(4):113-119.

Budyko, M.I. and L.T. Zubenok. 1961. Opredelenije Isparenija s Poverchnosti Sushi. Izd. AN SSSR. Ser. Geografula.

Bultot, F. and D. Gellens. 1989. Simulation of the Impact of Atmospheric CO_2 Doubling on Precipitation and Evaporation — Study of the Sensitivity to Various Hypotheses. Conference on Climate and Water. Helsinki, pp. 73-92.

Huzulák, J. and F. Matejka. 1994. Possible Impacts of Climatic Changes on Winter Wheat and Sugar Beet Evapotranspiration in Podunajská Nízina Lowland (in Slovak with extended English summary). Contributions of NCP SR, No. 2, Bratislava, MZP SR and SHMÚ, pp. 139-156.

Kalvová, J. and K. Vanícek. 1991. Present Predictions of Climate Changes in Relation to the Territory of CSFR (in Czech). *Meteorol. Zprávy*, 44(3):68-72.

Lapin, M. 1992. Possible Impacts of Expected Climate Changes on Water Balance in Slovakia (in Slovak with extended English summary). Contributions of NCP CSFR, Praha, CHMÚ, No. 7, pp. 51-87.

Lapin, M. 1994. Possible Impacts of Climate Change upon Water Balance in Central Europe. In: M. Desbois & F. Désalmand (eds.). *Global Precipitation and Climate Change.* NATO ASI Series, Springer-Verlag Berlin, Heidelberg, New York, London, Paris, Tokyo, Hong-Kong, Barcelona, Budapest, pp. 191-170.

Lapin, M. 1995. Possible Global Warming Impacts on Climatic Conditions Change in Slovakia, Monitoring and Climate Change Scenarios (in Slovak). NCP SR Summary Report No. 1, Bratislava, SHMÚ, 17 pp.

Lapin, M. and M. Šipocz. 1991. Possible Impacts of Expected Climate Changes on Water Balance in Lowlands of Slovakia (in Slovak). *Meteorol. Zprávy*, 44(3):79-83.

Lapin, M., P. Faško, and V. Zeman. 1994. Contribution to Analysis of Possible Global Warming Impacts upon Climate Change in Slovakia (in Slovak with extended English summary). Contributions of NCP SR, No. 2, MZP SR and SHMÚ, Bratislava pp. 35-77.

Majercáková, O. 1995. Possible Impact of Climate Changes on Hydrologic Cycle and Water Management in Slovakia and Mitigation Options Design (in Slovak). NCP SR Summary Report No. 2, Bratislava, SHMÚ, 16 pp.

Majercáková, O. and P. Šedík. 1994. The Runoff Change of the Slovak Rivers (in

Slovak with extended English summary). Contributions of NCP SR, No. 2, Bratislava, MZP SR and SHMÚ, pp. 107-137.

Mindáš, J. 1995. Possible Impacts of Climate Change on Forests in Slovakia and Mitigation Options Design (in Slovak). NCP SR Summary Report No. 3, Zvolen & Bratislava, LVU & SHMÚ, 9 pp.

Mindáš, J. and J. Škvarenina. 1994. Possible Impacts of Climate Change on Forest Ecosystems (in Slovak with extended English summary). Contributions of NCP SR, No. 1, Bratislava, MZP SR and SHMÚ, pp. 57-82.

Ministry of Forestry and Water Management of the Slovak Republic. 1975. *Master Water Management Plan of the Slovak Republic*. 2nd ed., 728 pp.

Moldan, B. et al. 1993. Risk of Climate Change and its Mitigation — Summary (in Czech with extended English summary). NCP CR, No. 10, Praha, CHMÚ, 175 pp.

Niedzwiedz, T.Z. 1993. Changes of Atmospheric Circulation (using the P, S, C, M Indices) in the Winter Season and Their Influence on Air Temperature in Cracow. Early Meteorological Instrumental Records in Europe — Methods and Results, Zeszity Naukowe Uj — Prace Geograficzne, Krakow, pp. 107-113.

Niedzwiedz, T. and Z. Ustrnul. 1994. Maximum and Minimum Temperatures in Poland and the Variability of Atmospheric Circulation. In: *Contemporary Climatology* (R. Brázdil and M. Kolar, eds.). Proceedings of the meeting of the Commission on Climatology of the IGU, Brno, Czech Republic, August 15-20, 1994. Litera, Brno, pp. 420-424.

Nieplová, E., M. Lapin, and P. Faško. 1994. Design of Climatological and Precipitation Station Network for Climate Changes and Variability Monitoring in Slovakia (in Slovak with extended English summary). Contributions of NCP SR No. 2, Bratislava, MZP SR and SHMÚ, pp. 5-33.

Scientific Assessment of Climate Change. 1990. Report prepared for IPCC. Geneva, WMO, 365 pp.

Špánik, F. 1995. Possible Impact of Climate Changes on Agriculture and Mitigation Options Design (in Slovak). NCP SR Summary Report No. 4, Nitra & Bratislava, VSP & SHMÚ, 9 pp.

Statistical Yearbook of the Slovak Republic 1993. 1994. Statistical Office of the Slovak Republic, Bratislava, 608 pp.

The Slovak First National Communication on Climate Change. 1995. The Slovak Ministry of Environment (draft in English, prepared for publishing in 1995).

Závodský, D. and E. Závodská. 1992. Air Quality and Climate Change (in Slovak with extended English summary). Contributions of NCP CSFR. Praha, CHMÚ, 7:4-50.

Závodský, D. and E. Závodská. 1994. The Effect of Global Warming and UVB Solar Radiation Increase on the Boundary Layer Chemistry in Slovakia (in Slovak with extended English summary). Contributions of NCP SR, No. 1, Bratislava, MZP SR & SHMÚ, pp. 18-53.

Chapter 15

Vulnerability and Adaptation Assessments for Ukraine

I. Trofimova,[1] S. Koublanov,[1] E. Shmurak,[2] I. Buksha,[3] V. Meshkova[3] A. Sidorov,[3] A. Shereshevsky,[4] L. Dmitrenko,[4] L. Sinitskaya,[4] P. Gozhik[5] Ya. Fedorin,[5] M. Yakimchuk,[5] V. Vernigorov,[5] I. Makridicheva[5]

[1] **State Committee of Ukraine for Hydrometeorology**
[2] **Ministry of Ecology and Nuclear Safety of Ukraine**
[3] **Ukrainian Research Institute for Forestry**
[4] **Ukrainian Research Hydrometeorological Institute**
[5] **Ukrainian Institute of Geological Sciences**

SUMMARY. This chapter presents the results of preliminary analyses of the vulnerability of the forest, water resources, and coastal resources sectors of Ukraine. For the forest sector, the Holdridge Life Zone Classification Model was used. Generally, the northern forests would remain unchanged, but total forested area would decrease by 39% to 47%. For the water resources sector, preliminary assessments were made of changes in the water supply at three main gauging sites of Dnieper water reservoirs under different climate change scenarios. Changes in runoff varied widely under the various scenarios. Water resources supply and the ratio of total water demand to total water supply were also estimated, holding demand constant because of the difficulty of predicting future demand given an economy in transition. The only cases in which water demand is estimated to exceed water supply are at one of three study sites under the Geophysical Fluid Dynamic Laboratory 1% transient and Max Planck transient scenarios. Water supply is predicted to increase under the Geophysical Fluid Dynamics Laboratory equilibrium and United Kingdom Meteorological Office scenarios in 2075, and decrease under the Geophysical Fluid Dynamics Laboratory 1% transient scenarios. For the coastal resources sector, dynamic investigations and onsite observations of the coastline were made, and land losses due to inundation and erosion were estimated to be 32 km² under a 0.5 m sea level rise and 370 km² under a 2.0 m sea level rise.

Introduction

Ukraine is located in central Europe in the southwest section of the Eastern European plain. Its area is 6.04 million km². The Ukrainian Carpathians are located in the western part of the country, and the Crimean Mountains are found along the southern

Note: This chapter does not contain a detailed description of the assessment methodologies that were recommended by the U.S. Country Studies Program. Refer to Chapter 2 for an overview of methodologies.

J. B. Smith et al. (eds.), Vulnerability and Adaptation to Climate Change, 313–333.

Figure 15.1. Map of Ukraine

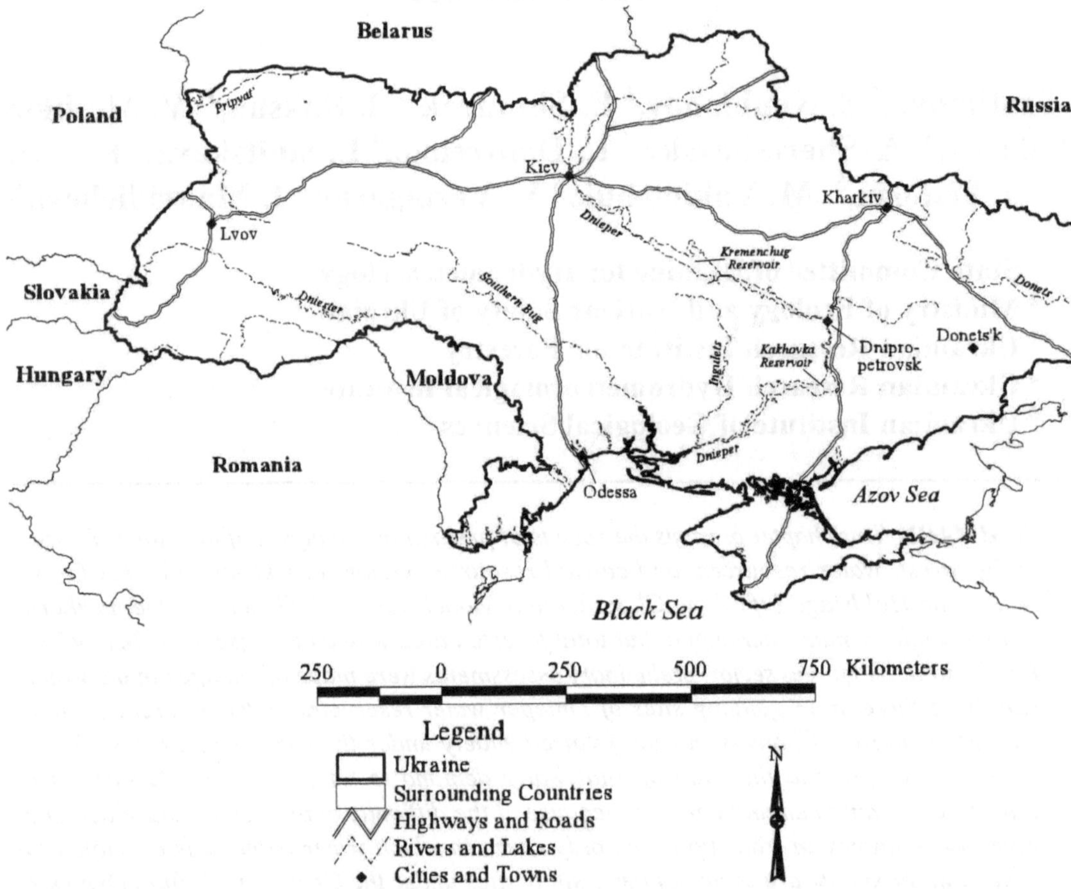

coast of the Crimean peninsula. The southern part of the country is bordered by the Black and Azov seas. The Ukrainian territory stretches north to south (from 52°22'54" to 44°23'18" latitude) for 900 km, and west to east (from 22°08'42" to 40°13'05" longitude) for 1,300 km (Figure 15.1).

Most of Ukraine (95%) is a plain of 5.74 million km^2. In the north are the lowland swamps of Polissya; the left bank of Dnieper is occupied by Dnieper lowlands (Pridnieprovskaya lowland). Spurs of the Central Russian mountains pass through northeastern Ukraine. In the southeast are the Donetsk Ridge and the Azov Mountains. In the south, the Black Sea lowlands (Prichernomorskaya lowland) are joined to the Northern Crimean plain by the Perekop neck. About 19% of the plain is a mixed forest zone in the north. A forest-steppe zone makes up 34% of the total area of the country and a steppe zone in the south comprises 40% of the territory.

Ukraine is dense with rivers, most of which drain to the Black and Azov seas. The largest river in Ukraine is the Dnieper, whose basin occupies one-half of the country's territory. The average flow of the Dnieper River is 53.5 km^3 per year.

Ukraine has a temperate continental climate, except along the southern Crimean coast where the climate is close to subtropical. The continental features of the climate increase from west to east, and the main meteorological elements change with latitude.

Fertile soils are the most important resource for agriculture in Ukraine. The most fertile soils in the Ukraine are black soils with a humus content from 3% to 7% and a humus layer 130 to 150 cm thick. More than 10% of Ukraine's total area is made up of these soils. Also, large areas of the country contain gray forest soils, in the forest-steppe zone, and chestnut soils, in the steppe zone, which have good fertility properties.

Given the country's unstable economy and critical ecological problems, the consequences of climate change in Ukraine could be serious. The results of scientific research during the last few years show that climate changes in Ukraine would have significant impacts on agriculture, forestry, water, and coastal resources. Crop yields could increase because of longer vegetation periods, but some changes in climate (such as drought frequency and duration, heat stress and cold weather effects) could have negative effects. As the climate of Ukraine changes, there could be changes in forest types, species, productivity, and stability. Coastal zone vulnerability is already a reality in Ukraine: the Black Sea is rising 1.5 mm per year (according to observation posts in Odessa, Sulini, and Primorsky).

All of Ukraine will be affected by climate change, but one can distinguish specific regions for each sector that will be most vulnerable to climate change:

- steppe zone — for agricultural crops and grasslands/livestock
- forest-steppe zones — for forests
- Dnieper river catchment — for water resources
- the Black Sea coast — for coastal resources.

Ukraine, as one of the 150 countries that signed the United Nations Framework Convention on Climate Change (FCCC), has an obligation to submit information on greenhouse gas (GHG) emissions within the country, intensify its scientific research on climate, create a national program to mitigate GHG emissions and to adapt to the effects of climate change, and assist in the development and implementation of educational programs and in increasing public understanding of climate change and its consequences.

The objectives of Ukraine's vulnerability and adaptation assessments are the following:

- development of baseline (socioeconomic, ecological) and climate change scenarios
- assessment of agriculture, forestry, water, and coastal resources sensitivity and vulnerability to climate changes
- identification of possible adaptation measures
- evaluation of both estimated losses from climate change impacts and estimated cost of adaptation measures in order to develop balanced response strategies

- development of recommendations for policy makers to create a national adaptation policy.

The data for the forest, water resources, and coastal resources sectors have been assembled, and are included here. In addition, the preliminary analysis has shown that climate change effects on the agriculture and grasslands/livestock sectors could also be considerable, but data for these sectors are still not available and are not included in this chapter.

SCENARIOS

Climate Change Scenarios

The modeling of climate systems has been successful, but current models cannot render realistic regional climate change processes. It is therefore expedient to consider a number of plausible climate factors in the future, that is, "climate scenarios." These scenarios can be obtained on the basis of calculations from general circulation models (GCMs).

For Ukraine, the following models were used: the Goddard Institute for Space Studies (GISS) model, the Geophysical Fluid Dynamics Laboratory (GFDL) equilibrium (GFD3) and 1% transient (GF01) models, the Canadian Climate Centre Model (CCCM), and the United Kingdom Meteorological Office 1987 (UKMO) model (Benioff et al., in press).

To choose the most suitable GCMs for Ukraine, the results for 1×CO_2 conditions were compared to observation data. Taking into account that the highest spatial resolution for GCMs is 250 km latitude, and considering their simplified topography, which does not consider local peculiarities, outputs from the GCMs were compared with:

- global climate data, presented in grid boxes, from the CLIM data (Taljaard et al., 1969; Crutcher and Meserve, 1970; Schutz and Gates, 1971) for most of Europe

- data of meteorological observations for 1951-1980 from 60 stations in Ukraine.

The results of these comparisons are presented in Figures 15.2 and 15.3.

In comparison with the CLIM temperature data for Europe, the GISS and CCCM models give the best description of annual temperature distribution, but the GISS model understates significantly the temperatures for spring and summer months. The UKMO model understates temperature for all seasons, and the GFD3 model understates temperatures for winter, spring, and autumn.

The comparison with Ukrainian average data reveals that the GISS model gives the best description of annual temperature distribution, but understates monthly temperatures. The CCCM overstates temperatures significantly for summer months, the GFD3 model overstates temperatures for June to September and understates them for winter months, and the UKMO model understates temperatures for all seasons.

In comparison with the CLIM precipitation data for Europe, no one model satisfactorily describes the annual precipitation distribution, and in particular the summer maximum. The best description for winter precipitation is provided by the CCCM; the

Figure 15.2. Comparison of 1×CO_2 GCM Results and CLIM Data for Precipitation and Temperature

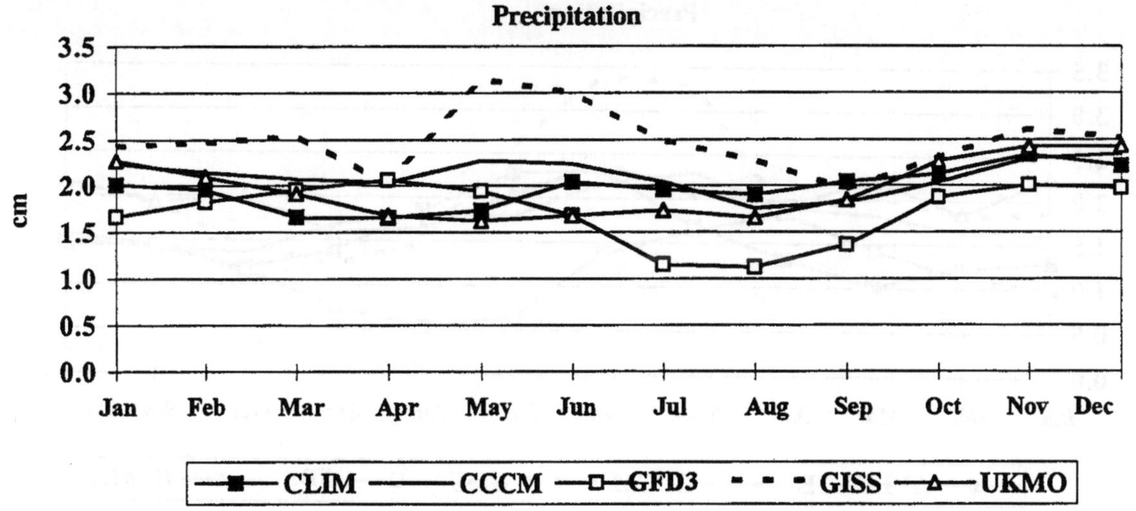

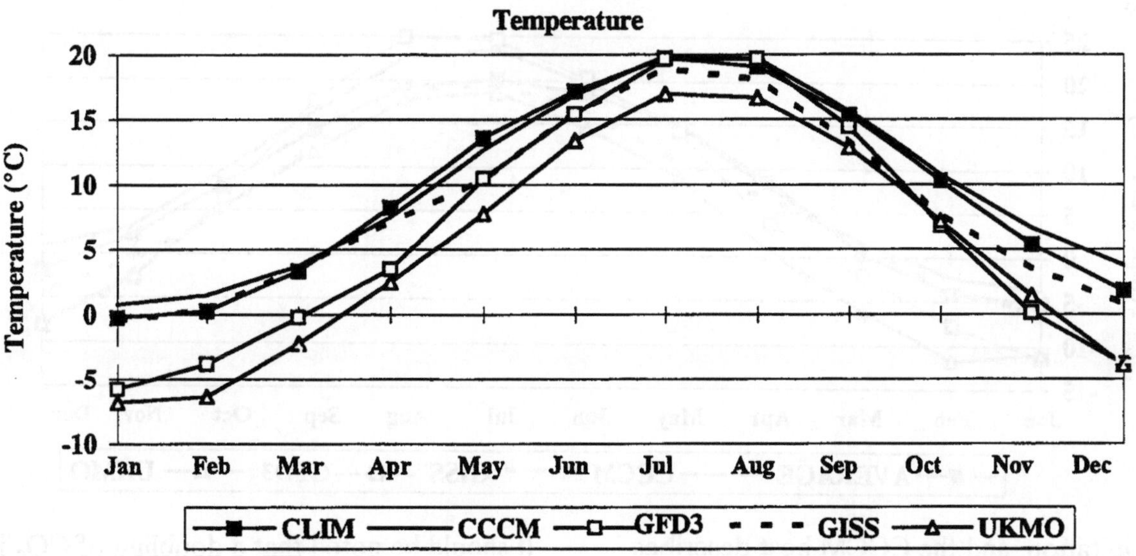

GISS and UKMO models overstate precipitation, and the GFD3 model understates precipitation significantly. Spring precipitation is best described by the CCCM and GFD3 models, and summer and autumn precipitation is best described by the GISS model. Precipitation from June to November is understated by all the models.

The results of comparison with Ukraine data show that no one model gives realistic precipitation distributions. The GFD3 model most realistically describes precipitation for winter and spring and the UKMO, GISS, and CCCM models significantly overstate precipitation for those seasons. The UKMO model gives the best description of summer

Figure 15.3. Comparison of 1×CO_2 GCM Results and Ukraine Area Averages for Precipitation and Temperature

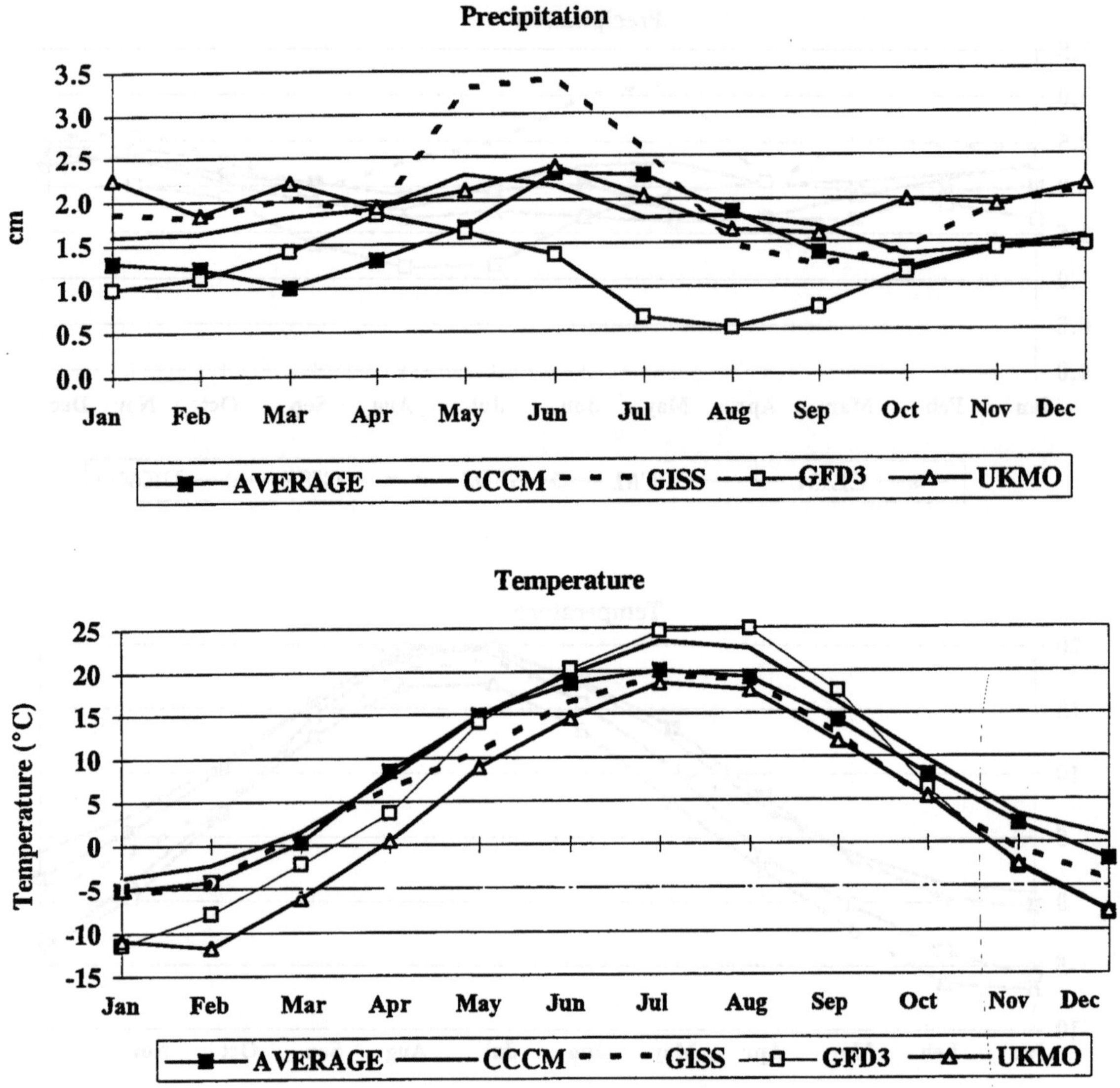

precipitation, and the CCCM best describes autumn precipitation.

The results show that no one GCM can best represent the climate of Ukraine. However, the GFD3 and CCCM models appear to best represent the climate of Ukraine.

The adjustment statistics for the CCCM, GISS, GFD3, and UKMO scenarios averaged for Ukraine are presented in Table 15.1.

It should be noted that a doubling of CO_2 in the atmosphere would increase the temperature in Ukraine according to all the scenarios for each season. Increased precipitation is also probable, except in the CCCM outputs for May to September.

Modeling climate response for transient conditions can provide information about climate change dynamics (rate and scales),

Table 15.1. Adjustment Statistics for the Difference between 2×CO_2 and Current (1×CO_2) GCM Scenarios Generated by General Circulation Models

Model	Jan	Feb	Mar	Apr	May	Jun	Jul	Aug	Sep	Oct	Nov	Dec	Annual
CCCM													
Temp. (C°)	6.52	6.04	6.62	4.35	3.43	3.46	3.78	3.29	3.40	2.85	3.45	3.30	4.21
Precip. Ratio	1.13	1.10	1.09	1.03	0.98	0.93	0.89	0.80	0.88	1.18	1.17	1.27	1.045
GISS													
Temp. (C°)	6.27	6.82	4.14	5.08	3.82	1.97	1.78	1.50	4.21	3.06	5.96	4.81	4.12
Precip. Ratio	1.20	1.28	1.29	1.19	1.17	1.07	1.07	1.53	0.97	1.32	1.20	1.07	1.175
GFD3													
Temp. (C°)	1.99	4.66	5.00	5.59	3.92	4.20	4.81	4.89	3.12	3.54	5.82	5.24	4.40
Precip. Ratio	1.09	1.38	1.20	1.01	1.05	1.08	1.24	1.12	1.55	1.10	1.21	1.04	1.14
UKMO													
Temp. (C°)	5.85	8.07	7.93	8.00	4.02	3.67	3.85	3.99	4.03	5.11	5.05	5.03	5.52
Precip. Ratio	1.05	1.33	1.01	1.04	1.10	1.00	1.15	1.29	1.02	0.88	1.36	1.15	1.087

Table 15.2. Adjustment Statistics for the GF01 General Circulation Model and Daily Weather Data from 1951-1980

Decade		Jan	Feb	Mar	Apr	May	Jun	Jul	Aug	Sep	Oct	Nov	Dec	Annual
Fourth (2000)	Temp. (C°)	1.17	1.80	3.41	1.45	-0.02	0.37	-0.60	0.39	2.12	1.91	0.78	1.23	1.17
	Precip. Ratio	0.85	1.18	1.06	1.08	1.17	1.30	1.44	1.09	1.03	0.76	1.19	0.97	1.059
Seventh (2030)	Temp. (C°)	3.79	2.16	3.71	1.99	1.12	1.87	0.63	2.25	2.42	2.56	1.68	2.53	2.22
	Precip. Ratio	0.95	1.15	1.25	0.89	1.20	1.27	1.48	0.96	0.80	0.95	1.20	0.93	1.07
Tenth (2070)	Temp. (C°)	6.05	3.09	4.46	3.32	2.65	3.21	3.88	4.72	3.74	4.26	3.93	5.06	4.03
	Precip. Ratio	1.02	1.40	1.29	1.25	1.30	1.34	1.10	0.97	1.07	0.87	1.24	1.05	1.15

therefore the Ukrainian assessment is also using outputs from the GF01 model.

Table 15.2 presents adjustment statistics averaged for Ukraine from the GF01 model.

In addition, a joint research effort between the Russian National Hydrological Institute and the Ukraine used the Max Planck Institute transient model (MPtran) to assess water resources vulnerability. An annual average temperature increase of 2.6°C under 2×CO_2 conditions is estimated using this model.

Socioeconomic Baseline Scenarios

The ongoing changes in the economy of Ukraine during its transition to a market economy do not allow accurate predictions of socioeconomic factors. Therefore, it is expedient in the preliminary assessments to use the Pepper et al. (1992) projections for a baseline socioeconomic scenario for Ukraine.

Pepper et al. (1992) does not take into account some peculiarities of individual countries. Specific socioeconomic and ecological factors for each country should be

Table 15.3. Forecast of GDP Growth in Ukraine

	1990	1995	2000	2025	2050	2075
Pepper et al. US$Bln	60.0	31.0	31.0	68.1	98.8	129.9
Kondrat'yev US$Bln	60.0	31.0	35.0	109.0	145.0	275.0

Table 15.4. Forecast of the Population Growth in Ukraine for 1990-2075

Year	Population (thousands)
1990	51,770
1995	51,728
2010	51,500
2030	51,410
2050	51,415
2070	51,412
2075	51,415

used to enhance the accuracy of the forecasts. Comparing forecasted GDP growth in Ukraine generated from Pepper et al.'s data for the European part of the former USSR with the forecast derived from the Kondrat'yev Heavy Waves method (Table 15.3), it is obvious that there are significant differences although general tendencies are maintained.

Forecasts of population growth were created based on 1995 population figures, as discrete models for the following periods: 1996-2029 and 2030-2075. The results are presented in Table 15.4.

The most recent 10-year cycle of birth, death, and net migration rates are assumed to continue during the next cycle from 1996 to 2029. The forecast for the period from 2030 to 2075 was derived from the theory of stable population and intrinsic rates.

Figure 15.4. Area of Forest Stands in Each Natural Zone

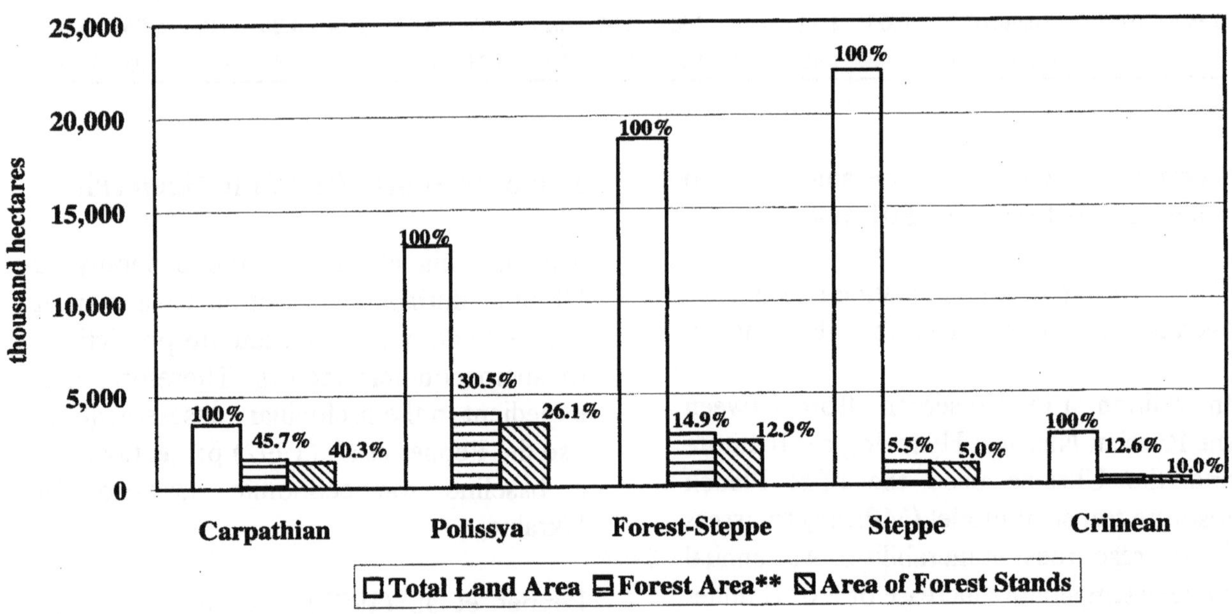

[1] Percentage of "Total Land Area" is shown next to the bar for each land type in each climate zone.

[2] Includes forest openings, glades, burned areas, etc.

Forestry Sector

Methods Used

Selection of Study Sites

The forested area of Ukraine is about 10 million ha. The distribution, growth, and condition of the forests are determined primarily by climate conditions that are generally continental temperate except for the narrow strip along the southern Crimean coast which has a subtropical climate. In the northern and western areas of Ukraine, forest land comprises 30% to 55% of the territory. In the central area, forest land comprises 10% to 20% of the territory and in the southern and eastern areas it comprises less than 10%.

Forests are present in each climate zone, but the forest area percentage (percent of area in zone covered by forest stands) differs from zone to zone. The highest forest area percentage is in the Carpathian region (40%). The average forest area percentage in Polissya, in the zone of mixed forest, is 26%; in the forest-steppe zone, it is 13%; in the Crimean region, it is 10%; and in the steppe zone, it is 5% (Figure 15.4).

Forest Ecosystem Models Used

The preliminary assessment of forest vulnerability used the Holdridge Life Zone Classification Model. Existing Holdridge zones were estimated in the Ukraine based on observed data, and the GFD3 and GISS scenarios were used in conjunction with the Holdridge model to estimate changes in the distribution of forests. This information will be used in future analyses to develop forest gap models for the Ukraine.

To verify the Holdridge model, the global climate database, applied at regular spatial grid, was used. The following Holdridge zones were defined within the Ukrainian territory:

- boreal moist forest — in the northern part of the Sumy region (eastward from Semyonovka and northward from Konotop)
- boreal wet forest — a small area at the southwestern part of Ukraine — Zakarpat'ye region, southern part of the Ivano-Frankovsk region, and western part of the Chernovtsy region
- cool temperature wet forest — Sumy, Chernigov, Zhitomir, Kiev, Vinnitsa, and all the western regions of Ukraine, except for the areas in the boreal wet forest zone
- cool temperature steppe — the northern border passes through the northern part of the Kharkov region, C. Gadyach, the northern part of the Cherkassy region, up to Borispol and down to the eastern boundary of Vinnitsa region.

Calculated from the global climate database, the Holdridge zones for Ukraine correspond with natural climatic zones obtained from physical-geographical Ukrainian maps, but the cool temperate steppe zone includes the forest-steppe, steppe, and Crimean Mountains, which have very different climates.

To identify the climate changes that may affect the forests of Ukraine, actual observation data for 1951-1980 from 18 meteorological stations of the National Committee on Meteorology of Ukraine, situated in

Polissya (mixed forest zone), forest-steppe, steppe, and Crimean zones, were used. Probable changes of climatic factors were determined using the GISS and GFD3 model outputs.

In future analyses, gap models will be used to forecast forest growth (changes in its species composition, structure, and productivity) for the forest land where the Holdridge life zones will undergo changes according to forecasts. The definition of input parameters for gap models will be somewhat difficult because Ukraine uses different assessment indices for plant response to climate change compared to those used in suggested models.

In addition to the methods described above, Ukraine national forestry methods of assessment of correlation between forests and climate were also used in this work. They are briefly described in the following text box.

The indices used as part of the Ukraine national forestry methods were calculated for all the indices for which there were factual meteorological data or data predicted on the basis of climate change scenarios (obtained from the State Committee on Hydrometeorology of Ukraine). For regions where climate changes occur according to scenarios, productivity and structure of vegetation are estimated by analogy methods.

Preliminary Results

Holdridge Model Results

The current climate of Ukraine is generally favorable for forest growth, but climate change could alter forest migration, and transform forest types, plant species of plantations, and their biological variety, productivity, and stability.

A map of the distribution of forests in Ukraine within the Holdridge life zones, obtained using meteorological observations from 18 meteorological stations, is presented in Figure 15.5. Data on changes of Holdridge life zones, derived from the results of two climatic scenarios (GISS and GFD3) are shown in Figures 15.6 and 15.7. Since this analysis was completed, work was undertaken to revise the analysis using data from 66 meteorological stations. It is possible that these results may change significantly.

The Holdridge life zones were analyzed to determine the areas that would undergo significantly changed climatic conditions. It is estimated that Polissya and the northern part of the forest-steppe zone would remain stable if the climate changed. As shown in Table 15.5, under the GFD3 scenario, total forest area would decrease by 39%(3.5 million ha). The lost forest area would contain 32% (443.7 million m^3) of total timber volume. Under the GISS scenario, a decrease of 47% (4.4 million ha) with 44% of timber volume (607.1 million m^3) is forecasted.

Discussion and Conclusions

The results presented here are preliminary and further research is under way to revise this analysis based on meteorological data from 66 stations. In addition, forest gap

Table 15.5. Estimated Changes in Forested Area Using the Holdridge Model

Scenario	Change in Forested Area (thousand ha)	% change
GFD3	3,499	-39
GISS	4,328	-47

Ukraine National Forestry Assessment Methods

Climate determines types of forest plots, and affects the structure and productivity of plantations. The typology of forest plots adopted in Ukraine is based on the classification of Pogrebnyak, which was further developed by Vorobyev (1961). It classifies forest plot type by soil fertility and moisture. There are four trophotopes or fertility groups defined according to degree of soil fertility (in ascending order): (A) sandy soils with mostly pine species, (B) more fertile sandy and subloamy soils, (C) subloamy soils, and (D) clay loamy soils.

There are six gygrotopes or moisture groups defined according to a degree of soil moisture: (0) excessively dry, (1) dry, (2) fresh, (3) moist, (4) damp, and (5) wet or marshy.

The formation of forest types is related to climatic factors: temperature, moisture, continentality of climate, prevailing winds, etc. The species of trees and other plants are considered as indicators of forest plot types under existing climatic conditions.

To assess forest response to climate change, the following indices are used:

- Sum of the positive monthly average temperatures (T) determines thermal zones, which correspond to the formation conditions of trophotopes (soil fertility groups). Many northern boundaries of natural habitats of tree and shrub species are related to total temperatures. Boundaries of zones are isotherms of index T with intervals of 20° beginning at the 4° isotherm.
- Climate moisture index according to Vorobyev: $W = R/T - 0.0286T$, where R is total precipitation over the months with average temperature above 0, and T is the sum of the monthly positive average temperatures, which defines climate moisture zones corresponding to soil moisture zones. Boundaries of zones are geographic isolines of index T with intervals of 1.4 beginning from the driest climate. Boundaries of natural habitats of nearly all species are related to moisture zones. Productivity of plantations and processes of natural regeneration are different in different zones, which is important for planning of forestry activities.
- Climate continental index as reflected in the difference between average temperatures for July and January (A), which determines continental zones separated at intervals of 5° beginning from 0°. Boundaries of these zones cross thermal zones and moisture zones, and correspond to average and maximum temperatures of the vegetation period, its duration and beginning date, average and minimum temperatures, etc. Boundaries of natural habitats of tree species are also related to this index.
- Yearly sum of temperatures above 10°C (T10).
- Total precipitation over months with average monthly temperatures above 10°C (R10).
- Hydrothermal factor according to Selyaninov, $10 \times R10/T10$.
- Yearly sum of temperatures above 0°C.
- Total annual precipitation.
- Average yearly temperature.

Figure 15.5. Distribution of Ukrainian Forests over Holdridge Zones on the Basis of Observed Data

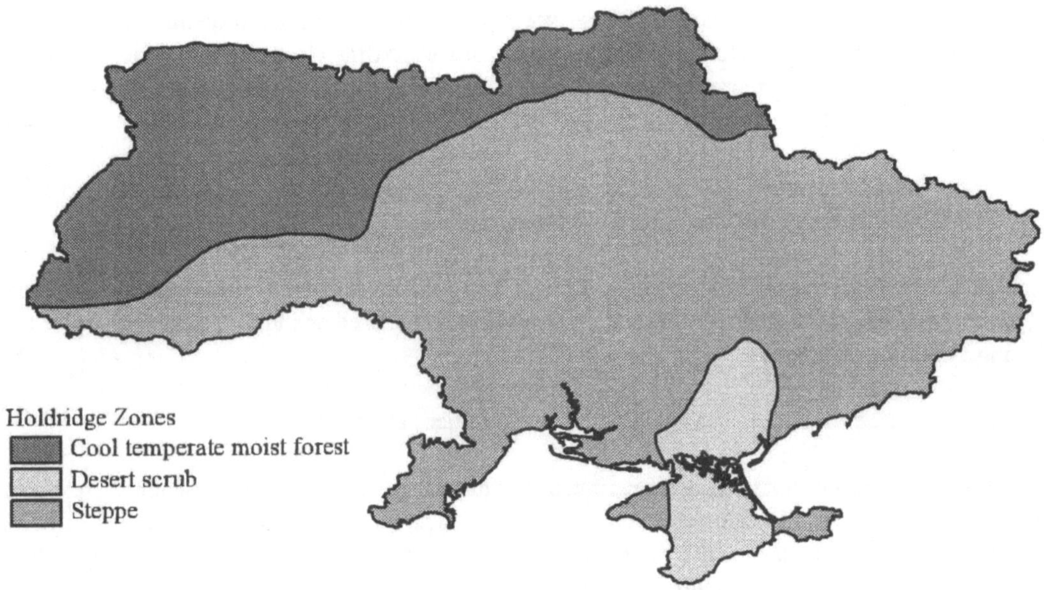

Figure 15.6. Distribution of Ukrainian Forests over Holdridge Zones on the Basis of GISS Model

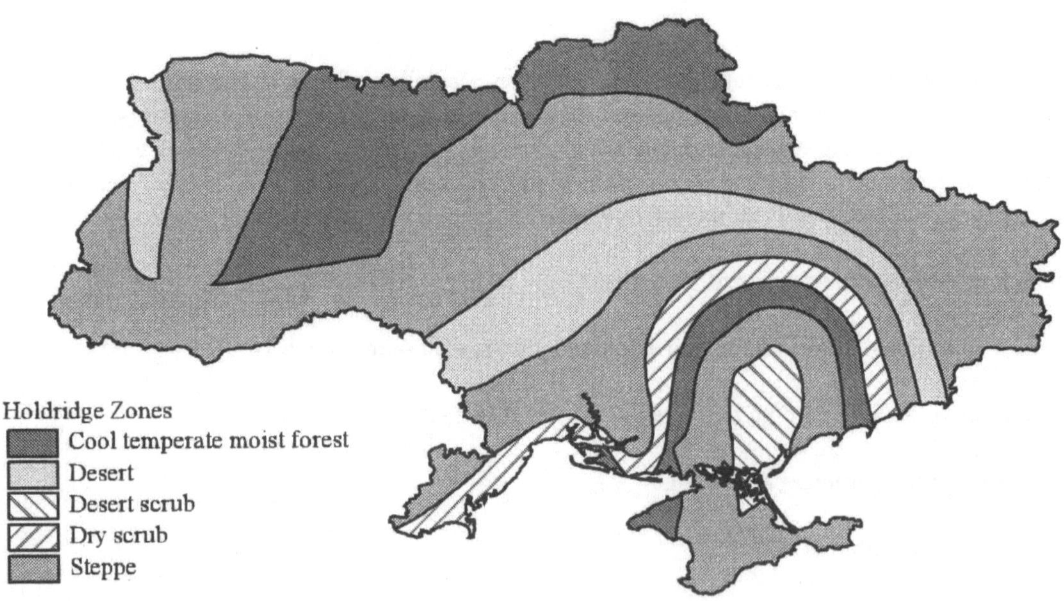

Figure 15.7. Distribution of Ukrainian Forests over Holdridge Zones on the Basis of GFD3 Model

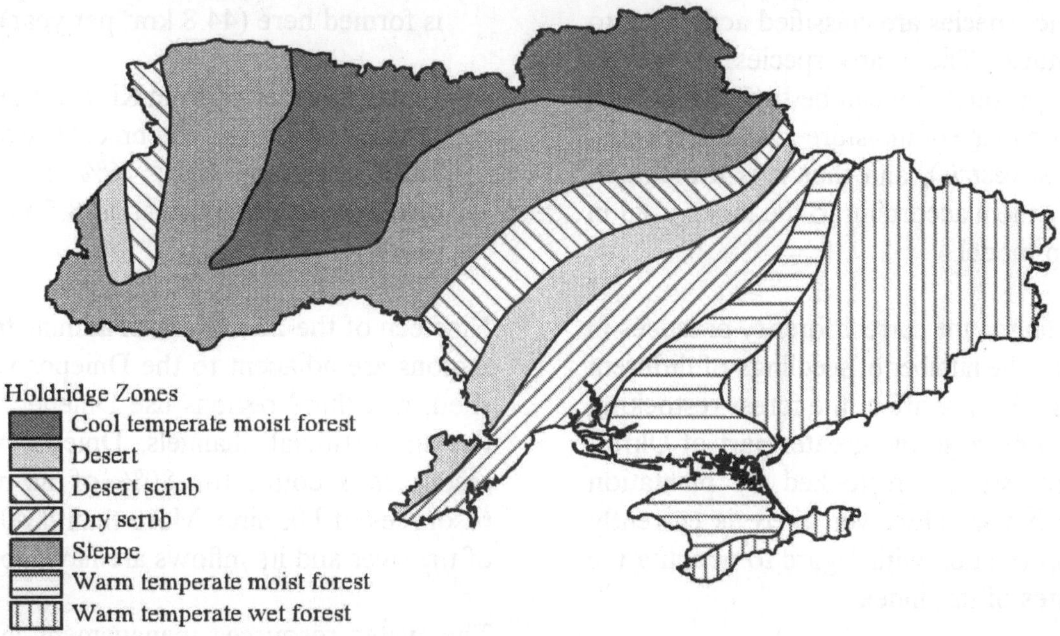

models will be used to further analyze the effect of climate on forests. However, there are several problems when adapting gap models to Ukrainian conditions.

Forest vulnerability includes the forests' abilities to resist droughts, changes in the length of the vegetation period, changes in soil moisture, and increased pests. Climate changes will shift the primary forest types.

Climate change could increase fire frequency, growth rate change, and mortality of different trees species. Changes in species competition and success will affect forest structure. Warmer winters could cause increases in parasite insects. It should be noted that there are uncertainties in how forests respond to regional climate changes and in what the potential role of forestry is in mitigating the effects of climate change.

Research on the biophysical and socioeconomic consequences of climate change for the forest sector is an important national task. The results will help improve long-term planning of future forestry.

For some forest-tree species, temperature indices of vegetation (minimum and maximum temperature sums) have not been established. The natural habitats of forest species spread beyond Ukrainian borders, and the temperature boundary in the south is virtually absent. For example, the spread of beech (*Fagus sylvatica*) and hornbeam (*Carpinus betalus*), whose eastern boundary lies in Ukraine, is checked by the temperature amplitudes of January and July (index of continentality) rather than by total temperature.

For the input data of gap models, drought (which is related to moisture availability) is calculated as the proportion of a growing season that a species can tolerate soil moisture below or at wilting point. In the Ukraine, species are classified according to this index. There are species, however, whose productivity can be sufficiently high in a wide range of moisture availability (e.g., *Pinus silvestris*), and thus it is difficult to classify them according to the index used in the gap model.

According to the current forestry practices in Ukraine, the number of seedlings of different species does not affect the forest restocking process because the greater part of Ukrainian forests are restocked by plantation establishment. However, there is currently little information with regard to quantitative estimates of this index.

WATER RESOURCES SECTOR

Methods Used

Selection of Study Sites

The Dnieper River is the largest river in Ukraine. Its length is 2,180 km, and its catchment area is 510.2 thousand m^2. The Dnieper runs through Russia, Belarus, and Ukraine before emptying into the Black Sea. Its average runoff equals 53.3 km^3 per year; its highest runoff was 71.5 km^3 in 1933, and its lowest was 19.0 km^3 in 1921. Its runoff regime is characterized by high freshets and comparatively low water from summer through winter. The Dnieper watershed in Ukraine occupies 291.4 thousand km^2 (48% of Ukrainian territory).

According to physico-geographical peculiarities the watershed can be divided into two parts:

- Upper Dnieper — from its source to Kiev, with a catchment area of 328 thousand km^2; 84% of annual runoff is formed here (44.8 km^3 per year)
- Lower Dnieper — from Kiev to the river mouth, with a catchment area of 182.2 thousand km^3; 16% of annual runoff is formed in this area (8.5 km^3 per year).

Nineteen of the 25 Ukrainian administrative regions are adjacent to the Dnieper watershed, and three regions use Dnieper water through artificial channels. Dnieper water resources account for 80% of all water resources of Ukraine. More than 8,000 km of the river and its inflows are navigable.

The water resources management system includes:

- 1,923 large and 18,493 small water courses with a total length of 104.6 thousand km
- a cascade of 6 large and 90 medium water reservoirs, and a water supply and sewerage system for 50 large cities and industrial centers
- 53 large irrigation systems with a watering area of 1.5 billion ha, 6 hydropower stations, 7 thermal power plants, 4 city power stations, 4 nuclear power plants, and 30 large fisheries.

The gauging sites of the Kiev, Kremenchug, and Kahovka hydrojunctions (groups of water engineering structures and related water bodies managed as a unified system) were examined in the work presented here.

Hydrologic and Water Resources Methodologies

To assess the effects of climate change and human activity on water resources, the following models were used:

- a model of natural flow recovery
- a model of river runoff hydrograph, based on meteorological data
- an assessment of the ratio of total water demand to total water supply
- a model of water reservoir management.

The natural flow recovery methodology was developed at the Ukrainian Research Hydrometeorological Institute in Kiev (Shereshevsky, 1986). This method is based on accounts of spring freshet waves moving along the main rivers of upper parts of the Dnieper river basin to Kiev, with further calculation of water movement along the main river gauging site to the river mouth, taking into account local inflows. The basis of spring runoff assessment is an engineering calculation of water discharge. For each site the water balance equation has being solved using channel capacity and lateral inflow change. Estimating the Dnieper's natural runoff for low-flow periods is based on calculations of flow rates at upper gauging sites rather than regulated flow sites (upstream from Kiev), taking into account lag time flow to gauging sites, lateral inflow, precipitation on water surface, evaporation of water, percolation, and irretrievable water use in reservoirs.

The model of river runoff hydrograph was developed by the Russian National Hydrological Institute (Georgievsky and Ezov, 1994). On the basis of decadal data on temperature, precipitation, and moisture, the main components of the hydrological cycle — evaporation, soil moisture, groundwater capacity, and surface and groundwater runoff — are calculated. The model also has components that account for the soil freezing effect, and divides the underground aquifer capacity into two parts: a lower semiconfined aquifer capacity with a stable water regime, and an upper semiconfined aquifer capacity with a dynamic water regime. The Russian National Hydrological Institute's model was used because it considers the local conditions of Dnieper runoff.

Because Ukraine is in the stage of transition from a centralized-planned economy to a market economy, it is not yet possible to obtain a reliable forecast of changes in water resources demand. However, the ratio of total water demand to total water supply was estimated, holding demand constant because of the difficulty in forecasting future demand.

A Dnieper water resources management model was developed at the Institute of Mathematical Machine and Systems Problems, Cybernetic Center of Ukrainian National Academy of Sciences, in Kiev (Zheleznyak et al., not dated). Because reservoir management practices will have to be adjusted in response to changes in climate conditions, future water management is estimated based on an optimizing model of reservoir cascade management. This model enables calculations of the operating regime under different water management or economic criteria. The model computes a reservoir's working regime to account for water resources management and ecological criteria. Time series of temperature and precipitation (1946-1991) constructed from data from 10 meteorological stations in the Dnieper river basin were used. The Ukrainian State Committee on Water Resources Management System provided data on water demand under current conditions.

Preliminary outputs of climate change scenarios from the GFDL (transient and equilibrium), UKMO (equilibrium), and the Max Planck Institute for Meteorology (MPtran; Cubasch et al., 1992) models were used.

Methodology for Economic Analyses
A preliminary estimate of hydropower energy losses due to changed water flow (according to the transient climate scenarios from GFDL and Max Planck Institute) was made by examining the relationship between energy production and annual runoff for six cascade power stations.

Preliminary Results

Hydrologic Vulnerability Assessment
An estimate of the hydrological vulnerability of Ukraine's water resources was developed using data from three main gauging sites in the Dnieper river basin and three climate change scenarios. The results are presented in Table 15.6. It should be noted that the percentage change in runoff determined by these scenarios differ significantly both in sign and in value. The distribution of the ratio of annual storage capacity to annual runoff is from 0.05 to 1.08.

The change to runoff at the gauging site of the Kahovka hydrojunction for observed conditions and the GFD3, UKMO, GF01, and MPtran climate change scenarios (ignoring changes in management practices) are presented in Figure 15.8.

Annual average values do not fully describe runoff changes for the climate change scenarios. The annual distribution for various climate change scenarios should also be considered. Figure 15.9 presents the dynamic changes in water supply across months at the Kahovka hydrojunction gauging site for an average water flow year for each scenario.

Table 15.6. Hydrologic Vulnerability Assessment for Dnieper River Basin

Basin Site	Percent Change in Precipitation	Change in Temperature (°C)	Percent Change in Runoff	Ratio of Annual Storage Capacity to Annual Runoff
Kiev Hydrojunction				
GFD3	+19.4%	+4.4	+31%	0.09
UKMO	+16.2%	+6.0	+128%	0.05
GF01 (2075)	—	—	-16%	0.14
Kremenchug Hydrojunction				
GFD3	+19.4%	+4.4	+27%	0.38
UKMO	+13.8%	+5.8	+107%	0.23
GF01 (2075)	—	—	-20%	0.58
Kakhovka Hydrojunction				
GFD3	+17.6%	+4.4	+25%	0.72
UKMO	+10.5%	+5.0	+101%	0.45
GF01 (2075)	—	—	-20%	1.08

Figure 15.8. Changes in Water Runoff for Gauging Site at Kakhovka Hydrojunction of Dnieper River

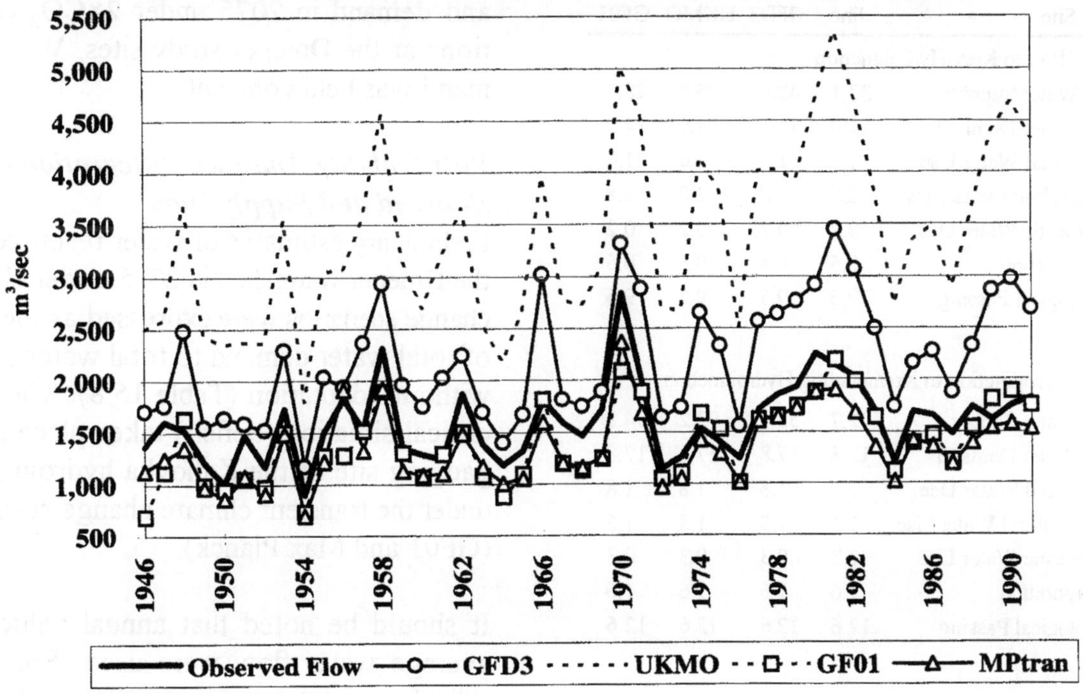

Figure 15.9. Annual Distribution of Water Runoff for Gauging Site at Kakhovka Hydrojunction of Dnieper River

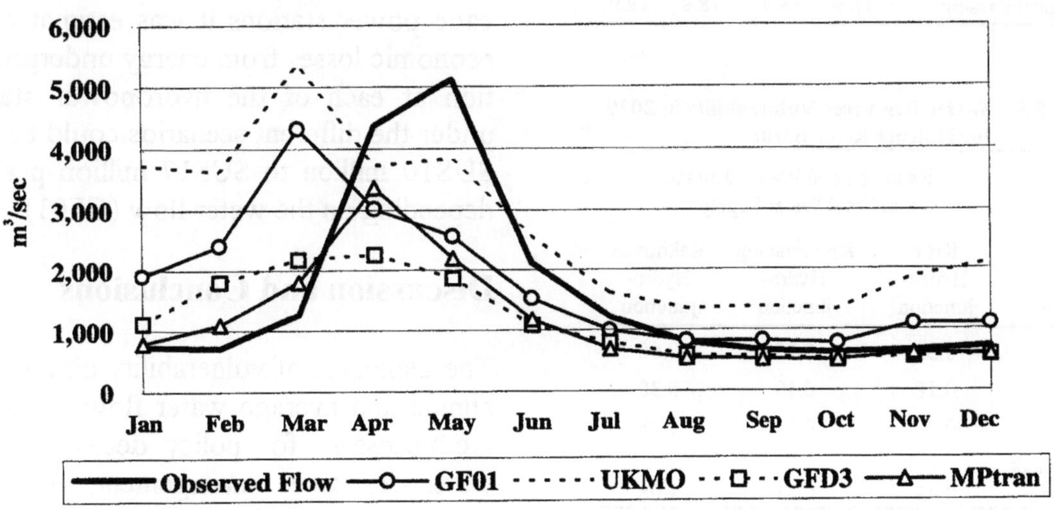

Table 15.7. Water Resource Supply and Demand in 2075 in km³/yr for Dnieper River Basin

River Site	Base-line	GFD3	UKMO	GF01
River Head to Kiev Hydrojunction				
Total Water Supply	33.1	42.6	75.7	28.5
Total Water Demand	14.0	14.0	14.0	14.0
Industrial Water Use	1.0	1.0	1.0	1.0
Agricultural Water Use	2.8	2.8	2.8	2.8
Domestic Water Use	0.2	0.2	0.2	0.2
Evaporation	0.5	0.5	0.5	0.5
Ecological Passing	9.5	9.5	9.5	9.5
Kiev Hydrojunction to Kremenchug Hydrojunction				
Total Water Supply	42.7	55.4	93.2	34.8
Total Water Demand	17.8	17.8	17.8	17.8
Industrial Water Use	1.8	1.8	1.8	1.8
Agricultural Water Use	1.5	1.5	1.5	1.5
Domestic Water Use	0.3	0.3	0.3	0.3
Evaporation	1.6	1.6	1.6	1.6
Ecological Passing	12.6	12.6	12.6	12.6
Kremenchug Hydrojunction to Kakhovka Hydrojunction				
Total Water Supply	41.3	54.3	92.8	32.8
Total Water Demand	35.6	35.6	35.6	35.6
Industrial Water Use	3.4	3.4	3.4	3.4
Agricultural Water Use	8.8	8.8	8.8	8.8
Domestic Water Use	2.3	2.3	2.3	2.3
Filtration	0.3	0.3	0.3	0.3
Ecological Passing	18.9	18.9	18.9	18.9

Table 15.8. Water Resources Vulnerability in 2075 for Dnieper River Basin

Scenario	Ratio of Total Water Demand to Total Water Supply: Kiev Hydro-junction	Kremenchug Hydro-junction	Kakhovka Hydro-junction
GFD3	0.32	0.32	0.66
UKMO	0.18	0.19	0.38
GF01	0.49	0.51	1.08
Max Planck Transient	0.49	0.51	1.08

Water Resources Supply and Demand Models by Climate Change Scenario

Table 15.7 presents water resource supply and demand in 2075 under 2×CO_2 conditions at the Dnieper study sites. Water demand was held constant.

Vulnerability Analyses: Integration of Demand and Supply Data

Preliminary estimates of water resources for the Dnieper watershed in 2075 under climate change scenarios were expressed as the ratio of total water demand to total water supply without adaptation (Table 15.8). The most critical situation (deficit) takes place at the gauging site of the Kahovka hydrojunction under the transient climate change scenarios (GF01 and Max Planck).

It should be noted that annual values for average water flow were used. For more reliable estimates, an investigation of water resource supply and demand ratios for various water flows throughout the year is planned.

Economic Analyses

Based on the energy production at six cascade power stations it was estimated that economic losses from energy underproduction at each of the hydropower stations under the different scenarios could be from $US10 million to $US17 million per year depending on the water flow (GF01).

Discussion and Conclusions

The estimates of vulnerability obtained for annual and average water flow conditions are necessary for policy decisions about water resources management. They are critical for assessing agricultural impacts, and for designing water resources management systems. In many cases it can be supposed that changes in extreme hydrological conditions would have more significant effects than changes in average hydrological conditions.

Further investigations are planned to analyze the effects of climate change and human activity on seasonal and monthly water resources estimates and to determine potential deficits under different water flow conditions at various Dnieper sites.

Coastal Resources Sector

Methods Used

Selection of Study Sites

The study sites include all of the Black and Azov sea coasts within the Ukrainian territory: the Black Sea lowland, the south Crimean coast, the Kerch peninsula, and the Azov seaside area. The entire shore line was considered because it is heterogeneous in geological, geomorphologic, and litodynamical terms. Investigations were carried out at sites where data have been collected for no less than 10 years.

Sea Level Rise Methodologies Used

Instrumental geomorphologic surveys of specific types of coastal zone sites (accumulative, erosion-landslide, erosion-crumbling) were used to estimate land losses. Data for 10 years collected at the beginning and end of summer were used. These data were interpreted based on the specific type of coastal zone at each observation site. The study included:

- the analysis of temporal coastal changes
- calculations of losses from erosion and abrasion.

During the final stage of this study, attention will be focused on scenarios with sea level rise of 0.5 m and 1.0 m. The calculations for these scenarios have not yet been completed.

Methodology for Assessment of Economic Losses

The only economic consequences of sea level rise estimated so far have been those of the flooding of agricultural lands. Losses from building destruction, loss of recreational areas, and soil salinization will be assessed in the final stage of this study.

Adaptation Alternatives and Evaluation Methods

A sea level rise of 2.0 m in the next century was considered and adaptation alternatives were evaluated using the Adaptation Strategy Evaluator (ASE) (Smith and Chu, 1994).

Preliminary Results

Data from the landslide mitigation stations of the Institute of Geological Sciences, Odessa University, at Odessa and Yalta were used to determine erosion rates for each coast type and soil loss volumes (for 1 m of coastal zone) at the current sea level rise of 1.8 mm per year. The cliff erosion rates of the main landslide sites are presented in Table 15.9. This table also shows that the erosion rates for bank types of coast are unequal and depend on bedrock composition and cliff height.

Table 15.9. Average Cliff Erosion Rates on Landslide Sites of the Black Sea

Northwestern Black Sea	Lat.	Long.	Erosion Rate
Burnasski	45°50'	30°10'3"	2.0
Balabanovski	45°53'5"	30°15'	0.5
Kurortnyi	45°56'	30°18'	1.1
Dniestrovski	46°08'	30°31'	0.7
Sanzheiski	46°14'	30°37'3"	1.4
Suholimanski	46°20'	30°41'3"	1.0
Bol'shefontanski	46°23'5"	30°45'4"	0.7
Kryzhanovski	46°33'5"	30°48'1"	0.8
Leskinski	46°34'	30°49'	1.7
Fontanski	46°33'5"	30°51'6"	0.9
Vapnyarski	46°34'	30°54'6"	0.7
Dophinovski	46°34'6"	30°55'2"	0.9
Gvardeiski	46°35'	30°57'1"	1.3
Chabanski	46°35'3"	30°58'7"	1.4
West-Grigor'evski	46°35'7"	31°06'	0.6
Grigor'evski	46°36'3"	31°2'3"	1.3
Stepnoi	46°36'7"	31°05'	1.7
Sychavski	46°37'	31°7'8"	1.8
Karabushski	46°36'6"	31°15'6"	1.1
Urozhaini	46°36'5"	31°18'7"	1.7
Adgiyaski	46°35'2"	31°20'9"	0.8
Chernomorski	46°37'7"	31°27'5"	2.2
Ochakovski	46°37'3"	31°31'8"	0.3
Kinburnski	46°30'7"	31°36"	0.4
Karkinitski	46°06'	31°27'	2.1
Perekopski	46°04'	33°38'	0.5
East-Stereguschenski	45°46'	33°19'	2.7
West-Stereguschenski	45°41'5"	33°07"	2.0
Tarhkankutski	45°20'6"	32°31'7"	0.1
Kisilyarski	45°01'2"	33°36'1"	2.4
Nikolayevski	44°57'2"	33°36'7"	0.2
Uchkuyevski			0.5
Lyubimovski	44°41'6"	33°35'4"	1.3
Balaklavski	44°34'8"	33°29'	0.02
Sarychski	44°23'4"	33°45'	0.2
Aidanil'ski			0.1
Sudakski	45°50'	35°02'	0.3
Primorski	45°7'3"	35°31'	0.2
Kiikatlomski	45°57'	35°23'4"	0.03
Chaudinski	44°59'9"	35°51'	0.05
Yakovenkovski	45°04'	36°22'	0.08
Zavetnovski	45°8'5"	36°20'3"	1.5
Geroeyvski	45°14'	36°24'5"	0.1
Perepravochni	45°22'	36°37'	1.4
Yurkinski	45°25'5"	36°34'	2.7

For this preliminary assessment, estimated flooding and soil salinization data were also used, but the primary losses are expected from erosion and submergence. For example, at the Voikovo collective farm, 12 ha of arable land were destroyed in 1981. Although erosion rates that year were unusually high, this part of the coast accounts for only 1% of the entire length of cliffs that are being eroded. A comparison of observed data from Odessa's site with the estimated rate of sea level rise shows that these values, although comparable, differ in that the current Black Sea rise lags behind ocean level rise. However, because of reduced river runoff and increased Mediterranean Sea inflow into the Black Sea, this difference is estimated to disappear. It has been forecasted from available data that the levels of the Black and Mediterranean Seas will be the same by the middle of the next century.

It is estimated that without protection wetlands loss for the Black Sea coastline would be 32 km^2 and 370 km^2, under 0.5 m and 2.0 m sea level rise scenarios, respectively.

Discussion and Conclusions

The information presented here is preliminary and does not cover all possible scenarios. A complication is that various factors influence erosion rate and the assessment of the role of sea level rise. In Ukraine's final country study report, this material will be included as well as geologico-geomorphic maps of the sea coast of Ukraine.

Assessments obtained for the entire Black Sea and Azov Sea coastlines will be the basis of state policy for both near- and long-term perspectives. Critical parts of the assessment will be a forecast of the scale of environmental changes, identification of the most sensitive sites, plans for coastal protection, and examination of the locations of planned municipal and hydrotechnic construction. Cartographic and prognostic results for two scenarios, average sea level rise and maximum sea level rise, will be presented in the final assessment.

Summary Conclusions

Future research efforts in the Ukraine will focus on obtaining reliable quantitative values describing the connection between potential biophysical impacts of climate change and the consequences for socioeconomic activities. This information will be important in developing adaptation strategies for the country.

References

Benioff, R., S. Guill, and J. Lee (eds.) (In press). *Vulnerability and Adaptation Assessments: An International Guidebook.* Dordrecht, The Netherlands: Kluwer Academic Publishers.

Crutcher, H.L. and J.M. Meserve. 1970. Selected Level Height, Temperatures and Dew Points for the Northern Hemisphere. NAVAIR 50-1C-52, Chief of Naval Operations. Washington, DC.

Cubasch, U., K. Hasselmann, H. Hock, E. Maier Reimer, U. Mikolajevicz, B.D. Santer, and R. Sausen. 1992. Time-Dependent Greenhouse Warming Computations with a Coupled Ocean-Atmosphere Model. *Climate Dyn.*, 8, 55-69.

Georgievsky, V. and A. Ezov. 1994. The Possible Changes of River Runoff and River's Regime under Different Scenarios of the Climate Warming. GGI, 32 pp.

Pepper, W. et al. 1992. Emission Scenarios for the IPCC: An Update. Assumptions, Methodology, and Results. Prepared for the Intergovernmental Panel on Climate Change, 115 pp.

Schutz, C. and W.L. Gates. 1971. Global Climate Data for Surface, 800 Millibars, 400 Millibars. R-915-ARPA, Rand Corporation.

Smith, A. and H. Quan Chu. 1994. *A Multicriteria Approach for Assessing Strategies for Anticipatory Adaptations to Climatic Change.* Washington, DC: Decision Focus, Inc.

Shereshevsky, A. and V. Voitsechovich. 1986. The Change of the Dnieper River Runoff under Influence of Human Activity. *Proc. UkrNIGMI*, Issue 211, pp. 22-36.

Taljaard, J.J., H. van Loon, H.L. Crutcher, and R.L. Jenne. 1969. Climate of the Upper Air: Southern Hemisphere.1: Temperatures, Dewpoints, and Heights at Selected Pressure Levels. NAVAIR50-1C-55, Chief of Naval Operations. Washington, DC.

Vorobyev. 1961. Forest Typical Classification of Climate. *Transactions of Kharkov Institute of Agriculture.* Kharkov, pp. 235-250.

Zheleznyak, M. et al. Not dated. Of Radionuclide Transport in the Set of Dnieper's Reservoirs, Intern. Seminar on Freshwater and Estuarine Radioekology. Lisbon, held 21-25, March.

Chapter 16

Vulnerability and Adaptation Assessments for Peru

Nestor Teves,[1] Gustavo Laos,[2] Carmen San Roman[1]
Sulma Carrasco,[3] Luis Clemente[1]

[1] **Universidad Nacional Federico Villarreal**
[2] **Dirección de Hidrografía y Navegación — Ministerio de Marina**
[3] **Instituto del Mar del Perú**

SUMMARY. The perception that the mostly cliff-lined Peruvian coast would escape any problems caused by a rise in sea level has changed. It is now thought that an increase in sea level caused by global warming could affect almost 30% of Peru's coastline. Some cities like Puerto Pizarro, Chimbote, Pisco, and Ilo could be in danger of flooding. In addition, almost 6 million people traveling to the coastline to enjoy the many summer resorts and recreational areas would be affected. No measures to adapt to the sea level rise have yet been taken because only limited information is available about global climatic change and the consequences of these changes. Some protection measures against marine erosion have been developed along the Lima coast, such as the breakwaters and groins to defend La Punta, a small city of 6,000 inhabitants that is 1.5 m above sea level. In addition, groins constructed at right angles to the coastline allow for the expansion of beaches along the Costa Verde.

INTRODUCTION

Peru has three natural regions: the coastal region (10.7% of the total area), the Andean region (31.8%), and the Amazonian region (57.5%). The area of Peru is 1,285,226 km^2, and the coastline is 3,079.5 km long (Figure 16.1).

The Peruvian coast is made up of alluvial and fluvioalluvial deposits and small percentages of marine, aeolian, and gravity deposits. Most of the coast is lined with cliffs, but some cities like Puerto Pizarro, Chimbote, Pisco, and Ilo occupy lowlands that would be in danger of flooding. In addition, summer resorts and recreational areas in these low-lying coastal areas could be flooded if the sea level rose.

The altitude of the coast ranges from 0 to 500 m above sea level. The coast is defined

Note: This chapter does not contain a detailed description of the assessment methodologies that were recommended by the U.S. Country Studies Program. Refer to Chapter 2 for on overview of methodologies.

J. B. Smith et al. (eds.), Vulnerability and Adaptation to Climate Change, 335–346.

Figure 16.1. Map of Peru

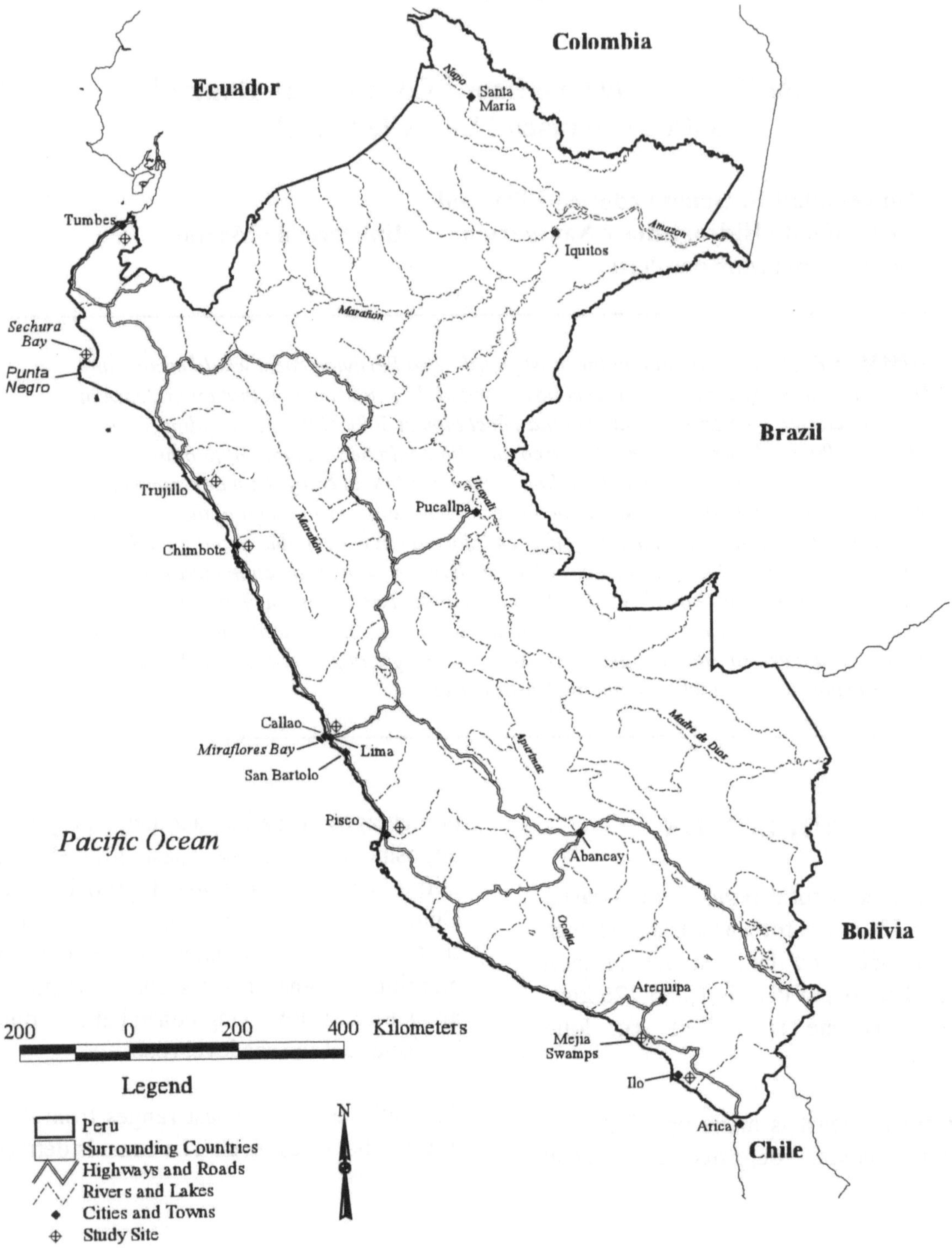

by the area from 03°23'S to 18°20'S latitude. There are 53 rivers that run through the coast and discharge into the Pacific Ocean. The main types of coastal landscapes are desert plains, fluvial and marine terraces, alluvial fans, isolated hills and rocky massifs, and dune fields (Teves, 1976). Most of the annual discharge from rivers (60% to 70%) occurs in a 3-4 month period and reaches 34.6 million m^3 (1.7% of the country's total discharge). The rivers' waters are mainly used for agriculture (12.0 million m^3) or for population needs (0.7 million m^3).

The coastal climate can be described as subtropical desert with minor variations. The temperature ranges from 10°C to 30°C with averages from 16°C to 20°C. Precipitation is generally low along the coast except in Tumbes, near the border with Ecuador. Along most of the coast, clouds are prevalent. The stratus cloud is near the surface and can be 800 m thick (Sanchez, 1976). The relative humidity is generally 98%, and the prevailing winds are from the southeast (Wirfel, 1980).

According to the 1993 census (INEI, 1993), the Peruvian population was 22,128,466 in the 24 political subdivisions (departments) and a constitutional province. The population in the coastal region was 11,558,204 (52%).

Some places along the coastline have been established by special laws as national sanctuaries, such as the Tumbes Mangroves (6,000 ha), the Virrila Estuary (1,500 ha), the National Reservation of Lachay (5,500 ha), the National Reservation of Paracas (335,000 ha), and the National Sanctuary of Mejia Swamps (690 ha).

To identify which sectors in Peru are vulnerable to climatic changes, we started with the assumption that the average global temperature is going to increase between 2.0°C and 3.0°C within the next century. The sectors that could be the most affected by global climate change, besides coastal resources, are agriculture, fisheries, and wildlife.

The coastal resources sector was selected for discussion in this chapter because a multidisciplinary group of scientists integrated through the Peruvian Group of Climatic Change of the Southeast Pacific Permanent Commission (CPPS) has been working on this sector since 1992. With the support of the U.S. government from 1993 to the present, the group has been studying the selected sites in danger of being flooded by sea level rise.

SCENARIOS

Sea Level Rise Scenarios

The scientific community is very concerned about the potential for sea level rise caused by global warming. A global sea level rise of 0.5 m in the next century is most probable, but uncertainties continue to exist about the real rate (Kjerfve, 1991). In accordance with the suggested criteria of the Intergovernmental Panel of Climatic Change (IPCC), Peru has used two scenarios to present this work: 0.3 m as a minimum and 1.0 m as a maximum (IPCC, 1990, 1991, 1992) (Table 16.1).

Table 16.1. Sea Level Rise Scenarios by 2100

Scenario	Sea Level Rise
Low (SLR1)	0.3 m
High (SLR2)	1.0 m

Baseline Socioeconomic Scenarios

Peru is undergoing the process of industrial conversion after many years of basing its hopes for economic development on the exportation of raw materials.

After the recession due to economic adjustment, stabilization and structural changes have taken place in the economy. Since 1993, the economy has begun to show some signs of economic revival. For the third consecutive year, the gross national product (GNP) has shown positive trends. However, the effect on employment is still negative, resulting in the paradox that as production has been increasing, the employment rate has been falling.

In 1995, Peru's GNP was calculated at US$35 billion. A growth of 7.5% in GNP is expected over the next 10 years. The rate of population growth estimated over the next 10 years is 2.2%. By 2060 to 2070, GNP growth is expected to slow to an increase of 1.6%, and population growth is expected to be 1.5% (Table 16.2). The Federation of Private Enterprise Institutions has estimated that Peru needs an infusion of US$100 billion throughout the next decade to reach the national objectives of increasing GNP growth through investment in tourism, mining, fisheries, industry, and services.

Table 16.2. Projected Growth in GNP and Population

	GNP		Population	
Year	(mill US $)	(Annual %)	(thousand)	(Annual %)
1995	35,000	7.5	23,030	2.2
2000	50,247	7.5	25,676	2.2
2005	72,136	5.5	28,628	2.0
2010	94,279	5.5	31,607	2.0
2015	123,219	3.5	34,897	1.9
2020	146,345	3.5	36,677	1.9
2025	173,812	2.5	40,296	1.8
2030	196,652	2.5	44,056	1.8
2035	222,494	2.5	48,166	1.8
2040	251,732	2.0	52,660	1.7
2050	306,860	2.0	62,329	1.6
2060	374,060	1.6	73,052	1.5
2070	438,408		84,779	

Source: Federation of Private Enterprise Institutions.

The production sectors of the coastal zone are assumed to account for 83% of GNP because the most important economic activities (agriculture, fishing, mining, industry, oil, housing, transport, ports and piers) are all found in this area.

Global warming and a subsequent sea level rise would affect fishing and agriculture primarily because of species substitution; industry because of the flooding of fish powder plants located near the coastline; tourism because of the flooding of beaches; oil platforms producing from the shallow shelf and directed wells near the coastline; and ports and piers because of variations in the water level. In addition, national sanctuaries like Tumbes Mangroves and Mejia Swamp could disappear altogether because of flooding.

COASTAL RESOURCES SECTOR

Methods Used

Selection of Study Sites

The following eight areas were selected as study sites:

- Lima-Callao
- Ilo
- Mejia Swamps
- Sechura Bay
- Tumbes River Delta
- Chimbote
- Trujillo
- Pisco.

These sites were selected through an initial evaluation of vulnerability based on a report presented by the Peruvian Group of Climatic Change/CPPS titled "The Effects of Climatic Changes in Coastal and Marine Ecosystems in the Southeast Pacific" (Teves et al.,1992).These sites include some conservation areas like the mangrove swamps in the Tumbes Delta; the national sanctuaries of Mejia Swamps and Villa (near Lima) where migratory birds rest on their flight from the Northern Hemisphere to the Antarctic Circle; the national sanctuary of Pisco, which has pre-Incan ruins famous for their textiles and multicolored pottery; and centers of population, education, commerce, tourism, and fishing like Lima-Callao, Chimbote, Pisco, Trujillo, and Ilo.

In general, the flooded areas discussed here were identified using data on the marine dynamics of each selected area. These data identified the importance of the ecological systems, both marine and littoral, to economic and social activities and structures. In addition, the data provided information that could be used in (1) planning for the expansion of urban centers along the Peruvian coast, (2) proposing response

Coastal Resources Effects of El Niño

The effects of the El Niño phenomenon on the coastal areas of Colombia, Ecuador, Peru, and Chile could be used to help predict the future impact of global climatic change and variations in sea level (Quispe, 1986).

The primary effects of El Niño are the invasion of warm oceanic waters into temperate coastal waters and excessive rainfall along the coast. During an El Niño event, the sea level rises from 0.40 to 0.50 m along the Peruvian coast, and mainly in the northern zone. El Niño has caused great changes in the morphology of South America's west coast, including mass movements, overflowing of rivers, floods, droughts in certain areas, an increase in fluvial solids transport, sedimentation along the coastline and on the shelf, and an increase in vegetation in arid zones. These processes have had a harmful impact on several economic sectors, including agriculture, transportation, housing, industry, and ports and piers. They have also had harmful impacts on the social systems of the coastline inhabitants, e.g., public health, education, and archeological sites.

The warming of surface waters in the ocean diminishes the stock of some species, mainly anchovy (*Engraulis ringes*), and results in species substitution, for example, the substitution of anchovy by sardine. El Niño is also responsible for the unusual appearance of the "titi" crawfish species and the concha de abanico (*Argopecten purpuratus*). In addition, El Niño impacts are associated with high mortality rates in species such as guano birds that are closely linked to anchovy abundance.

measures to the impacts due to sea level rise, and (3) disseminating among the coastal populations (mainly authorities) information about climate changes at local, regional, and global levels and the estimated impacts on coastal zones.

Sea Level Rise Methodology

The vulnerability assessment of the coastal resources sector was based on assumptions about the potential impacts of a rising sea level, including: (1) accelerated erosion of beaches and the littoral would take place, (2) sea water would penetrate the mangroves and lagoons, (3) the coastal areas would be flooded more frequently, and (4) natural vegetation would be destroyed and other species adapted to salt waters would be substituted. Besides these four environmental impacts, coastal infrastructure damage and socioeconomic impacts are possible, and the magnitude of these impacts has been estimated (Fifita and Mimura, 1992).

To determine the flooding zones, the following data were calculated or assumed:

- average observed high tide in a month (one value per month)
- average maximum high tides in a day (30 values per month)
- height of storm waves
- increase of 0.3 m and 1.0 m in mean sea level.

Values were determined for observed high tide or the value of the highest tide during a month, as well as the maximum upper high tide or the average of highest tides each day in a month. The data for these variables have been collected from 1942 to the present. This information was measured by a mareograph that had been installed along the Peruvian littoral under the Dirección de Hidrografía y Navegación. In cases where a mareograph had not been installed, interpolation based on the harmonic characteristics of the referred area in the tides tables was used to obtain estimates of the data for high tides.

To determine the height of normal and storm waves, statistical measures of waves were taken by the Fisheries Ministry for one year. In the zones where we did not have wave measures but where information was available from a nearby area, we used numeric methods to estimate the height and period of waves in deep water. This information plus the bathymetry of the site was used to project the height of waves heading toward the coastline using the orthogonal method and refraction diagrams. Wave height was calculated using different coefficients of refraction and diffraction in shallow waters (Table 16.3).

In other areas where there was no wave parameter information even in nearby areas, visual observations were made during a three-day period at different hours of the day. These observations were compared to accounts by fishermen and local inhabitants. Both the observations and the accounts coincided (KUROIWA, 1986).

After determining the maximum flooding level and to establish the extent of flooding, topographic charts were developed for the different sites. The charts were from the National Geographic Institute and from some municipalities in the study areas.

Table 16.3. Estimated High Tides and Storm Surges for Study Sites under Two Sea Level Rise Scenarios

Site	Current Condition (m)	SLR1 (0.30m)	SLR2 (1.00 m)
Lima			
Miraflores Bay			
High Tide Level[1]	1.24	1.54	2.24
Storm Surge[2]	2.66	2.96	3.66
Callao Bay			
High Tide Level	1.13	1.43	2.13
Storm Surge	2.40	2.70	3.40
Mejia Swamps			
High Tide Level	1.19	1.49	2.19
Storm Surge	2.27	2.57	3.27
Ilo			
High Tide Level	1.34	1.64	2.34
Storm Surge	2.05	2.35	3.05
Sechura Bay			
High Tide Level	0.98	1.28	1.98
Storm Surge	1.70	2.00	2.70
Tumbes Delta			
High Tide Level	1.16	1.46	2.16
Storm Surge	2.25	2.55	3.25
Chimbote			
High Tide Level	0.70	1.00	1.70
Storm Surge	1.99	2.29	2.99
Trujillo			
High Tide Level	1.34	1.64	2.34
Storm Surge	2.49	2.79	3.49
Pisco			
High Tide Level	1.03	1.33	2.03
Storm Surge	1.81	2.11	2.81

[1] High tide levels correspond to ordinary conditions.
[2] Storm surge levels correspond to extreme events.

In general, these charts were made by the Water Supply and Drainage Service (SEDAPAL).

In most cases, the chart scale was inadequate for obtaining topographic lines 1 m equidistant. Therefore, field work was undertaken to develop adequate littoral charts to plot the maximum level of flooding and to determine the total area of flooding.

After the flooded area was defined, we developed inventories of the infrastructures, population, and uses of the terrain in the areas that may be flooded.

Methodology for Economic Analyses

The infrastructures and production centers that may be threatened by sea level rise have been identified, and we are in the process of estimating the losses due to potential flooding. We are also obtaining estimates of activities that are secondary to the identified industries affected. In addition, using data from the Industry, Tourism, and Integration Ministry and from the private sector, we are estimating the value of future activities or economic projects that would have to be abandoned or not developed. The analysis will also estimate the cost of protecting vulnerable areas in cases where that is economically feasible. The results section of this chapter includes some examples from the preliminary economic evaluation.

Adaptation Alternatives and Evaluation Methods

The IPCC Working Group of Response Strategies suggested three key types of adaptation strategies: retreat, accommodate, and protect. These strategies have been discussed briefly for some of the study sites and will be developed more fully in future analyses.

Results

In a scenario where the sea level rises 1.0 m, less than 30% of the Peruvian coastline would be affected and in general only small parts of some coastal cities and towns. Only the 6,000 inhabitants in La Punta (near Lima), which is 1.5 m above sea level, would be seriously affected (Teves, 1990). Both normal and storm waves could destroy this town under a 1.0 m sea level rise scenario.

The population traveling to beaches and recreational areas (about 6 million) would also be affected. In addition, many ports would be seriously affected because it would be necessary to establish new levels for piers, and relocate activities such as fishing, unloading cargo, and fish processing plants. At this higher sea level, severe storm waves would increase coastal erosion in areas that are already experiencing problems such as the destruction of the infrastructure for coastal mariculture (crawfish) and the intertidal hydrobiological communities. Protected areas such as the Tumbes Mangroves, Mejia Swamps, and Paracas National Reserve would also be affected by this level of sea rise.

Potential Land Loss

Estimates of land lost in the study sites due to inundation from a 1.0 m sea level rise are shown in Table 16.4. Peru's vulnerability and adaptation assessment includes information indicating that in some places like La Punta we already have substantial problems due to erosion (Teves, 1989). La Punta has been partially protected on the southern side, but if the sea level rose 1.0 m it would be necessary to elevate the present breakwater (as has been partially done this year), and construct a groin parallel to the coastline on the northern side of La Punta, as well as a levee.

In La Costa Verde, south of La Punta, the construction of groins has allowed the formation of beaches where in the summer more than 2 million come for vacations (Teves, 1991). These beaches would disappear under a 1.0 m sea level rise. We recommend elongating the groins and constructing longitudinal seawalls like the municipality of Lima is doing at present, to reconstruct the road in La Costa Verde.

Table 16.4. Land Loss due to a 1.0 m Sea Level Rise by 2100

Site	Land Lost (km^2)
Lima	
Miraflores Bay	2.60
Callao Bay	8.47
La Chira-Pucusana	12.58
Ilo	2.63
Mejia Swamps	6.90
Sechura Bay	29.24
Tumbes Delta	4.54
Chimbote	0.84
Trujillo	8.23
Pisco	2.29

Potential Economic Losses

The following are some examples from our preliminary economic evaluation. In the Pisco zone, several fish powder and cannery plants are near the coastline at an altitude of 2.0 m. Relocating these plants would cost an estimated US$80 million. The cost of repositioning eroded beaches in Pisco and Paracas Bay is estimated at US$11 million.

The fishing plants planned north of the mouth of the Pisco River would have to be changed to an inland location. The same would be true for a main road to Lima that is now planned near the coastline.

In the Lima-Callao zone, the cost of reconstructing eroded beaches is estimated at US$118 million. The cost of relocating clubs, recreational areas, restaurants, etc., is estimated at US$100 million. La Punta would be completely flooded, resulting in an estimated cost of US$68 million, to reposition all of the city's infrastructure. In La Costa Verde, plans to develop buildings, hotels, clubs, and recreational areas are estimated at US$5 billion. At La Herradura and La Chira beaches, there are plans to invest US$300 million in each. The cost of the coastal protection work for La Costa Verde and La Punta is estimated at US$50 million.

Adaptation Alternatives

In cities where the urban infrastructure is near the coastline (as in La Punta), we recommend protection based on preliminary evaluations (Teves et al., 1994). To control present-day erosion in La Punta, it has been necessary to construct two breakwaters, three groins in T construction, and seawalls, all of which have provided acceptable results.

In other cases where the existing ocean-land intersection is fundamental to the natural character of the area, we recommend retreat and accommodation. In the Tumbes Delta, for example, protection is not possible for the mangrove areas. The same would be true for the Mejia Swamps.

We think that when possible, ocean flooding should be allowed to take place and we should retreat to allow coastal ecosystems to migrate. Measures we are evaluating include using sand to form protective dunes or constructing breakwaters to diminish wave height, primarily during storms.

An overview of the potential vulnerability and (in most cases) recommended adaptation option for each of the study sites is provided below.

- South of Lima, the beaches of Conchan, Arica, Punta Negra, San Bartolo, and Santa Maria are growing. Conchan Beach has a longitudinal beach ridge, which is limiting the growth of an ancient, filled lagoon. Some summer resorts are being constructed along Conchan Beach, but they would also be in danger from the increased level of storm waves. If the sea level were to increase, these beaches would be covered and eroded by sea water, affecting more than 500,000 tourists during the summer. Constructing groins or elongating them to favor the accretion of sand would be recommended as would the construction of seawalls on the upper part of the beach ridge to protect the urban area and the Villa Swamps.

- Ilo is slated for development to aid the southern part of the country. By developing this port, increases in loading and unloading of cargo from Brazil, Bolivia, Argentina, and Peru will be possible. A rise in sea level could affect this development because beach flooding would reduce the number of tourists coming from Brazil, Bolivia, and northern Argentina, as well as the tourism of Peruvian people coming from the mountains. Construction of groins and breakwaters may be necessary in this area and some alteration to the coastal infrastructure (i.e., raising piers) may be necessary.

- The Mejia Swamps would almost disappear if the sea level were to rise, affecting the natural flora and fauna that are important to migrating birds. Retreat is the only adaptation recommended for the Mejia Swamps.

- In Tumbes Delta, the mangroves area would be totally flooded, including some infrastructure for mariculture (ponds and buildings) where crawfish are being raised. In this case, we recommend not protecting the mangroves, but protecting the crawfish ponds, if possible. An alternative would be to construct groins that would allow sand to accumulate, because large amounts of marine sediment will be transported to the zone.

- A scenario similar to the one described for the Tumbes River Delta would be considered for Sechura Bay.

- Chimbote Bay is currently protected by a seawall; protection would only be necessary in the southern part of the bay.

- Trujillo is already suffering coastal erosion and sand accumulation along its littoral. Additional construction of groins and breakwaters could reduce loss of the coast. Delicias, a summer resort, has been affected by substantial erosion: two blocks of houses were destroyed in recent years. Construction of a breakwater would be necessary to protect Las Delicias.

- Pisco and Paracas bays are currently undergoing extensive sedimentation, but if the sea level increased, strong erosion from storm waves could occur. Some hotels and summer resorts located on the inner part of the bays could be affected. In addition, problems could occur for the road to San Martin Port, the fish powder plants, the fishing town of San Andres, and the beaches in Pisco. Because of the development in this area a number of adaptation options, construction of groins, construction of a dam, closing Paracas Bay, or retreat, should be considered. Economic analyses of these options will be needed.

If the sea level were to rise, some small fishing towns and summer resorts would most likely disappear. For these smaller locations, it would be necessary to more accurately assess the costs of retreat or protection. In the near future, these preliminary evaluations should be supplemented with detailed studies and evaluations to plan for the protection of areas considered to be threatened by a potential sea level rise.

Discussion and Conclusions

In Peru today, some projects for the management of coastal areas are being developed. In some places, specialists are working to evaluate and identify all existing sectors potentially affected by sea level rise (e.g., industrial, educational, health, civil defense, population, commerce, tourism) and their characteristics, including the areas threatened by erosion and contamination. All of this information, gathered by field work, air photographs, and remote sensors, is being classified in a system of geographic and georeference information. With this system in place, we will know what we have, what we need, and what we must correct to maintain a sustainable coastal management plan.

These studies will make it possible to apply the measures, recommendations, and policies that are necessary for the integrated sustainable development of coastal zones.

Peru's vulnerability and adaptation assessment, dealing with climate changes, sea level rise, and subsequent impacts, will provide some insights on potential problems with the coastal sector that cannot be resolved at this point in time, but which we must be prepared to solve in order to minimize the adverse impacts of climate change.

In the near future, the areas we have identified through this preliminary assessment as vulnerable to a sea level rise will be included in the planning involved in the integrated management and sustainable use of coastal zones.

SUMMARY CONCLUSIONS

Vulnerability and adaptation assessments for the Peruvian coast are being developed for eight selected sites. The vulnerability assessment is in intermediate stages, but the adaptation assessment is only in preliminary stages. We have also performed other work in the coastal resources sector (NOAA, 1976). In the future, vulnerability and adaptation assessments will be completed for the coastal resources in the selected sites, and studies of new coastal sites should be undertaken. In addition, we should be collaborating to a great extent with government departments to disseminate our conclusions and recommendations for the adaptation measures (Sorensen et al., 1992).

REFERENCES

Fifita, P. and N. Mimura. 1992. Assessment of the Vulnerability to Sea Level Rise for the Kingdom of Tonga. Report of Vulnerability Assessment for Tonga submitted to the IPCC Coastal Zone Management Subgroup Workshop.

INEI. 1993. Instituto Nacional de Estadistica. Censode 1993 Lima.

IPCC. 1990. Strategies for Adaptation to Sea Level Rise. Response Strategies Working Group. Intergovernmental Panel on Climate Change.

IPCC. 1991. The Seven Steps to the Assessment of the Vulnerability of Coastal Areas to Sea Level Rise. Metología Común. Intergovernmental Panel on Climate Change.

IPCC. 1992. Global Climate Change and the Rising Challenge of the Sea. Report of the Coastal Zone Management Subgroup. Intergovernmental Panel on Climate Change.

Kjerfve, B. 1991. Perspectiva de Cambio en el Nivel del Mar. Boletin Internacional de Ciencias del Mar. UNESCO 57/58.

KUROIWA. 1986. Efectos de los Tsunamis en la Costa del Perú entre Ancón y Pucusana. Universidad Nacional de Ingeniería.

NOAA. 1976. The Economic Benefits of Coastal Zone Management and Overview. Washington, DC.

Quispe, J. 1986. Las Bravezas de Mar a lo Largo de la Costa Peruana y su Relación con el Fenómeno de "El Niño." Dirección de Hidrografía y Navegación.

Sanchez, W. 1976. La Nubosidad a lo Largo de la Costa Occidental de Sud América entre 0° a 30° Sur. *Cuaderno de Física y Meteorología.* Vol. I. No. 1.

Sorensen, J.C., S.T. McCreary, and A. Brandani. 1992. Arreglos Institucionales para Manejar Ambientes Costeros. CRC, University of Rhode Island. USAID, Office of Environmental and Natural Resources.

Teves, N. 1976. Neotectonics in Peruvian Litoral. IV Intercongress Pacific Science Association. Valparaiso, Chile.

Teves, N. 1989. Análisis de los Problemas de Erosión en la Costa Peruana. CPPS/ PNUMA (OCA)-PSE-COI WG1/1, Bogota.

Teves, N. 1990. Cambio Climático Global y Variaciones del Nivel del Mar. Rev. Ingeniería — CIP Año 1, Nro. 2, Lima.

Teves, N. 1991. Las Variaciones de las Orillas en el Litoral Peruano. VII Congreso, Perú Geol. Anales T-II.

Teves, N., R. Calienes, J. Zuzunaga, R. Guevara, C. Del Carmen, E. Culqui, G. Laos, and E. Perez. 1992. Efectos de los Cambios Climáticos en los Ecosistemas Costeros y Marinos del Pacífico Sudeste. Informe de Avance del Grupo Peruano de Cambios Climáticos. CPPS, PNUMA.

Teves, N., G. Laos, S. Carrasco, L. Pizarro, C. San Roman, V. Moreno, E. Jaimes, S. Barrantes, and A. Romero. 1994. Implicancias Socio Económicas del Incremento del Nivel del Mar en Lima Metropolitana y el Callao como Consecuencia del Calentamiento Global. Estudio de Caso. Informe GNCC/CPPS/PNUMA.

Wirfel, J. 1980. Estudio de la Atmósfera sobre la Costa Peruana. Trabajo de Investigación.

Chapter 17

Vulnerability and Adaptation Assessments for Venezuela

Martha Perdomo,[1] Maria de Lourdes Olivo[1]
Yamil Bonduki,[1] Luis Jose Mata[2]

[1] **Ministry of Environment and Renewable Natural Resources**
Ministry of Energy and Mines
[2] **Institute of Fluid Mechanics FI-UCV**

SUMMARY. This chapter presents a summary of the preliminary results on vulnerability and adaptations assessments in the forest and coastal resources sectors of Venezuela. The Holdridge Life Zone Classification Model was used, and the entire country of Venezuela was selected as study site. The results of the vulnerability assessment show very clearly that, in Venezuela, vegetation will suffer from drier climate patterns. Adaptation strategies in Venezuela to reduce possible forest damage from climate change are being analyzed for forest plantations and for the natural reserve forests based on the life zones identified by the Holdridge model. The IPCC common methodology for assessing the vulnerability of coastal areas to sea level rise proposed by the Intergovernmental Panel on Climate Change was used to study five Venezuelan areas. The most vulnerable areas for both erosion and flooding are the eastern coast of Maracaibo Lake, the Barlovento region, and the eastern coast of Falcon State. Under a 0.5 m sea level rise scenario, the total land lost by erosion would be 32.91 km², and under a 1.0 m sea level rise scenario, total land lost would be 47.00 km². Land lost from flooding would be 52.63 km² and 77.74 km², respectively.

INTRODUCTION

Venezuela is located in the northern part of South America, between 60° and 73°W and 2° and 13°N. It has a total area of 916,445 km². It is divided in 22 states and 1 federal district, plus 311 small islands in the Caribbean Sea (Figure 17.1).

The population growth rate is high, although it has decreased in recent years. Mortality rates are low and birth rates are still high. It is a nation with just over 20 million inhabitants (OCEI, 1990), with a population density of 21 inhabitants/km². The population is considered to be in a state of transition between rapid growth and slow growth. A high percentage of the population, approximately 71%, is located in the central-north

Note: This chapter does not contain a detailed description of the assessment methodologies that were recommended by the U.S. Country Studies Program. Refer to Chapter 2 for on overview of methodologies

J. B. Smith et al. (eds.), Vulnerability and Adaptation to Climate Change, 347–366.

Figure 17.1. Map of Venezuela

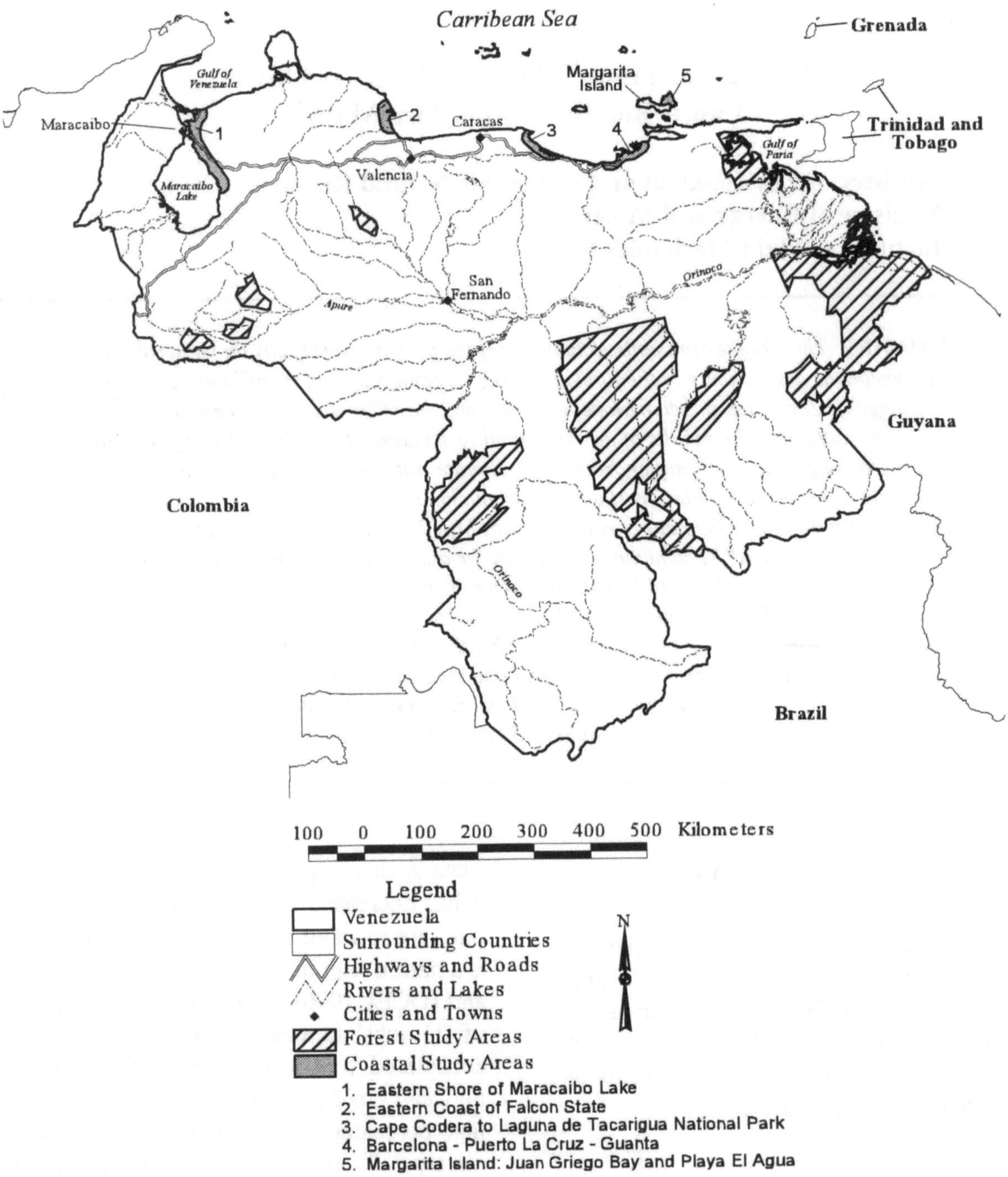

coastal region, where most of the economic activities have been developed.

The government of Venezuela signed the Framework Convention on Climate Change. Because the Convention requires all parties to develop and publish national inventories of anthropogenic greenhouse gas emissions, as well as national plans to reduce or control these emissions, the Ministry of Environment and Renewable Natural Resources and the Ministry of Energy and Mines developed the Country Study to Address Climate Change.

A team of experts from several Venezuelan institutions were in charge of conducting the study, which had the following objectives:

- develop a national inventory of anthropogenic emissions
- predict future greenhouse gas emissions under various economic development scenarios
- identify, analyze, and rank abatement strategies by forming a national plan to mitigate greenhouse gas emissions and enhance reservoirs and sinks
- assess the potential impacts generated by a sea level rise on Margarita Island and the Venezuelan coastal zones and outline possible adaptation responses
- assess the potential impacts and adaptation strategies for the Venezuelan forests within the context of a climate change.

The forest area of the country is approximately 58 million ha, which represents more than 60% of the national territory. About 70% of the forest land is in the southern part of the country, in the Amazon Basin.

The annual rate of forest clearing in Venezuela has not been consistently documented. Data from Pepper et al. (1992) was used to estimate an average deforestation of approximately 517,000 ha per year (excluding the southern region of Venezuela), which represents a deforestation rate of less than 1% per year.

Most of the agricultural activities in Venezuela are related to extensive cattle raising, which does not involve land tilling. However, some important crops have been established on open savannas, especially during the 1984-1989 period, when government subsidies resulted in a substantial increase in agricultural production.

Venezuela's coastline has approximately 4,000 km^2, 52% on the Caribbean Sea and 48% on the Atlantic Ocean. There are many types of coastal environments: sandy marshland, beaches, cliffs, deltaic plains, coastal lagoons, estuaries, barrier islands, and bays.

The majority of the coast is classified as a dry-warm climate with temperatures above 25°C. Trade winds from the east and northeast prevail, from December through March. From April to September, the prevailing winds are from the south and southeast, with periods of equatorial calm. Precipitation shows a bimodal pattern with a dry season from October through April and a rainy season from May through September. The coastal zone is semi-arid and precipitation is less than 500 mm/yr, except adjacent to mountainous regions where trade winds are forced to rise, causing high precipitation.

The deltaic coast of the Orinoco is a wet-warm climate, with more than 500 mm of precipitation (Roa-Morales, 1991). The coastal zone of the Federal District (Caracas, the national capital, is located about 50 km inland and 800 m above sea level) is the most developed part of the coastal area. Narrow coastal plains, no more than 1 km wide, are highly developed with tourist facilities, harbors, and industrial infrastructures.

Venezuela has developed land use plans for all coastal states at the regional, subregional, and local levels based upon its Organic Law for Territorial Ordination. The law was passed to reorganize land use in response to environmental pressure caused by overpopulation in urban areas and increasing concern about pollution and irresponsible resource use. In addition to those concerns, natural and human-induced risks are to be taken into account while developing and implementing these plans. Therefore, sea level rise is one of the risks that needs to be addressed within the context of integrated coastal zone management. Integrated coastal zone management (ICZM) is seen as a component of the Organic Law. Formalizing an ICZM program is a commitment of the Venezuelan Ministry of Environment and will require new legislation and institutional arrangements.

SCENARIOS

Climate Change Scenarios

The models used to simulate the future climate of Venezuela were: the Geophysical Fluid Dynamics Laboratory equilibrium run (GFD3) model, the Goddard Institute for Space Studies (GISS) model, the Oregon State University (OSU) model, and the United Kingdom Meteorological Office 1987 (UKMO) model (Benioff et al., in press). From each of these models, average annual temperature and precipitation were obtained for the Venezuelan region under both $1\times CO_2$ and $2\times CO_2$ conditions.

In addition, an analysis of the observed average annual temperature and precipitation was made using IIASA's database of global terrestrial climate, which has a resolution of 0.5 by 0.5 degrees (Leemans and Cramer, 1990). The data provided by the Venezuelan Meteorological Stations were compared to the data obtained from the IIASA database to create maps of precipitation for the observed data (1951-1980). As expected the agreement is quite remarkable.

The results of the general circulation model (GCM) simulations of current conditions and climate change conditions were compared with the observed data provided by the Meteorological Venezuelan Database. Climate change scenarios were defined for four

Table 17.1. Adjustment Statistics for the Difference between $2\times CO_2$ and Current ($1\times CO_2$) Generated by the GFD3 Model

	Jan	Feb	Mar	Apr	May	Jun	Jul	Aug	Sep	Oct	Nov	Dec	Annual
Temperature (ΔC°)	3.09	2.69	2.81	3.27	2.69	2.21	2.75	3.45	2.43	2.16	2.58	2.29	1.12
Precipitation Ratio	0.79	1.03	1.07	0.99	1.34	1.26	0.73	1.12	1.43	1.20	0.82	0.92	1.06

GCMs. To provide values at the same spatial resolution as the IIASA database, the data from each GCM had to be interpolated to a 0.5 × 0.5 degrees resolution.

Table 17.1 presents adjustment statistics for temperature and precipitation for GFD3. The temperatures are expressed as absolute difference ($2\times CO_2 - 1\times CO_2$) and the precipitation as a ratio ($2\times CO_2 : 1\times CO_2$).

It is important to emphasize that the results presented in Table 17.1 are only an example of data from a particular GCM grid. The Holdridge Life Zone Classification Model analysis that is presented later in this chapter analyzes the entire country of Venezuela.

Table 17.2 presents the results from the transient, coupled ocean-atmosphere GCM climate change experiments conducted at the UKMO. These results were not used as input for the assessments presented in this chapter.

Sea Level Rise Scenarios

Both a 0.5 m and a 1.0 m sea level rise scenario were used for Venezuela's sea level rise analysis.

Baseline Socioeconomic Scenarios

The baseline socioeconomic scenarios have not yet been fully incorporated into the assessment, but are presented here to provide a context for the assessment work that has been completed.

Table 17.3 represents baseline growth in gross national product (GNP in US$ billion). The GNP shows a trend toward growth between 1990 (US$2.8 billion) and 2075 (US$11.5 billion). The average annual growth rate for GNP also shows a slowly increasing trend (BCV, 1994). These data indicate that notwithstanding the country's current economic difficulties (foreign debt, fiscal deficit, exchange control, currency devaluation, unemployment), productivity in the different economic activities will tend to grow.

Table 17.4 represents the projected percentage growth in the population of Venezuela. The projected population for 2075 (64.4 million) will practically triple the recorded population for 1990 (19.7 million). However, average annual population growth percentages in the extrapolation for 2075 show a decreasing trend (OCEI, 1991), thus indicating that the population growth rate will slow in the coming years.

Table 17.2. Transient Scenario from the UKMO Model and Daily Weather Data from 1951-1980

Decade		Jan	Feb	Mar	Apr	May	Jun	Jul	Aug	Sep	Oct	Nov	Dec	Annual
Second (2000)	Temperature (Δ°C)	2.65	2.84	1.69	0.10	0.61	0.90	1.43	1.26	1.50	1.92	1.98	1.94	1.06
	Precipitation Ratio	1.62	0.70	2.81	3.74	1.24	1.19	0.90	1.00	0.80	0.69	0.97	0.65	0.97
Third (2030)	Temperature (Δ°C)	5.06	5.52	3.93	2.61	2.38	2.87	2.60	2.60	3.06	3.78	3.12	3.60	1.13
	Precipitation Ratio	0.27	0.12	1.30	0.79	0.76	0.81	0.90	0.95	0.64	0.39	0.98	0.45	0.75

Table 17.3. Baseline Growth in GNP (average annual percentage growth)

Period	Growth Rate
1990-2000	0.9%
2000-2025	1.2%
2025-2050	1.9%
2050-2075	3.3%
Year	**GNP (US$ billions)**
1990	$2.8
2075	$11.5

Source: BCV, 1994.

Table 17.4. Baseline Growth in Population (average annual percentage growth)

Period	Growth Rate
1990-2000	2.2%
2000-2025	1.6%
2025-2050	1.1%
2050-2075	0.8%
Year	**Population (millions)**
1990	19.7
2075	64.4

Source: OCEI, 1991.

In addition, the population distribution pattern indicates a high concentration in the northern area of the country near the coast. This trend will increase due to the location of oil fields, oil industry, and supporting activities in the northeast and northwest regions of the country. Population growth in coastal areas is also occurring due to the large concentration of installations for the extraction of petroleum and its byproducts in Zulia, as well as infrastructures for processing petroleum and its byproducts located near the coast of the states of Zulia, Falcón, and Anzoátegui.

The socioeconomic scenario described above also anticipates an increase in conflicting pressures because the wealth of coastal, island, and continental natural resources offers attractive opportunities for many different types of investors. One particular situation already beginning to be obvious in the coastal area included in this study is a shift from agricultural uses to tourist/recreational uses. If the necessary controls are not applied by coastal zone administrators, there will be an ever-increasing reduction of agricultural activities in future years.

FORESTRY SECTOR

The importance of studying the impacts of global change on tropical forests is widely recognized because forests are highly sensitive to climatic change. The impact of climate change on tropical forests has not been as well studied as it has for other types of forests: boreal, warm temperate, and cool temperate forests.

Methods Used

Selection of Study Sites

The analysis presented here addresses the entire country of Venezuela. However, forest reserves such as Sipapo (1,215,500 ha), Caura (5,134,000 ha), and Imataca (3,203,250 ha) have been identified as potential areas that could be affected by climate change. The Imataca Forest has been very extensively inventoried for wood production. It is a specific area that should be studied. There are also other areas where plantations are planned for development that need to be studied to evaluate climate change conditions.

Forest Ecosystem Models Used

A Holdridge model was used to examine broad patterns of vegetation and climate and define to what extent climate changes could influence the suitability of a region to support different vegetation and forest types.

The close correlation between climate and vegetation formation has long been recognized by scientists. This has led to the practice of using vegetation maps to create climate maps and vice versa (Thornthwaite, 1948; Holdridge, 1967; Walter, 1985). A database of potential life zones can be mapped, providing a base map for the entire country of Venezuela.

To validate the Holdridge model for Venezuela, the generated map for current conditions was compared to existing maps of vegetation.

Maps of Venezuelan life zones under 2×CO_2 conditions were also obtained for each of the GCM simulations evaluated in this assessment (GFD3, GISS, OSU, and UKMO).

It should be noted, however, that the Holdridge model intends to provide a global picture of vegetation cover and, consequently, these results should be viewed with caution. More specific studies on ecosystem dynamics at local levels are needed to support the outcomes of this exercise, considering the highly complex nature of tropical ecosystems and the lack of knowledge on crucial issues, which constrains our ability to predict such changes.

Results

Estimates of the current distribution of life zones for the nine largest life zones in Venezuela are presented in Table 17.5. The Holdridge model and the IIASA current climate database were used to generate these results.

Table 17.6 presents the changes in land area associated with different vegetative cover under 2×CO_2 climate conditions as estimated by the UKMO model. The change in land area indicates the difference between the

Table 17.5. Estimated Area of Current Vegetative Cover[1]

Life Zone	Area (km^2)
Subtropical Moist Forest	281,946
Tropical Dry Forest	241,401
Tropical Moist Forest	125,957
Subtropical Dry Forest	104,272
Subtropical Wet Forest	82,249
Tropical Very Dry Forest	51,926
Warm Temperate Wet Forest	12,690
Tropical Thorn Woodland	12,239
Subtropical Thorn Woodland	6,855

[1] Results generated using the Holdridge model and the IIASA database.

Table 17.6. Changes in Vegetative Cover under 2×CO_2 Climate Change Conditions[1]

Vegetative Cover	Change in Land Area (km^2)	Percent Increase/Decrease
Subtropical Moist Forest	-274,188	-97.2%
Tropical Dry Forest	+33,079	+13.7%
Tropical Moist Forest	-125,957	-100.0%
Subtropical Wet Forest	-82,248	-100.0%
Tropical Very Dry Forest	+419,859	+806.6%

[1] Climate change conditions estimated using the UKMO model and the Holdridge model.

Figure 17.2. Map of Venezuela Showing Holdridge Life Zones for Present Day and UKMO Scenarios

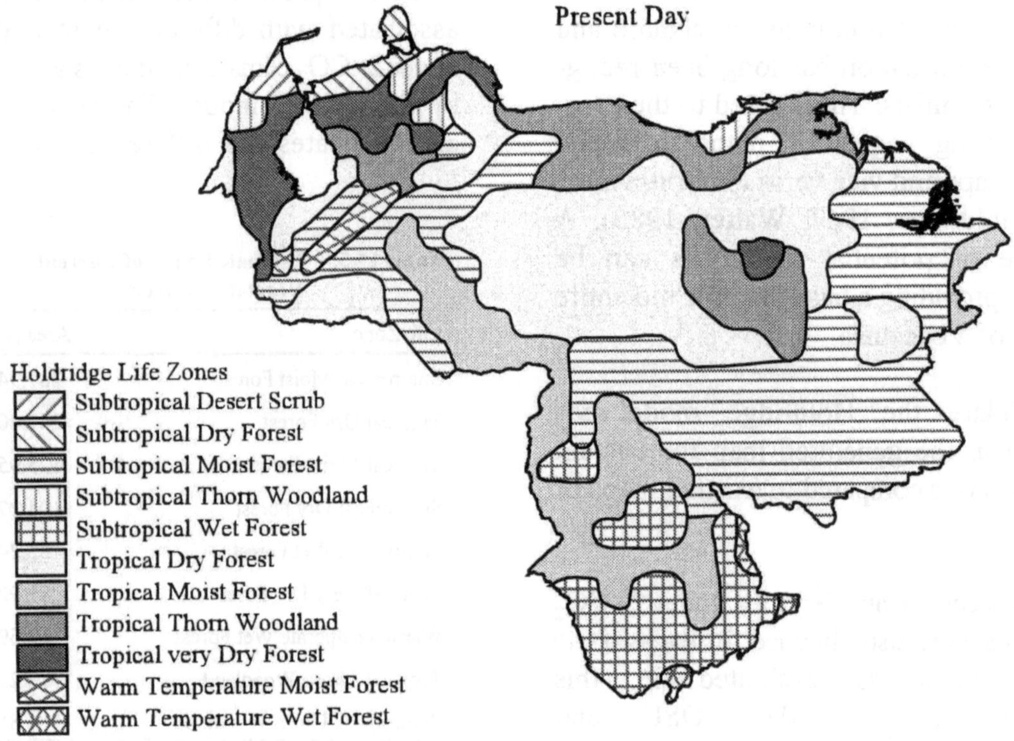

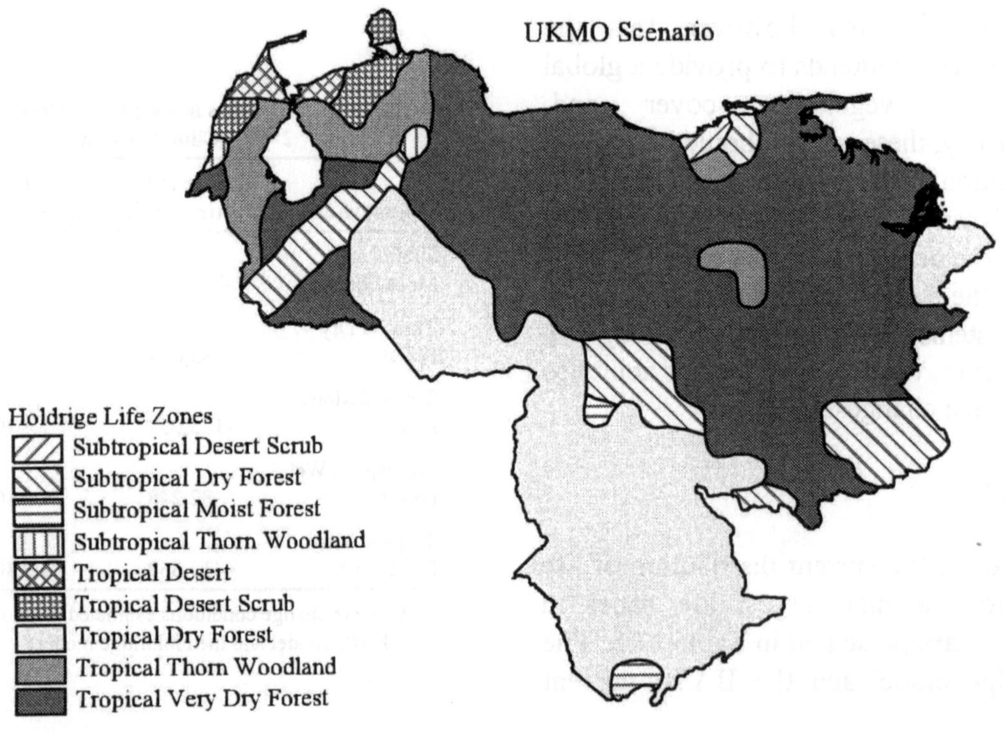

Table 17.7. Land Area for the Tropical Very Dry Forest Life Zone under Climate Change Scenarios

Model	Area (km^2)
Baseline	51,926
GFD3	383,468
GISS	426,909
OSU	223,236
UKMO	470,785

estimated current life zones in Venezuela (IIASA data using the Holdridge model) and the area estimated using the UKMO model. The life zones under present conditions and the UKMO simulation are presented in Figure 17.2.

The life zones predicted for the four GCMs indicate a tendency to drier patterns. Table 17.7 shows the land area in square kilometers for the tropical very dry life zone as predicted using the different GCM simulations.

Discussions and Conclusions

The application of the Holdridge model to the Venezuelan region shows a clear tendency to a drier pattern over all the country. Most of Venezuela's natural forest reserve will be affected. It is estimated that tropical very dry forests will cover most of the northern region currently covered by tropical dry forest. The models also suggest that subtropical wet forests in southern Venezuela will be replaced by tropical dry forest.

Some preliminary conclusions can be reached from this vulnerability assessment. The Holdridge model, which was used to develop a classification map under Venezuela's current conditions, provides a good representation of the distribution of life zones within the country. The four GCMs predict, in general, the same change in vegetation toward drier patterns under 2×CO_2 climate conditions.

Several critical zones were identified. The model results indicate that the subtropical wet forest in the southern part of Venezuela would no longer be sustainable under changed climate conditions, and that this life zone would be replaced by tropical dry forest (in this study, the definition of subtropical zone is not based on latitude but on temperature characteristics). It is estimated that tropical very dry forest would cover most of the northern region currently covered by tropical dry forest; thus all nature reserve forests in Venezuela would be affected. Work is currently underway to build on these results through forest gap model analysis of adaptation options.

Coastal Resources Sector

Methods Used

Selection of Study Sites

Five study sites were selected according to their urban, industrial, tourist, or physical-natural characteristics. The five areas are the eastern coast of Maracaibo Lake, the eastern coast of Falcon State, the Barlovento Region, Barcelona-Puerto La Cruz-Guanta, and two sections of Margarita Island (Bahia de Juan Griego and Playa El Agua).

The five selected areas were chosen based on a study of the entire Venezuelan coast completed by Volonté and Arismendi (1995) at a scale of 1:100,000.

They actually proposed eight study areas but because of restricted financial resources, accessibility, and time limitations, this number was reduced to five. The final site selections were made for the following reasons.

The eastern shore of Lake Maracaibo. This site was selected because of the presence of the petroleum industry, the country's most important source of foreign currency. In this area, towns and petroleum industry infrastructures would be affected by subsidence; in addition, there is an important wildlife refuge in this area.

The eastern coast of the State of Falcón. This site was selected for its moderate but growing tourist development, and the existence of two protected areas, a wildlife refuge and a national park, offering a variety of recreational scenarios.

Barlovento region (Cape Codera-Laguna de Tacarigua National Park). This site was selected because of the presence of a coastal lagoon with great diversity of fauna, which is protected as a national park, and because of a recent tourist-recreational development. This site is a highly attractive area for the population due to its proximity to the capital, Caracas.

Barcelona-Puerto La Cruz-Guanta. This site was selected because of the extensive tourist and recreational development with a high investment level, important fishing, and industrial (oil) activities; in addition, there is a national park located here.

Margarita Island. The two locations on Margarita Island were selected for their great scenic beauty and important tourist developments.

Sea Level Rise Methodologies Used

Since 1990, two research studies on the potential consequences of a rise in sea level have been conducted in Venezuela. Roa-Morales (1991) conducted a study on "Coastal Morphology and Sea Level Rise Consequences in Venezuela," which gave the first warning of the consequences of this phenomenon for the Venezuelan coast. Recently, Volonté and Arismendi (1995) applied the aerial videotape-assisted vulnerability analysis (AVVA) to develop a preliminary assessment of the vulnerability of the Venezuelan coast to a sea level rise.

The results determined that the low coastal lands of the Falcon, Anzoátegui, and Sucre states, as well as some areas of the Federal District, are the most vulnerable areas in the country (Orinoco Delta and Margarita Island were not included in this study).

These two research studies resulted in new information on the vulnerability of the Venezuelan coast, but they did not follow the common methodology suggested by the IPCC (1992).

The assessment presented in this chapter is a quantitative assessment of the vulnerability of the Venezuelan coast to the impacts of sea level rise and it follows the IPCC common methodology for scenarios of a 0.5 m and a 1.0 m sea level rise by 2100.

Four methods were used to obtain data on coastal geomorphology:

- aerial photographs when available
- review of related bibliographies, particularly the National Land Inventory carried out by COPLANARH (1970)

- review of aerial video maps made by Volonté and Arismendi (1995)
- field visits to update and obtain new information at study sites.

Two scenarios of eustatic sea level rise for 2100 were considered: 0.5 m and 1.0 m. In accordance with convention, the 1.0 m scenario was considered in most detail (IPCC, 1990; 1992).

The eustatic scenarios were applied directly to the study sites as relative sea level rise scenarios since the gauge station records did not go back in time far enough for a meaningful analysis of sea level trends. Each sea level rise scenario produced a new coastline and, consequently, a new area of land loss. The retraction model was applied to the five study sites. An analysis related to the subsidence pattern of the eastern coast of the Maracaibo Lake was also included.

Based on coastal characteristics, it was determined which of the following mechanisms would be in force:

- erosion (sandy beaches and soft cliffs)
- floods (low-lying land, particularly wetlands, marshes, swamps, etc.)
- no land loss (hard, high rocky shores).

From the analysis of all the information, an estimate was made of the amount of land loss from erosion by applying the Bruun Rule as modified by Hands (1983). This formula was applied as indicated in Nicholls and Leatherman (1994) and Leatherman et al. (1994). It should be noted that since there are no well-shaped dune formations on Venezuelan shores, dune height was measured as the height from the mean sea level to the nearest infrastructure or vegetation formation. Because of the absence of continuous records of sediment transport and coastal wave systems, bathymetric maps were used to estimate the closing depth for present-day conditions.

To estimate land loss from inundation, the direct inundation concept was applied, assuming that the land lost consisted of wetlands and other coastal lowlands (estuaries, deltas, marshes, swamps, etc.). These zones were estimated by measuring their surface and overflow areas on topographical maps with a planimeter. A constant gradient of slope was assumed.

Some difficulties were experienced in gathering information related to the inventory of physical, natural, and socioeconomic features of the sites under study. One problem was that the information was dispersed among numerous public and private agencies. Bureaucratic procedures to obtain the information were extremely slow. The data supplied were not sufficiently detailed for application of the methodology according to the scale of the work. Series of records across the significant periods of time were not available, and in many cases, no updated records of climatological, hydraulic, or oceanographic variables existed.

The sources of climatological information were the Year Books and Statistical Reports published by the Ministry of Environment and Renewable Natural Resources (MARNR, 1994) and the Venezuelan Air Force (FAV, 1994). The climatological stations that provided the information comply with the universal rules established by the World Meteorological Organization.

Methodology for Economic Analyses

The type and number of buildings in coastal urban centers and the type of land use along the coast was estimated by means of ground surveys and sea travel; the 1991 population census and slums census (OCEI, 1991); and aerial photography. For each sea level rise scenario, buildings seaward of the predicted shoreline location for that scenario were considered at risk and their value was recorded.

The methodology was modified by substituting an estimate of the number of fishing ports and industries with an estimate of the surface area of fishing and industry. An inventory was made of the 1995 market value of land and buildings (tourism, urban, and industrial). Based on this information, capital value of the lands, including the buildings on them, was estimated.

Three types of maps were used to estimate the size, pattern, and value of development: the study of Urbanistic Regulation Maps, scale 1:10,000 and 1:5,000 prepared by the Urban Development Ministry; the study of State Land Regulation Maps (scale 1:250,000); and the Coastal Belt Regulation Maps (scale 1:25,000) prepared by the Ministry of the Environment and Renewable Natural Resources. The area devoted to each type of development was obtained from these maps by using a planimeter to calculate the surface area covered.

To establish the value for urban developments, the cost of municipal land was determined using the real estate registry divisions of Municipal Councils. The value of privately owned land and buildings was investigated through real estate agencies. The investment in tourist developments was investigated through the Venezuelan Tourism Corporation (CORPOTURISMO). For agricultural developments, the volume and value of agricultural and fishing production was obtained from the Ministry of Agriculture and Livestock. Last, representatives of the various oil industry subsidiary companies were contacted to obtain investment values and maintenance costs for the different coastal areas.

It should be mentioned that the oil industry, the country's main income-producing activity, was excluded from the economic analysis described above. It is difficult to obtain economic information on this industry, which is considered to be strategic for Venezuela. However, in the discussion of results, the oil industry is included in the estimated replacement cost for the eastern coast of Maracaibo Lake. A large proportion of the oil industry is located here and 65% of the national oil is produced here (1,600,000 barrels per day).

An estimate of the cost of the coastal infrastructure that will be affected was also included for the eastern coast of Falcón State and the eastern region.

The economic value of coastal lagoons, mangrove vegetation, and coral reefs as contributors to fisheries was not considered.

The value of urban land was assumed to be included in the value of urban buildings. For rural land, an average land value based on actual sale prices was used. Changes in market values were not considered. Based on these data, an estimate of financial loss was made.

The average number of residents per house taken from population information and the number of houses at the state and municipal level (OCEI, 1991) were estimated. These data were combined with the number of buildings at risk to arrive at an estimate of the population at risk for the 1.0 m sea level rise scenario. Since most at-risk buildings consisted of secondary housing used for recreation and tourism during holidays and weekends, this methodology tends to overestimate the resident population at risk. The at-risk rural population was not taken into consideration due to a lack of information on rural population density. These results are not presented in this chapter.

Adaptation Alternatives and Evaluation Methodologies

Some adaptation alternatives are currently being analyzed. Among others, the strategies being considered are retreat, accommodation, and protection, as identified by the IPCC Coastal Zone Management Subgroup. These strategies are being evaluated for each of the study sites. The ones that are ultimately selected must be appropriate for the industrial, tourism/recreational, urban development, and natural characteristics of each area. For the analysis, some data were introduced into the Adaptation Strategy Evaluator for Coastal Zone Planning (Smith, 1994).

Results

Of the five study areas, the ones estimated to be most vulnerable to the loss of land caused by flooding are the eastern coast of Falcón, the eastern coast of the Maracaibo Lake, and the Barlovento Coast. The most vulnerable areas to the loss of land by erosion are the eastern coast of Maracaibo Lake first, followed by Cabo Codera-Parque Nacional Laguna de Tacarigua, and the eastern coast of Falcón.

Land Loss Analysis

The most relevant results obtained for the land loss analysis are summarized in Tables 17.8 and 17.9. These results are discussed below in conjunction with the estimates of development threatened by sea level rise.

The estimated area at risk of both erosion and inundation at the five study sites in 2100 will increase significantly over the estimated area currently at risk (Table 17.10).

Table 17.11 presents the results obtained thus far for shoreline recession by 2100 for 643.58 km of coastline. These estimates were derived using the Bruun Rule and direct inundation for 0.5 m and 1.0 m scenarios with no protection for both land lost and the percentage of wetland area lost. As Venezuela's study progresses, information for protection alternatives and costs of protection will be developed.

Although the local subsidence analysis has not been completed, it will be very important to take this factor into account, especially in oil-producing areas where land subsidence is already a problem and where ample information on this topic is available. It has been proposed that tide records could be used to derive an estimate for the areas affected by subsidence.

Table 17.8. Low and High Estimates of Land Loss due to Inundation and Erosion (0.5 m and 1.0 m sea level rise scenarios)

	Scenarios					
	Erosion (km^2)					
	Lower Bound Estimates[1]		Upper Bound Estimates[2]		Inundation (km^2)	
Study Site	0.5 m	1.0 m	0.5 m	1.0 m	0.5 m	1.0 m
Eastern Shore of Lake Maracaibo	11.20	17.19	17.84	18.58	13.19	21.52
Eastern Coast of Falcón	3.34	6.69	5.47	10.86	22.62	32.76
Cabo Codera-Laguna de Tacarigua National Park	5.22	9.59	8.38	15.43	15.43	21.18
Barcelona-Puerto La Cruz-Guanta	0.24	0.65	0.38	1.05	1.35	2.21
Margarita Island						
Juan Griego	0.03	0.06	0.74	0.80	0.04	0.07
Playa El Agua	0.04	0.06	0.10	0.28	—	—

[1] $d_{\Delta,1}$ - Actual.
[2] $d_{\Delta,100}$ - 100.

Table 17.9. Estimated Land Loss due to Inundation and Erosion (1.0 m sea level rise scenario)

Study Site	Total Land Loss (km^2)	Erosion (km^2)	Inundation (km^2)	No Impact (km^2)
Eastern Shore of Lake Maracaibo	40.10	18.58	21.52	0.00
Eastern Coast of Falcón	47.30	10.86	32.76	3.68
Cabo Codera-Laguna de Tacarigua National Park	37.90	15.43	21.18	1.29
Barcelona-Puerto La Cruz-Guanta	3.53	1.05	2.21	0.27
Margarita Island				
Juan Griego	0.90	0.80	0.07	0.03
Playa El Agua	1.40	0.28	0.00	0.12

Estimates of the Development Threatened by Sea Level Rise

Eastern coast of Lake Maracaibo. This study site is located in the western part of the country, skirting the eastern coast of the largest lake (12,870 km^2) in Venezuela where the country's largest quantities of light crude are found. A large infrastructure has been developed in this area to extract and process oil, which accounts for a significant amount of the national GDP (23.3%). Industries associated with oil extraction and processing occupy approximately 6,427 km^2. In addition, since the area is adjacent to the coast, it has become attractive for tourism and recreation. This fact has promoted the expansion of facilities to support tourism and recreation, which currently occupy 7,436 km^2. The gradual development of these two activities has led to the development of several populated centers whose residential and commercial use areas occupy approximately 5,415 km^2. It should also be mentioned that in the northern part of this study site, a wildlife refuge occupies 242 km^2 (Ciénaga de Los Olivitos).

Table 17.10. Area at Risk of Both Erosion and Inundation at All Five Study Sites (km²)

Sea Level Rise (m)	1995	2100
Low (0.5)	73.36	86.20
High (1.0)	121.37	142.67

Table 17.11. Shoreline Recession by 2100 for 643.58 km of Venezuela's Coastline Estimated Using the Bruun Rule and Direct Inundation

Sea Level Rise Scenarios	Land Lost Due to Sea Level Rise (km²)	Land Loss (% of total)	Wetlands Loss (% of total)
0.5 m	32.91	0.25	3.97
1.0 m	55.54	0.42	5.10

All these activities will be affected by a loss of land caused by erosion. A sea level rise of 0.5 m would cause erosion of 11.2 km² to 17.84 km². A rise of 1.0 m would cause potential erosion of between 17.19 km² and 18.58 km². These results put this area at high risk to the potential impact of sea level rise, since the above mentioned erosion impacts would occur in unprotected extensions near populated centers and oil industry developments.

Similarly, this sector is highly vulnerable to floods caused by a mean sea level rise. Values of areas subject to floods ranging from 13.19 km² to 21.52 km² have been obtained for the 0.5 m and 1.0 m scenarios, respectively. This, taken together with the subsidence problem found in Maracaibo Lake, would increase the risk of flooding in the area.

The serious subsidence problem in the eastern coast of Maracaibo Lake caused by oil and gas extraction should be taken into consideration in addition to the potential erosion and flooding effects caused by sea level rise. It has been reported that a land area of 1,224.4 km² could be affected, as well as a lake area of 761.6 km² (MARAVEN, 1995).

The potential land loss in 100 years under a 1.0 m sea level rise scenario would total 40.10 km² with a coastline regression of 246.43 km. For a 0.5 m sea level rise scenario, estimated land loss would total 31.03 km².

Based on the above study results and on the oil industry's replacement value estimate of US$10 billion, this study site is regarded as high risk.

Eastern coast of the State of Falcón. This study site located in the western part of the country has high potential for development of tourism/recreational activity due to its beautiful beaches and the scientific and recreational interest of the Cuare Wildlife Retreat and the Morrocoy National Park that has an approximate area of 4,391.5 km². Because of the proximity of this study site to several of the most important cities in the country (Maracaibo, Barquisimeto, Maracay, Valencia, Caracas, and San Felipe) and because of the availability of a roadway infrastructure allowing access, the development of several facilities supporting tourism/recreational activities has been expanded. This expansion of approximately 189.9 km² has taken place without taking into consideration any urban development criteria. At the present, no drainage system has been developed due to the extremely flat topography of the land, which in turn makes drainage difficult and increases the cost of its implementation.

Because of its flat topography, this area has a high potential for sea level rise-caused flooding. Potential flooding values obtained in this study ranged from 22.62 km² to 32.76 km²; therefore, this area could be classified as the most vulnerable study site to flooding. Potential erosion would affect between 6.69 km² and 10.86 km² for the 1.0 m scenario. For the 0.5 m scenario, potential erosion would range between 3.34 km² and 5.47 km². Therefore, this study site could also be regarded as highly risky for erosion.

Estimated total land loss would be 43.62 km² for a 1.0 m sea level rise scenario in 100 years, with an approximate coastline regression of 50.84 km. For the 0.5 m sea level rise scenario, estimated land loss would total 28.09 km².

In economic terms, approximately US$6.1 million would be at risk. This amount includes a first- class tourism infrastructure (resorts, buildings, and beach houses).

Barlovento region (Cabo Codera-Parque Nacional Laguna de Tacarigua). This study site is located in the north central part of the country, very close to the capital. It has recently become a highly attractive tourism/recreational area for most of the Caracas population. The development of hotel and resort infrastructures has been encouraged because of the importance of tourism to the country's economy. Currently, land dedicated to tourism and recreational uses comprises an area of approximately 9.3 km². The attraction of this area is due not only to its wonderful beaches but also to the Laguna de Tacarigua National Park, which covers 15.8 km² and which has been the center of several research studies because of its great ecological diversity.

In addition, there are several population centers in this area occupying approximately 10.3 km² of land dedicated to residential use. These population centers first developed because of the boom in agricultural activity consisting of cocoa and coffee plantations. More recently, agriculture has been replaced by tourism and recreation. This study site should be taken into account when plans for the management of coastal areas are implemented in the near future.

For the 0.5 m scenario, estimated land loss due to erosion ranged from 5.22 km² to 8.38 km². For the 1.0 m scenario, land loss due to erosion was estimated between 9.59 km² and 15.43 km². These results classify this study area as one of medium risk to erosion. However, this area is at high risk for flooding, since the estimated values for land loss at the 0.5 m and 1.0 m scenarios were between 15.43 km² and 21.18 km², respectively, near Laguna de Tacarigua.

Total potential land loss for the 1.0 m sea level rise scenario would be 36.61 km² in 100 years; coastline regression would be approximately 61.48 km. In the 0.5 m sea level rise scenario, total land loss would be 23.81 km².

The value of at-risk structures and land would be approximately US$7.4 million. Of this amount, the tourism infrastructure component would be the most significant.

Barcelona-Puerto La Cruz-Guanta. This area is located on the northeastern coast of the country. It has been extensively developed for the processing, storage, and transportation of oil; for tourism, recreation, and residential activities; for port and airport installations; and for important roadways. This study site's economic development has significant importance at the regional and national level because the country's economic development currently depends on oil and tourism. Therefore, any event that even slightly affects infrastructures could result in severe economic losses for the country. According to study results, the area that could be affected by erosion in a potential 0.5 m sea level rise would be between 0.24 km² and 0.38 km². For a 1.0 m rise, it could be between 0.65 km² and 1.05 km². The areas potentially subject to flooding for the 0.5 m and 1.0 m scenarios would be 1.35 km² and 2.21 km², respectively.

Estimated total land loss for the 1.0 m sea level rise scenario would be 3.26 km² in 100 years, with a coastline regression of 116 km. In the 0.5 m sea level rise scenario, total land loss would be 1.73 km².

Since the value of structures and land is high (US$138.5 million), this study site could be classified as high risk. First-class tourism infrastructures and some oil structures would be at risk.

Margarita Island. Margarita Island, located in the northeastern part of Venezuela, has both natural beauty and historical value that made it an important tourism and recreation center. Its importance became established after it became a "free zone." A very important economic dynamic for the country was initiated at that time, consisting of investments in hotels, resorts, and commercial infrastructures (mostly in the southern part of the island — Porlamar, La Asunción, Pampatar) and transportation facilities (the International Airport and port terminals for ferry boats and speed boats), which connected this beautiful island with the mainland and the rest of the world.

The study sites located in the northwestern part of the island (Juan Griego Bay) and the northeastern part (Playa El Agua) have not been subject to large investments when compared with the southern part of the island, which explains the limited area of 0.584 km² for residential use (existing only in Juan Griego Bay), 0.316 km² for tourism in the bay, and 1.4 km² for tourism in Playa El Agua. These study sites are shown as areas where investments for the development of tourism could be increased.

In Juan Griego Bay, erosion values in the coastline caused by a sea level rise of 0.5 m comprise between 0.03 km² and 0.74 km²; for a 1.0 m rise, there would be an eroded surface of 0.06 km² and 0.8 km². For the 0.5 m and 1.0 m scenario, the flooding that this area would be between 0.04 km² and 0.07 km². Estimated total land loss for the 1.0 m sea level rise scenario in 100 years would be 0.87 km², which represents a coastline regression of 13.56 km.

The Playa El Agua study site would experience potential erosion between 0.04 km² and 0.10 km² along the lower coast for a 0.5 m scenario; for the 1.0 m scenario, potential erosion would be between 0.06 km² and 0.28 km². The study site is not vulnerable to inundation because it does not have rivers or wetlands. Estimated total land loss for the 1.0 m sea level rise scenario in 100 years would be 0.28 km², with a coastline regression of 29.48 km.

The value of the infrastructure at risk in these two study sites on Margarita Island would be approximately US$0.67 million and result primarily from tourism infrastructure. In view of the above results, Margarita Island is regarded as an area at high risk from sea level rise.

In the analysis presented above, developed areas were defined at those that had basic services allowing for the establishment of industrial infrastructures, residential and tourism installations, airport facilities, etc. Likewise, those land areas associated with production (agricultural or forestry) important to the local, regional, or national economy were considered to be developed.

Adaptation Alternatives

The results obtained from the Adaptation Strategy Evaluator for Coastal Zone Planning seem very appropriate because land use planning, i.e., the assignment of different land uses — urban, tourist, industrial, agricultural — to coincide with the features and potentialities of existing natural resources, was the most favored strategy for all the sectors. More work will be done utilizing this model.

Discussion and Conclusions

The areas most vulnerable to land loss by erosion are the eastern coast of Maracaibo Lake, the Cabo Codera — Laguna de Tacarigua National Park, and the eastern coast of Falcón State. On the other hand, the most vulnerable areas to land loss by flooding are the eastern coast of Falcón State, Cabo Codera — Laguna de Tacarigua National Park, and the eastern coast of Maracaibo Lake.

All five study sites have a total area of 131.43 km². In aggregating the results from these study sites, land loss by erosion (lower bound estimate) under a 0.5 m sea level rise is estimated to be 34.24 km² and the land loss by flooding is estimated to be 52.63 km². Under a 1.0 m sea level rise land loss erosion (upper bound estimate) is estimated to be 47.0 km² and land loss by flooding is estimated to be 77.74 km².

The greatest population at risk is located in the eastern coast of Falcón State and in the eastern coast of Maracaibo Lake. The highest capital value was obtained for the eastern coast of Maracaibo Lake and the Barcelona-Puerto La Cruz-Guanta sectors, where most of the oil and gas infrastructure in the country are located.

Assuming a nonprotection response, and a 1.0 m sea level rise, the land area lost (upper bound estimate) would be 131.13 km², while with the option of protecting developed areas, the land area lost would be 86.16 km² (US$30,017). With adequate planning and coastal management, the future vulnerability of these coastal areas could be reduced.

Summary Conclusions

A vulnerability and adaptation assessment is important for Venezuela, whose current national plans include programs of industrial development, an increase in the public service sector, and expansion of petroleum industry activities. These developed plans affect forest areas extensively because a wide range of Venezuela's economic activities has traditionally been linked to land clearing. In addition, most of the coastal activities are related to oil extraction, tourism, or fishing. All of these activities make the coastal zone a very important part of the national economy. Thus, it is important to assess the potential effects of climate change on Venezuela's forests and coastal zone.

The information presented in this chapter provides preliminary information. However, further assessments and efforts to implement policies based on these findings should be undertaken in the future. For example, a proposal for an integrated coastal zone management program should be developed.

References

BCV. 1994. *Informe Banco Central de Venezuela.* Caracas, 145 pp.

Benioff, R., S. Guill, and J. Lee (eds.) (In press). *Vulnerability and Adaptation Assessments: An International Guidebook.* Dordrecht, The Netherlands: Kluwer Academic Publishers.

COPLANARH. 1970. Inventario de Suelos y Tierra. Ministerio de Obras Públicas. Caracas. Venezuela.

FAV. 1994. *Anuarios de Información y Estadística.* Fuerza Aérea. Fuerzas Armadas Venezolanas. Venezuela, 125 pp.

Hands, E.B. 1983. *Handbook of Coastal Processes and Erosion.* Boca Raton, FL, USA: C.R.C. Press, 167 pp.

Holdridge, L.R. 1967. *Life Zone Ecology.* Tropical Science Center, San Jose, Costa Rica.

IPCC. 1990. Strategies for Adaptation to Sea Level Rise. Report of the Coastal Zone Management Subgroup. IPCC Working Group III, Rijkswaterstaat, The Netherlands: Ministry of Transport, Public Works and Water Management, 122 pp.

IPCC. 1992. Global Climate Change and the Rising Challenge. Report of the IPCC Coastal Zone Management Subgroup. Working Group III, Rijkswaterstaat, The Netherlands: Ministry of Transport, Public Works and Water Management, 123 pp.

Leatherman, S.P., R.J. Nicholls, and K.C. Dennis. 1994. Aerial Videotape-Assisted Vulnerability Analysis: A Cost-Effective Approach to Assess Sea Level Rise. *Journal of Coastal Research,* Special Issue, 14:15-25.

Leemans, R. and W. Cramer. 1990. The IIASA Climate Database for Land Area on a Grid of 0.5 Resolution, WP-41, International Institute for Applied System Analysis, Laxenburg, Austria.

MARAVEN. 1995. Gerencia de Diques y Drenaje. Personal Communication. Lagunillas. Estado Zulia.

MARNR. 1994. *Anuarios de Información y Estadística.* Ministerio del Ambiente y de los Recursos Naturales Renovables. Caracas, Venezuela, 234 pp.

Nicholls, R.J. and S.P. Leatherman. 1994. Sea Level Rise and Coastal Management. In: Mc Gregor, D. and D. Thompson (ed.), *Geomorphology and Land Management in a Changing Environment,* UK: John Wiley and Sons.

OCEI. 1990. Oficina Central de Estadísticas e Información. Anuarios. no. 2. Caracas, Venezuela: OCEI.

OCEI. 1991. Oficina Central de Estadísticas e Información. Tiempo de Resultados no. 1. Caracas, Venezuela: OCEI.

Pepper et al. 1992. *Emission Scenarios for the IPCC: An Update: Assumptions, Methodology, and Results.* Prepared for the Intergovernmental Panel on Climate Change, Working Group 1, May.

Roa-Morales, P. 1991. Coastal Morphology and Sea Level Rise Consequences in Venezuela. International Sea Level Rise Studies Project Institute of Marine and Coastal Sciences Rutgers. New Brunswick, The State University of New Jersey, 24 pp.

Smith, A. and H. Quan Chu. 1994. A Multi-Criteria Approach for Assessing Strategies for Anticipatory Adaptation to Climate Change. Washington, DC: Decision Focus, Inc.

Thornthwaite, C.W. 1948. An Approach Toward a Rational Classification of Climate. *Geogr. Review*, 38:55-89

Volonté, C. and J. Arismendi. 1995. Sea Level Rise and Venezuela: Potential Impacts and Responses. *Journal of Coastal Research*, 14:285-302.

Walter, H. 1985. *Vegetation of the Earth and Ecological Systems of the Geo-Biosphere*, 3d ed., Springer-Verlag, Berlin, Germany.

Environmental Science and Technology Library

1. A. Caetano, M.N. De Pinho, E. Drioli and H. Muntau (eds.), *Membrane Technology: Applications to Industrial Wastewater Treatment.* 1995 ISBN 0-7923-3209-1
2. Z. Zlatev: *Computer Treatment of Large Air Pollution Models.* 1995 ISBN 0-7923-3328-4
3. J. Lemons and D.A. Brown (eds.): *Sustainable Development: Science, Ethics, and Public Policy.* 1995 ISBN 0-7923-3500-7
4. A.V. Gheorghe and M. Nicolet-Monnier: *Integrated Regional Risk Assessment.* Volume I: Continuous and Non-Point Source Emissions: Air, Water, Soil. 1995 ISBN 0-7923-3717-4
   Volume II: Consequence Assessment of Accidental Releases. 1995 ISBN 0-7923-3718-2
   Set: ISBN 0-7923-3719-0
5. L. Westra and J. Lemons (eds.): *Perspectives on Ecological Integrity.* 1995 ISBN 0-7923-3734-4
6. J. Sathaye and S. Meyers: *Greenhouse Gas Mitigation Assessment: A Guidebook.* 1995 ISBN 0-7923-3781-6
7. R. Benioff, S. Guill and J. Lee (eds.): *Vulnerability and Adaptation Assessments.* An International Handbook. 1996 ISBN 0-7923-4140-6
8. J.B. Smith, S. Huq, S. Lenhart, L.J. Mata, I. Nemošová and S. Toure (eds.): *Vulnerability and Adaptation to Climate Change.* Interim Results from the U.S. Country Studies Program. 1996 ISBN 0-7923-4141-4
9. B.V. Braatz, B.P. Jallow, S. Molnár, D. Murdiyarso, M. Perdomo and J.F. Fitzgerald (eds.): *Greenhouse Gas Emission Inventories.* Interim Results from the U.S. Country Studies Program. 1996 ISBN 0-7923-4142-2

KLUWER ACADEMIC PUBLISHERS – DORDRECHT / BOSTON / LONDON

MIX
Papier aus verantwortungsvollen Quellen
Paper from responsible sources
FSC® C105338

If you have any concerns about our products,
you can contact us on
ProductSafety@springernature.com

In case Publisher is established outside the EU,
the EU authorized representative is:
Springer Nature Customer Service Center GmbH
Europaplatz 3, 69115 Heidelberg, Germany

Printed by Libri Plureos GmbH
in Hamburg, Germany